NATO ASI Series

Advanced Science Institutes Series

A series presenting the results of activities sponsored by the NATO Science Committee, which aims at the dissemination of advanced scientific and technological knowledge, with a view to strengthening links between scientific communities.

The Series is published by an international board of publishers in conjunction with the NATO Scientific Affairs Division

A Life Sciences B Physics	Plenum Publishing Corporation London and New York
C Mathematical and Physical Sciences D Behavioural and Social Sciences E Applied Sciences	Kluwer Academic Publishers Dordrecht, Boston and London
F Computer and Systems Sciences G Ecological Sciences H Cell Biology I Global Environmental Change	Springer-Verlag Berlin Heidelberg New York London Paris Tokyo Hong Kong Barcelona Budapest

PARTNERSHIP SUB-SERIES

1. Disarmament Technologies	Kluwer Academic Publishers
2. Environment	Springer-Verlag
3. High Technology	Kluwer Academic Publishers
4. Science and Technology Policy	Kluwer Academic Publishers
5. Computer Networking	Kluwer Academic Publishers

The Partnership Sub-Series incorporates activities undertaken in collaboration with NATO's Cooperation Partners, the countries of the CIS and Central and Eastern Europe, in Priority Areas of concern to those countries.

NATO-PCO DATABASE

The electronic index to the NATO ASI Series provides full bibliographical references (with keywords and/or abstracts) to about 50000 contributions from international scientists published in all sections of the NATO ASI Series. Access to the NATO-PCO DATABASE compiled by the NATO Publication Coordination Office is possible in two ways:

- via online FILE 128 (NATO-PCO DATABASE) hosted by ESRIN, Via Galileo Galilei, I-00044 Frascati, Italy.
- via CD-ROM "NATO Science & Technology Disk" with user-friendly retrieval software in English, French and German (© WTV GmbH and DATAWARE Technologies Inc. 1992).

The CD-ROM can be ordered through any member of the Board of Publishers or through NATO-PCO, Overijse, Belgium.

Series I: Global Environmental Change, Vol. 31

Springer
Berlin
Heidelberg
New York
Barcelona
Budapest
Hong Kong
London
Milan
Paris
Tokyo

The Role of Water and the Hydrological Cycle in Global Change

Edited by

Howard R. Oliver

Institute of Hydrology
Wallingford OX10 8BB, U.K.

Sylvia A.Oliver

Chalgrove
Oxford OX44 7SS, U.K.

Springer

Published in cooperation with NATO Scientific Affairs Division

Proceedings of the NATO Advanced Study Institute on the role of water and the hydrological cycle in global change, held in Il Ciocco, Lucca, Italy, May 27–June 6, 1994

ISBN 3-540-59429-9 Springer-Verlag Berlin Heidelberg New York

Printed in Germany

Typesetting: Camera ready by authors/editors
Printed on acid-free paper
SPIN 10134974 31/3136 - 5 4 3 2 1 0

Preface

Water is an extremely important factor in global environmental change. Water influences the processes causing change. The human and economic consequences of changes in the water system can be very significant.

The aim of this NATO Advanced Study Institute was to present a state-of-the-art assessment of the role of water in global change, ranging from a consideration of atmospheric processes to the social and political impacts of changes in water resources. Many initiatives have recently been developed, such as international conferences and research programmes in particular themes, but there was still a need for information from these diverse activities to be brought together. One of the aims of the ASI was to encourage cross-fertilization between the various disciplines looking at water in the global system.

This book contains the expanded written versions of the lectures presented during the ASI held in Italy in May/June 1994. It falls into two basic parts.
The first twelve chapters cover the role of water in the climate system and climate modelling. Various areas of the water balance including global budgets, the effect of each element of the water balance on regional and global climates, and procedures used to model hydrological processes at all scales are discussed. Precipitation, ice, lake, groundwater, land surface and atmospheric considerations are included together with hydrological process linkage to climate models. Ocean effects were not covered as they were considered to be outside the scope of this particular ASI.

The second part of the book is concerned with the impacts of climate change and variability on hydrological and water resource systems. These last four chapters review methodologies used for estimating impacts and creating climate change scenarios as well as discussing impacts on hydrological and water resource systems together with social and political implications.

The ASI was attended by post-doctoral level students mainly from Europe and North America. As well as the formal lecture sessions the students were given opportunity to show posters of their own work which covered a very wide range of related topics. The need for such inter-disciplinary meetings to aid in the development of coordinated approaches to the study of global problems came out very clearly throughout the meeting.

Thanks are of course due to many people. Firstly to the Scientific Affairs Division of NATO and to participants' organisations for provision of funding for lecturers' and students' expenses. To the lecturers themselves who gave their time to produce and present their lectures and book chapters as well as taking part in the discussions during the meeting. To Sylvia Oliver for book editing and indexing as well as indispensable administrative work throughout. To Margaret Howarth for the financial management. To Ezio Todini for help with visa applications. To Sally Austin for word processing support and to Nigel Arnell and Max Beran who initially proposed this ASI. Finally the ASI could not have been the success it proved to be without the efficient and friendly help of Bruno Giannasi and his staff at the Il Ciocco Centre, Castelvecchio Pascoli near Pisa, Italy who hosted the meeting.

Howard Oliver (ASI Director)

Contents

THE ROLE OF WATER IN GLOBAL ENVIRONMENTAL CHANGE PROCESSES

Max Beran
TIGER Programme
Institute of Hydrology
Wallingford, OX10 8BB
UK

OUTLINE

Global Environmental Change (GEC) is understood as a set of human-induced environmental changes which have the capacity to perturb the Earth System. This paper ranges widely, if superficially, over the ways in which water is implicated in GEC processes. We first consider GEC itself - what is it, what causes it, is there a science of it, and what is the role of water? Earth System Science provides answers in terms of exchanges, cycles and feedbacks of energy, momentum, water and other biogeochemicals. GECs of current concern - ozone depletion, the greenhouse effect, reduction of biological diversity, and soil and atmosphere degradation - are driven by human interventions in these cycles.

On close inspection we discover that man's direct intervention in the water cycle is not large by comparison with that in the other element cycles, and water's importance lies in the way it mediates the global cycles of energy, carbon, nitrogen, phosphorus and sulphur. This paper goes on to present an "organic" view of water-related processes in the terrestrial biosphere. Examples then follow of the problems of trying to forecast how GEC will influence the water cycle, and this leads us finally to consider international scientific programmes of which the NATO Special Programme on the Science of Global Environmental Change is one example.

GLOBAL ENVIRONMENTAL CHANGE

Definitions: Global Environmental Change is a term that has come to apply to a range of man-induced environmental problems, notably climate change and stratospheric ozone depletion. There is great confusion about how GEC should be defined, centring on the

NATO ASI Series, Vol. I 31
The Role of Water and the Hydrological Cycle
in Global Change
Edited by Howard R. Oliver and Sylvia A. Oliver

significance of the adjective "global". The two cases just quoted exemplify this confusion. While climate change is unarguably global - it affects the entire globe - ozone depletion is merely zonal. Indeed there are several environmental problems which are not conventionally regarded as global changes, like river pollution and acid rain, and which are at least as widespread. What then is it about a GEC that merits it being termed "global"?

A litmus test that can be used to determine whether an environmental issue qualifies as a genuine Global Environmental Change is to ask: *"is the change to the environment capable of perturbing global stability?"* If the answer is *"yes"*, then the issue qualifies as a Global Environmental Change.

To understand this test it helps to go back to the origin of the subject - Global Change science - the branch of geology that concerns the evolution of oceans, landmasses, deserts, mountains, ice ages, *etc*. One intriguing question thrown up by Global Change science is how the rather narrow conditions that permit life to persist have remained unbroken in the face of massive external changes like continental drift, meteoric impact, vulcanism, orbital changes, glaciations and increased insolation? Another asks how an atmosphere which consists of a "cocktail" of reactive gases in concentrations far from chemical equilibrium can persist?

Answers to these problems became possible when systems theory was applied to Global Change problems. This blend, called Earth System Science, views the Earth as made up of four "spheres" - hydrosphere, geosphere, atmosphere and biosphere. Though simplistic, the concept has been extraordinarily fruitful in explaining how the Earth's surface temperature, and the atmosphere's chemical composition have remained within limits suited to life. Our present understanding of the Earth System reveals this not to be a matter of chance or due to the precise balance between opposing forces, but as the outcome of homoeostatic adjustments, *ie* negative feedbacks.

These feedbacks operate on the exchanges and cycling of energy, momentum, water, carbon and other entities that take place between the four "spheres", such that if any part of the system begins to drift then this initiates a chain of events which acts to counteract it (see for example Lovelock's and other chapters in Schneider and Boston, 1991). In the Gaia hypothesis (Lovelock, 1979) biotic links are emphasised, though many abiotic links are also present in feedback chains. An example from the hydrological cycle illustrates this through the chain of geochemical reactions thought to be responsible for stabilizing atmospheric CO_2 concentration (Berner and Lasaga, 1989). In their scheme a drift towards increased concentration of carbon dioxide in the atmosphere, perhaps due to a period of vulcanism, lowers the pH of rainwater and acidifies the soil water in contact with rocks. Acid conditions at the soil-rock interface favour erosion and the transport of carbonates and silicates to the ocean. Ocean alkalinity derived from the silicates increases the rate of carbon burial so reducing net efflux of carbon dioxide to the atmosphere and finally restoring the *status quo*.

Berner's feedback, which involves the cycling of carbon between all four Earth System "spheres" to resist **Global Change,** operates over the million-year time scale. By contrast **Global Environmental Changes** caused by man's activities operate at a rate and are of a magnitude that do not permit restorative feedbacks by the Earth System, at least not over the decade-to-century time scale. Table 1 shows the GECs and the cycles that are perturbed.

Table 1 - Global Environmental Changes and key associated biogeochemical cycles (C-Carbon, N-Nitrogen, Cl-Chlorine, P-Phosphorus, K-Potassium, S-Sulphur, O_3-Ozone)

Greenhouse effect and climatic change	C N
Stratospheric ozone depletion	Cl N
Reduction in biological diversity	N P
Acidification and soil degradation	C N S
Eutrophication of freshwaters and oceans	P
Desertification and dryland degradation	C N
Reduction of atmospheric clarity	S O_3

We will be returning to these cycles in subsequent sections, but at this point we note that the justification for the word "global" arises through the potentiality for an environmental change to be **globally destabilizing** rather than because it is global in its impact.

Causes and Effects: Over the decadal time scale of GEC concern, feedbacks responsible for thermostatic and chemistatic control of the Earth's radiation budget are most prone to destabilization. The hydrosphere and biosphere are deeply implicated in these feedbacks via (a) biophysical processes - *ie* changes to surface energy and momentum exchange, and (b) biogeochemical cycling - *ie* emissions and uptake of radiatively active biogenic gases. Massive conversions of land cover, and releases of eutrophying and soil degrading chemicals by man, play havoc with the natural exchanges at the land-atmosphere interface. Industrial and agricultural emissions alter the composition of the atmosphere by injecting greenhouse gases, ærosols, and ozone destroying substances, which influence the radiation budget.

We need to be more precise about the concept of "destabilization". Variability exists in unperturbed environments on many time scales, *eg* interannual variability of weather patterns and streamflow, and the swings in relative abundances of species comprising ecosystems. The interior dynamics of the hydrological cycle appears to be capable of generating long-term, say decadal, fluctuations such as Sahel drought (Entekhabi *et al*, 1992). What "destabilization" means is failure to contain variability within the preset bounds that permitted the affected system to perform its role within the Earth System.

Let us also distinguish GEC *processes* from their *causes* (drivers) and *consequences* (impacts). Land-cover changes like deforestation and drainage, and emissions of gases like CFCs and CO_2 drive global environmental changes along different paths. Global environmental changes have consequences such as those discussed in chapters 12 to 16 for water resources. GEC consequences need not all be negative in socio-economic terms, for example a moderate warming in some northern latitude regions would enhance agricultural productivity, and elevated carbon dioxide acts as a fertilizer by enhancing plant growth and reducing plant water use (chapter 6). These trade-offs add to the complexity of framing policy responses to GEC; but like consequences, *responses* too are distinct matters and not to be confused with the GEC *processes* themselves. Though Global Environmental Research encompasses in its totality GEC's causes, impacts and responses, this paper focuses on the water-related processes of change.

QUANTIFYING HUMAN INTERVENTIONS

In this Section we develop a numerical index of the scale of human interventions in biogeochemical cycles. For this we employ the most basic mathematical tool of the Earth System scientist: the conservation of mass applied to a reservoir with steady state inputs and outputs. Figure 1 illustrates the principle where the arrows signify the flux of some entity

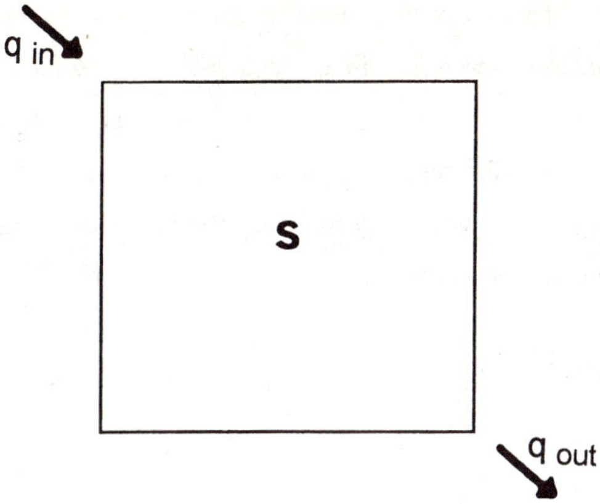

Figure 1: Reservoir and mass balance concepts

into and out of one of the components of the earth system. A fuller specification would require a rate law to describe the dependence of q_{out} on the magnitude of S and other factors, the simplest being q_{out} = kS which leads to an exponential time decay for S. Budget calculations concern changes in the magnitude of S, hence differences between q_{in} and q_{out}. Most real-world applications also require the model to be spatially disaggregated, *eg* separate treatment of land and ocean. However, for present purposes the basic, steady-state, global form of Figure 1 suffices.

Consider water in the atmosphere. In this case S=13000 km^3 of water vapour (chapter 8) from which there is an annual flux q_{out}=500,000 km^3 falling as precipitation (chapter 2). A much-used parameter in Earth System science is the turnover time (alternatively mean residence time) which is defined as the time taken to fill or empty the reservoir. In the current example S/q_{out} is .026 yr or 9.5 days. There are two main reasons for the importance of residence time. Firstly it is the e-folding time for reservoir depletion in the linear q_{out}=kS case. Secondly, in many instances one has prior information or intuition about its magnitude, perhaps more so than S or q. In this regard we may note that 9.5 days conforms reasonably to an intuitive guess at the interval between significant (ie atmosphere emptying) rainstorms.

Returning to the matter of indexing the scale of human intervention, this is expressed as the contribution of the anthropogenic flux in q_{in} or $q_{out.}$ to the natural background flux. The volume of water exploited by man by diversion into temporary storages and passed through subsidiary loops between aquifer, soil, plant, industry and consumption (Shiklomanov, 1989), is of the order of 5,000 km^3, which is small in relation to the natural land atmosphere exchange. Round-number values for water and the other main cycles are shown in Table 2.

Table 2 Magnitudes of human interventions in Earth System cycles

Entity/element	Human intervention in cycle (%)
Energy	0.1
Water	1
Carbon	5
Nitrogen	50
Sulphur	50
Phosphorus	100

The low value for water in Table 2 contrasts with our expectation based upon its undoubted importance for terrestrial life and human society. However this is a separate issue which would be more appropriately indexed through seasonal volumes in populated regions. Table 2 confirms that man's direct intervention in the hydrological cycle cannot play more than a minor role in creating Global Environmental Change through, for example, disturbing poleward transports of latent heat. The real importance of water lies in its indirect role in the other cycles of Table 2 which are considered in following Sections.

PERVASIVE ROLE OF WATER

Figure 2 illustrates water's involvement in earth system processes. The following paragraphs emphasise those that take place over land and are capable of human interference. We start at the molecular level before exploring the individual water-related biosphere processes from Figure 2.

Water as a molecule: Water exhibits many anomalous physical and chemical properties which underlie its pervasive role in the earth system. Its melting and boiling points are unexpectedly high for a material of its molecular weight, so solid, liquid and gaseous phases can co-exist

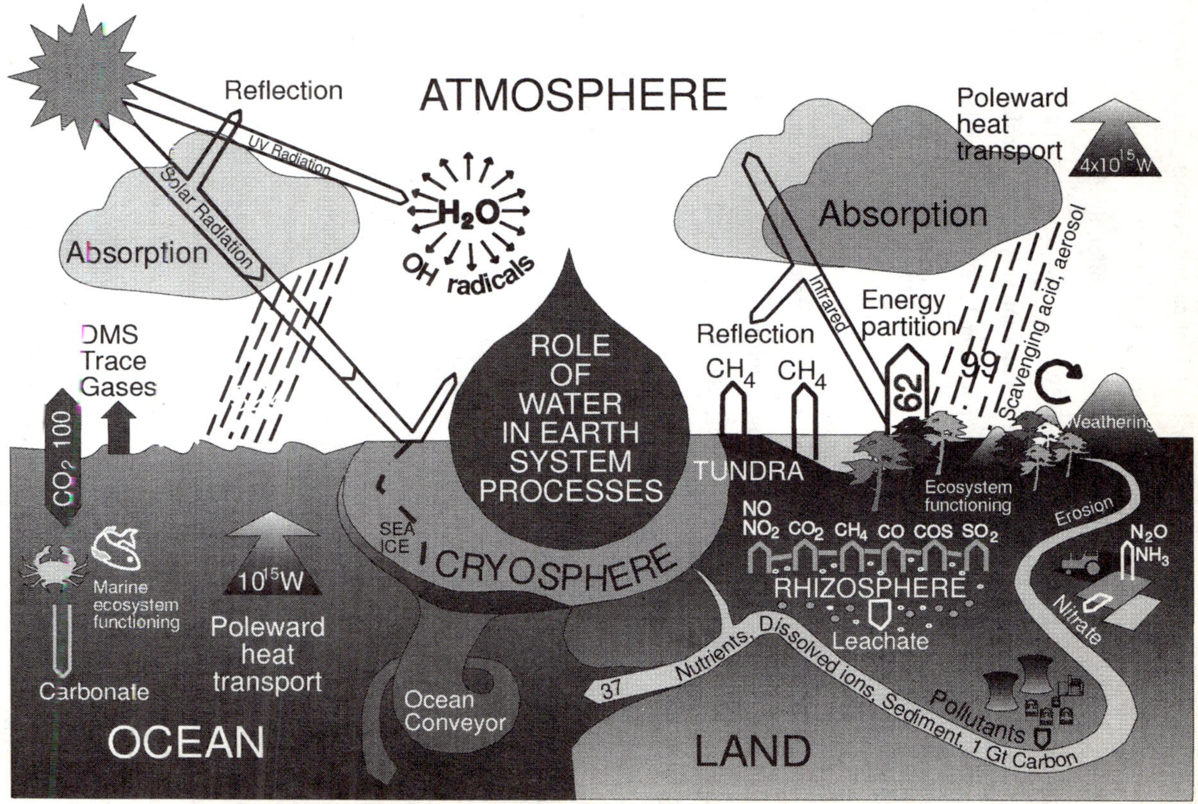

Figure 2 The role of water in earth system processes

at ambient temperatures. It has an unusually large heat capacity so there is a release of heat on ice formation and on condensation, balanced by high uptakes of heat on melting and evaporation. Unlike other liquids it does not contract uniformly on cooling, but has a maximum density at 4 °C and a solid form, ice, which is less dense than the liquid. Perhaps liquid water's most important property in the current context is its remarkable solvent power which permits it to be the reaction and transport medium for almost all biochemical processes. These properties are not independent as they can all be ascribed to hydrogen bonds and water's polymeric nature.

Biota consist of elaborate systems of organic macromolecules in an aqueous medium, organized into cells which depend on insoluble membranes for their integrity. Most organisms comprise more than 60 per cent water - juvenile humans, up to 80 per cent! The biosphere (living biota plus their supporting medium) is fuelled by shortwave energy from the sun of which much less than one per cent is used in photosynthesis. This complex of processes involves a light reaction with chlorophyll that splits water into its components, and which energises the following reduction of CO_2 to carbohydrates. Cellular processes then work on the carbohydrates to build the macromolecules. Post-photosynthetic products fuel links in food chains which can be viewed chemically as an oxidation reaction chain taking the reduced forms back to CO_2. Important features of these reactions are that they occur in small steps and that they involve enzymes which reduce the activation energies, but which require nitrogen and phosphorus for their own activation.

The biosphere processes nearly one-half of the incoming shortwave energy abiotically, mostly in evaporation (see chapters 4 and 5). The second abiotic use is to warm surfaces which then re-radiate in the infra-red. Molecules of certain gases in the atmosphere, principally water vapour and carbon dioxide, absorb and re-radiate this infra-red, *ie* the *natural* greenhouse effect. Any addition to the atmospheric burden of CO_2 intensifies the heat exchange in the lower atmosphere and leads to global warming; around 1K for a doubling of CO_2. This increases the moisture holding capacity of the atmosphere, leading to a positive feedback which augments the basic warming, *ie* the *GEC* greenhouse effect.

Biophysical properties of ecosystems: At the ecosystem level, interest centres on the surface radiation budget, gas exchange, heat storage of soil and vegetation, and interactions with the planetary boundary layer *eg* through surface roughness (Garratt, 1993; de Bruin & Jacobs, 1993). These biophysical properties have special significance to the climate system (Beran, 1991 and chapters 3 and 5 of this volume).

One of the ways that water enters this picture is through its control on the type of vegetation that can survive at any location. Historically, broad-scale classification of vegetation - ie biome definition - has been predicated on this control (*eg* Holdridge, 1947). Temperature and moisture-based zonations such as tropical-wet, cool temperate-dry, boreal-moist *etc* find correspondences in growth-habit based categories like rainforest, savanna, taiga *etc*. However, this level of classification is of little direct use for earth system modelling for which it is necessary to represent the individual properties of vegetation that influence the energy and water budgets - canopy architecture, albedo, leaf area and conductance (chapter 3). Of these, leaf area index, canopy architecture and biomass are predictable from the hydrological budget, and life form and physiognomy from drought and other seasonal temperature factors (Woodward, 1987).

As explained in chapters 5 and 11, it is not sufficient to consider a surface element of vegetation in isolation. The above-ground vegetation has to be set within a soil-vegetation-atmosphere framework in which the soil's water- and heat-holding capacities, and water transport mechanisms are accounted for. The horizontal scale - the way that heterogeneous patches build into landscape units and that lateral transports of water and nutrients occur - is important for understanding GEC processes. Research takes the form of experiments and field campaigns such as the HAPEX series and FIFE, and modelling. This area provides much of the inspiration for international programmes (Henderson-Sellers *et al*, 1993).

Vegetation as an active participant: While modellers have incorporated biophysical properties with increasing sophistication, they tend to neglect the most distinctive property of the biota, that it is *alive*. Plants and ecosystems ought not to be regarded as passive participants whose energy, water and nutrient capture are under the one-way control of environmental conditions. This is most clearly seen with transpiration. Evaporation from a free water surface can be expressed correctly as a function of vapour pressure difference and the capacity of the receiving medium to accept the newly gasified molecules. Plant transpiration does not follow this inert pattern but imposes a "donor-side" demand on the behaviour of the system. This fact has long been recognized through the observed decline in afternoon transpiration, somewhat different from the pattern of energy supply and atmospheric demand which are more symmetric about local noon. Vegetation clearly follows a different agenda based on its own embedded rhythm.

Many examples of demand-side control have appeared in the literature and operate at all time and space scales. Biological control on the daytime planetary boundary layer is most evident in rainforest areas where the boundary layer can rise to 2km. The climate within the layer is also very much under biological control (Shuttleworth, 1988, Salati and Nobre, 1991). Possibly up to 10 per cent of the rain at the forest floor emanates from transpired moisture condensing within the canopy on biogenic condensation nuclei, *ie* an internal cycle at patch scale. Of the remainder, up to two-thirds can be locally derived as "rainout" from upwind evaporation. The residue, which is from long-distance advection, is the fraction which receives most attention in climate models such as those discussed in Chapter 3 (Brubaker *et al*, 1993).

Monteith (*eg* in 1988) discusses a belowground example of an active biosphere: hydrologists tend to think in terms of a water front approaching roots and develop models accordingly. However there is a matching root front "searching" out the water. Other examples occur with the spacing of plants in communities using mechanisms based upon light signalling, and on nutrient harvesting. This latter is well known in semi-arid environments where runoff takes place from the bare soil between vegetated patches (Schlesinger, 1991). Other mechanisms derive from canopy-induced turbulent transfer (Hollinger *et al*, 1994) and litter-mediated partitioning of water between runoff and percolation. At the biome scale the "green-wave" that pushes polewards each spring immediately influences its climate as the wave-front passes (Schwartz, 1992).

Indeed it is hard, when describing these phenomena, to refrain from using metaphors imputing intelligence to vegetation. New discoveries to extend this list are likely: the combination of signals that are used by the plant to recognize drought stress are yet to be worked out; biogenic cloud condensation nuclei of terrestrial origin may be of substantial importance (Novakov and Penner, 1993); the ecological role of biogenic emissions like isoprene and terpene is unknown (Hewitt and Street, 1992). This family of hydrocarbon gases is important in tropospheric chemistry and costly for the plant to produce so it is improbable that it will not be shown to be capable of some biological service. For a selection of cases where vegetation appears to adjust its phenology community interactions, life history or structure in order to engineer an environment suitable for its own survival see *inter alia* Lashof (1989) and chapters by Waring, Bazzaz and Chapin in Ehleringer and Field (1993).

Soil water and biogeochemical cycling: Earlier paragraphs have emphasised aboveground matters and here we consider the role of water in chemical fluxes that take place belowground. Carbon is the most abundant biogeochemical belowground and one of its GEC functions in relation to the control of soil water acidity has already been raised. Soil is the largest reservoir of carbon in the terrestrial biosphere, equivalent to the sum of the aboveground and atmospheric reservoirs, and its residence time (Figure 1) is important to the global cycle (Gorshkov *et al*, 1994). This is partly under control of post-photosynthetic processes which translocate carbon within the plant and determine the proportion that is in refractory and hence long-lived forms; structural parts of woody biomass such as lignin have turnover times orders of magnitude longer than leaf litter and fine roots. But soil water also plays a key role as bacterial decay is inhibited when oxygen is excluded, which is why bogs and peatlands are localities of net carbon accumulation. This has been a very partial introduction to a large and complex topic. The interested reader is guided to Schlesinger (1991), Bouwman (1990) or Rounsevell and Loveland (1995).

Apart from carbon, the chemical elements of most importance to earth system and GEC feedbacks are nitrogen, phosphorus and sulphur. Though there are large reservoirs of the first two within living biota, cycling of all these elements is mediated through soil processes, microbial action in the cases of nitrogen and sulphur (Andreae & Shimel, 1989). These processes and subsequent transports - back to the biota, to the atmosphere, and to rivers and ocean - cannot take place in the absence of soil water.

For example, the cycling of nitrogen within the belowground part of the biosphere is closely linked with the hydrological cycle because water is its primary mode of transport - ionically via plant roots or as leachate to groundwater. Man's intervention in the natural nitrogen cycle is an order of magnitude larger than in the carbon cycle (Table 2) and the vast bulk is applied at the soil surface.

N species are implicated in several global environmental changes - eutrophication, acid rain, desertification, ozone depletion - as well as the greenhouse effect, where nitrous oxide (N_2O) accounts for six per cent of current global heating potential. Much of the anthropogenic N is applied as fertilizer to agricultural soils, the rest via fossil fuel and biomass burning. Soil moisture status determines whether N_2O is produced by the nitrification (oxidation of ammonium) or denitrification (reduction of nitrate) pathways. Moist conditions favour the latter (Bandibas *et al*, 1994; Davidson, 1993).

Anthropogenic additions of phosphorus to its global cycle, like nitrogen, are largely as an agricultural fertilizer. There are no biogenic gaseous forms of phosphorus so it plays no direct role in GEC processes related to the atmosphere. However, its indirect role is important, stemming from a requirement by all living organisms for building protein. Adequate C/P and N/P ratios are necessary before green plants and microbes respectively can flourish. Unlike nitrogen and sulphur, phosphorus cycling does not depend on soil bacteria but is taken up efficiently by specialized fungi living in association with plant roots. Plants exude substances from their roots into the surrounding soil that enhance this process. Because phosphorus is readily soluble it is easily leached and is the key element in the widespread eutrophication of watercourses, potentially even remote oceans.

Ocean biosphere processes are more important than terrestrial ones for the global cycling of sulphur. However the repercussions are global due to sulphur's function in the climate system - for cloud and rainfall formation and atmospheric aerosol. Dimethyl Sulphide (DMS) is indicated on Figure 2 as a marine emission where it is produced by the bacterial decay of marine plants. DMS is transported over land and reacts to provide an important source of cloud condensation nuclei. Carbonyl Sulphide (COS), the sulphur species derived from soil microbial processes, is more important in the stratosphere where it forms an aerosol layer which regulates radiation input. Additions by man to the sulphur cycle largely derive from fossil fuel burning and industrial processes.

The propensity for sulphur compounds to condense to acid, particulate forms explains their implication in GECs like acidification and reduced atmospheric transparency. In the absence of the enhanced greenhouse effect it appears likely that the added sulphur burden would have given rise to a cooling trend.

A potential GEC issue resides in moist soils which currently sequester many types of pollutants and nutrient residues, but which may be released into the atmosphere, rivers and aquifers if the hydrological conditions change as a consequence of the greenhouse effect. Stigliani *et al* (1991) has termed this the "chemical time bomb".

Water and methane production from wetlands: Decomposition of organic matter normally concludes with the return of CO_2 to the atmosphere, though methane is an alternative end-product. It is important to understand the conditions that favour the two pathways because methane is about 20 times more powerful a greenhouse gas, molecule for molecule (IPCC, 1994), and is a sink for beneficial atmospheric oxidants which ultimately degrade it to CO_2 in the atmosphere. The regional hydrological balance is the critical factor as flooded soils provide the anaerobic conditions necessary for the methane producing soil bacteria. However much of this production can be oxidized in the upper aerobic layers, which is under water table control. Boreal wetlands and tropical swamps contribute around 20 per cent of the global emissions of methane. A strong driver for research in this area is the concern that warming will lead to much increased emissions of methane, a concern which has some support from palaeological studies that show how temperature and methane have risen and fallen in step over the last ice age (Prinn, 1994).

Water in the atmosphere: Once in the atmosphere water participates in processes which are mostly abiotic, and hence fall outside the scope of this note. However there is a link with the biosphere through the CH_4-CO-$OH^\cdot$ cycle. As indicated in Figure 2, $OH^\cdot$, the hydroxyl radical, is formed in a photochemical reaction of water vapour and ozone. The main sink for methane is oxidation by the hydroxyl radical. There are regional variations in the reactions such that, in polluted areas like Europe, high-NO_x restores the $OH^\cdot$ concentration after oxidation - one of many examples of opposing global environmental changes.

Methane oxidation to CO_2 and H_2O in the stratosphere is the main source of water vapour there. This is significant because of the ozone depleting reactions which take place on the surface of ice particles in stratospheric clouds (Graedel and Crutzen, 1993).

Another biospheric link is with the atmospheric phase of the sulphur cycle. It was stated earlier that biogenic sulphur is a source of ærosol, and IPCC (1994) stresses the role of sulphur ærosol as a cooling influence. It also draws attention to its spatial variability such that radiative forcing is reduced in polluted regions such as Europe.

As with other key processes there is much interest in how GEC will affect the atmosphere's oxidation capacity. Because the water holding capacity of the atmosphere increases by six per cent per °C, the first order response to global warming is for an increase in hydroxyl radical concentration and a reduction in the atmospheric lifetime of methane. The actual situation is likely to be much more complex than this because of interactions with other atmospheric oxidants and competition for oxidation from biogenic carbon monoxide and other hydrocarbons (Thompson, 1992).

CASE STUDIES OF GLOBAL ENVIRONMENTAL CHANGE

As has already been stated at several points, GEC issues, especially climate change, have provided new impetus for research. If a particular GEC, eg global warming, increases the rate at which a greenhouse gas is produced or reduces its sink capacity, then this will lead to a positive feedback that amplifies the initial forcing (Lashof, 1989). Some feedbacks have been outlined in the previous Section; the following paragraphs concern others where the hydrological cycle is involved.

Elevated CO_2 and plant-water relations: Early findings that atmospheric CO_2 enrichment lowers stomatal conductance, hence reduces transpiration and increases water use efficiency (WUE), has been quoted as a potential benefit of this GEC. However, to quote from Eamus' (1991) review.

In earlier short-term studies, large increases in WUE were observed. The response was the product of a 30-40% decrease in stomatal conductivity (g_s) and a substantial (but highly variable) increase in assimilation. In more recent, longer-term studies, it has been found that the large increase in assimilation observed at the start of the experiment is often not maintained and "ecosystem homeostasis" appears to operate. In addition, although elevated atmospheric partial pressure of CO_2 $[CO_2]_a$ does appear to reduce g_s (there is no evidence of acclimation of g_s to elevated $[CO_2]_a$) there is debate as to whether this decline will have a significant impact upon transpiration from entire canopies.

A second review (Tyree and Alexander, 1993) stresses the need for knowledge of the implications of the altered root to shoot ratio that is exhibited by vegetation growing in elevated CO_2. There are obvious hydrological implications of a shift in internal flow pathways from above to below-ground structures.

Some very fundamental questions remain unanswered: how do plants sense high CO_2; how will ecosystems (as opposed to individual plants) respond? The first question is not just of theoretical importance as knowledge of physiological processes allow predictions to be made of plants' response to simultaneous stresses such as drought, pollution or elevated UV-B. There are indications that several biochemical pathways may be involved in the signalling process (Mott, 1991 and Mansfield *et al*, 1990); indeed it is not impossible that it may originate in the root zone and be mediated by soil water.

Technical advances now allow ecosystem-level experiments. FACE (Free-air Carbon Dioxide Enrichment) systems have been constructed in USA and Switzerland (Hendrey and Kimball, 1994), and a mobile version is under development in the UK. A FACE system consists of an array or ring of pipes through which CO_2 is delivered to the experimental area to maintain a target concentration. FACE overcomes many of the drawbacks of conventional experimental techniques. The first two years operation of a US FACE system with a cotton crop supports all the findings mentioned above. All indicators of biomass showed an enhancement by CO_2 fumigation; transpiration remained unchanged despite higher leaf areas; root to shoot ratio was considerably reduced (Prior *et al*, 1994).

How may the effect of elevated CO_2 on vegetation influence the atmospheric phase of the hydrological cycle? We have already discussed the biophysical role of vegetation in the climate system. The philosophy applied by the climate modelling community to test whether any given biophysical process is likely to be climatically significant is to model it first in a simplified version of a General Circulation Model (GCM), for example a single column model, parameterised to represent a sensitive biome. The model may be run for current and future conditions, and with the process "switched on" and "switched off". Such trial evaluation is necessary because fully interactive models are expensive to run and to parameterise. Any additional computational burden has to be justified in the light of the other approximations of a climate model.

A recent TIGER programme simulation of rainforest vegetation clearly indicated the importance of incorporating stomatal feedbacks (Friend and Cox, 1995). In a double-CO_2 simulation they found a reversal of sign from 10 per cent more transpiration with fixed vegetation to 9 per cent less when stomatal opening was free to respond to environmental conditions. The chain of consequences predicted by the model included a very clear impact on cloudiness under a high-CO_2 regime. We can anticipate similar surprises from other biophysical interactions such as soil moisture changes and their surface albedo consequences or vegetation canopy changes leading to altered momentum transfer.

Climate change, biome shift, and water: Predicting biome shifts that may be forced by GEC presents even greater difficulties than forecasts at the organism level because we cannot fully explain why, even under current conditions, a given ecosystem should occupy any given location (as opposed to another that appears equally fitted). Nor have we yet the tools to predict reliably the time-trajectory from one system to its replacement (ie succession at biome scale).

However, for purposes of climate modelling it is vegetation's functional properties which are significant, not the details of the botany. Because ecosystem function is strongly tied to water and radiation this simplifies the problem a little, but not a lot (Woodward and Diament, 1991)! Also this redeeming feature does not carry over to other GEC such as biodiversity where the individual members of ecosystems have importance (Naeem *et al*, 1994). Initial experiments by Henderson-Sellers (1990, 1993) are based on the Holdridge Life scheme in which characteristic vegetation is deemed to be controlled by annual precipitation and temperature. Martin (1993) has reviewed and classified alternatives to the Holdridge scheme based upon climatic controls on vegetation. All schemes retain the strong link between vegetation type and the hydrological cycle. Another key feature that will be required for full dynamic interaction will be the ability to model the time course of vegetation change during a period of rapid climate change (Prentice *et al*, 1992).

Drought, soil respiration and transient climates: The advent of transient climate runs by recent GCMs has highlighted the importance of time lags. Rind et al (1990) predicted that thermal buffering in the oceans will cause rainfall increases to lag continental temperature increases. Thus there will be a period of aggravated droughtiness which could give rise to further loss of soil carbon as well as obvious hydrological effects. There is support for the basic supposition of a mid-continent lag from the UK Meteorological Office's GCM run in transient mode (Murphy, 1992 and 1995) which displays extensive regions of mid-latitude continents experiencing reduced precipitation resulting from diminished local evaporation and soil moisture. A concern from this scenario would be a positive feedback from a carbon pulse, conceivably up to half of current fossil fuel emissions, due to vegetation dieback during the establishment of a new equilibrium state (King and Neilson, 1992).

Increase in UV-B radiation: The threat from increased UV-B stems from the depletion of stratospheric ozone that has been observed in northern mid-latitude. Recent estimates, suggesting that UV-B radiation could be one-third higher by the end of the century than in 1980, exceed estimates quoted only two years ago (UNEP, 1991). UV-B is known to be damaging to plant cells and research into the issue is only now beginning to gather momentum. There is a strong link to problems of water through the extreme sensitivity of aquatic ecosystems, freshwater as well as marine. There have also been experiments on combined CO_2 and UV-B effects on land plants which have shown that the improvement to WUE is nullified. As with CO_2 effects, the major gap concerns field information.

INTERNATIONAL RESEARCH FRAMEWORK

The interface between global climate change processes, the hydrological cycle and ecosystems falls within the remit of the International Geosphere Biosphere Programme, in particular its core project "Biospheric Aspects of the Hydrological Cycle" (BAHC). IGBP-BAHC is structured into four foci:

Focus 1: Water, energy and carbon transfer between soil, vegetation and the atmosphere at patch scales.

Focus 2: Regional scale studies of land surface properties, experiments, interpretation and modelling.

Focus 3: Interactions between the biosphere, water resources and climate - region to continental scale.

Focus 4: The weather generator project.

Other core projects which include some reference to water cycle issues are the Global Change and Terrestrial Ecosystems (GCTE) and International Global Atmospheric Change (IGAC).

The hallmark of IGBP projects is its concern with the biosphere; the physical-system counterpart is the WMO/ICSU programmes GEWEX and GCIP, parts of the World Climate Programme. These also have strong needs for data and knowledge on the soil/plant system. Institutionally this should be arranged by ensuring appropriate data are made available to the Global Climate Observing System, GCOS, the Global Terrestrial Observing System, GTOS and equivalent IGBP structures, DIS. The data types that have principally engaged operational hydrologists are river discharge and rainfall. Water-related ecological aspects of climate change highlight the importance of the soil, vegetation and atmospheric phases, and on dissolved materials involved in biogeochemical cycling. As well as rivers, other types of aquatic system need to be monitored, especially wetlands. New bases for the assembly of data become necessary in order to understand the properties of biomes and ecosystems.

NATO and the European Science Foundation fund workshops, visits and collaboration though not usually research. The environment line in the EU's fourth framework programme is designed to support participation in the international programmes, WCP and IGBP. SCOPE also organizes global environmental change programmes.

REFERENCES

Andreae, M.O. and Shimel, D.S. (eds) 1989. Exchange of trace gases between terrestrial ecosystems and the atmosphere. Dahlem Workshop Reports. Wiley, Chichester, 347p.

Bandibas, J., Vermoesen, A., de Groot, C.J. and van Cleemput, O. 1994. The effect of different moisture regimes and soil characteristics on nitrous oxide emissions and consumption by different soils. Soil Science, **158**, 106-114.

Beran, M.A. 1991. The climate system and hydrological cycle. pp 57-81 in R W Corell and P A Anderson (eds), Global environmental change. NATO ASI Series, Vol I1. Springer-Verlag, Berlin. 264p.

Berner, R.A. and Lasaga, A.C. 1989. Modeling the geochemical carbon cycle. Scientific American, March 1989, pp 54-61.

Bouwman, A.F. (ed) 1990. Soils and the greenhouse effect. Wiley, London. 575p.

de Bruin, H.A.R. and Jacobs, C.M.J. 1993. Impact of CO_2 enrichment on the regional evapotranspiration of agro-ecosystems, a theoretical and numerical modelling study. Vegetatio **104/105**, 307-318.

Brubaker, K.Y., Entekhabi, D. and Eagleson, P.S. 1993. Estimation of continental precipitation recycling. Journ Clim. **6**, 1077-1089.

Davidson, E.A. 1993. Soil water content and the ratio of nitrous oxide to nitric oxide emitted from soil. pp 369-386. In R.S. Oremland. Biogeochemistry of global change. Chapman and Hall, New York. 879p.

Eamus, D. 1991. The interaction of rising CO_2 and temperatures with water use efficiency. Plant Cell and Environment **14**, 843-852.

Ehleringer, J.R. & Field, C.B. 1993. Scaling physiological processes: leaf to globe. Academic Press, San Diego. 388p.

Entekhabi, D., Rodriguez-Iturbe, I. and Bras, R.L. 1992. Variability in large-scale water balance with land-surface-atmosphere interaction. Journ Clim. **5**, 798-813.

Friend, A.D. and Cox., P.M. 1995. Modelling the effects of atmospheric CO_2 on vegetation - atmosphere interactions. Ag Forest Meteor. (in press)

Garratt, J.R. 1993. Sensitivity of climate simulations to land-surface and atmospheric boundary-layer treatments - A review. Journ Clim **6**, 419-449.

Gorshkov, V.G., Kondratyev, K.Ya. and Losev, K.S. 1994. The natural biological regulation of the environment. Journ Ecol Chem. **3**, 85-90

Graedel, T.E. and Crutzen, P.. 1993. Atmospheric change: An earth system perspective. Freeman, New York. 446p.

Henderson-Sellers, A. 1990. Predicting generalized ecosystem groups with the NCAR GCM: First steps towards an interactive biosphere. Journ Clim **3,** 917-940.

Henderson-Sellers, A. 1993. Continental vegetation as a dynamic component of a global climate model: a preliminary assessment. Climatic Change **23,** 337-377.

Henderson-Sellers, A., Yang, Z-L, and Dickinson, R.E. 1993. The project for intercomparison of land surface parameterization schemes. Bull Amer Met Soc **74** (7), 1335-1349.

Hendrey, G.R. and Kimball, B.A. 1994. The FACE programme. Ag Forest Meteor. **70**, 3-14.

Hewitt, C.M. and Street, R.A. 1992. A qualitative assessment of the emission of non-methane hydrocarbon compounds from the biosphere to the atmosphere in the UK: Present knowledge and uncertainties. Atmos Env. **26A**, 3069-3077.

Holdridge, L.R. 1947. Determination of world plant formations from simple climatic data. Science **105**, 367-368.

Hollinger, D.Y., Kelliher, F.M., Schulze, E.-D., and Köstner, B.M.M. 1994. Coupling of tree transpiration to atmospheric turbulence. Nature, **371**, 60-62.

IPCC. 1994. Radiative forcing of climate. WMO/UNEP Intergovernmental Panel on Climate Change. Geneva.

King, G.A. and Neilson, R.P. 1992. The transient response of vegetation to climate change: a potential source of CO_2 to the atmosphere. Water Air and Soil Pollution **64** (1/2), 365-383.

Lashof, D. 1989. The dynamic greenhouse: Feedback processes that may influence future concentrations of atmospheric trace gases and climatic change. Climatic Change **14,** 213-242.

Lovelock, J.E. 1979. Gaia: a new look at life on earth. Oxford, Oxford University Press. 157p.

Mansfield, T.A., Hetherington, A.M., and Atkinson, C.J. 1990. Some current aspects of stomatal physiology. Annual Review of Plant Physiology and Plant Molecular Biology, **41**, 55-75.

Martin, P. 1993. Vegetation responses and feedbacks to climate: a review of models and processes. Climate Dynamics **8**, 201-210.

Monteith, J.L. 1988. Does transpiration limit the growth of vegetation or vice versa? Journ. Hydrol. **100**, 57-68.

Mott, K.A. 1991. Sensing of atmospheric CO_2 by plants. Plant Cell and Environment **13**, 731-737.

Murphy, J.M. 1992. A prediction of the transient response of climate. Climate Research Technical Note 32, Hadley Centre, Bracknell, UK. 27p.

Murphy, J.M. 1995. Transient response of the Hadley Centre coupled ocean-atmosphere model to increasing carbon dioxide. Part I: Control climate and flux adjustment. Journ Clim **8**, 36-56. Part II: Spatial and temporal structure of response. pp 57-80.

Naeem, S., Thompson, L.J., Lawler, S.O., Lawton, J.H. and Woodfin, R.M. 1994. Declining biodiversity can alter the performance of ecosystems. Nature **368**, 734-737.

Novakov, T. and Penner, J.E. 1993. Large contribution of organic ærosols to cloud-condensation nuclei concentrations. Nature 365, 823-826.

Prentice, I.C., Cramer, W., Harrison, S.O., Monserud, R.A., and Solomon, A.M. 1992. A global biome model based on plant physiology and dominance, soil properties and climate. Journ Biogeography **19**, 117-134.

Prinn, R.G. 1994. The interactive atmosphere: Global atmospheric-biospheric chemistry. Ambio **23**, 50-61.

Prior, S.A., Rogers, H.H., Runion, G.B. and Mauney, J.R. 1994. Effect of free-air CO_2-enrichment on cotton root growth. Ag. and Forest Meteor. **70**, 69-86.

Rind, D., Goldberg, R., Hansen, J., Rosenzweig, C. and Rudy, R. 1990. Potential evapotranspiration and the likelihood of future drought. Journ. Geophys. Res. **95**, 9983-10004.

Rounsevell, M.D.A. and Loveland, P.J. (eds) 1995. Soil responses to climate change. NATO ASI Series I Vol 26. Springer, Berlin (in press).

Salati, E. and Nobre, C. A. 1991. Possible climatic impacts of tropical deforestation. Climate Change **19**, 177-196.

Schlesinger, W.H. 1991. Biogeochemistry; an analysis of global change. Academic Press, San Diego, 443p.

Schlesinger, W.H., Reynolds, J.F., Cunningham, G.L., Huennekc, L.F., Jarrell, W.M., Virginia, R.A. and Whitford, W.G. 1990. Biological feedbacks in global desertification. Science **247**, 1043-1048.

Schneider, S.H. and Boston, P. J. (1991) Scientists on Gaia. MIT press, Boston USA. 433p. (Proceedings of AGU Chapman conference, March 1988).

Schwartz, M.D. 1992. Phenology and springtime surface-layer change. Journ Clim. **12**, 2570-2578.

Shiklomanov, I.A. (1989) Climate and water resources. Hydrological Sciences Bulletin **34**, 495-529.

Shuttleworth, W.J. 1988. Evaporation from Amazon rainforest. Proc Roy Soc London **233B**, 321-346.

Stigliani, W.M., Doelman, P., Salomons, W., Schulin, R., Smidt, G.R.B., and van der Zee, S.E.A.T 1991. Chemical time bombs: predicting the unpredictable. Environment **33**, 4-30.

Thompson, A.M. 1992. The oxidizing capacity of the earth's atmosphere: probable past and future changes. Science **256**, 1157-1165.

Tyree, M.T. and Alexander, J.D. 1993. Plant water relations and the effects of elevated CO_2: a review and suggestions for future research. Vegetatio **104/105**, 47-62.

UNEP, 1991. Environmental effects of ozone depletion: 1991 update. Nairobi, 52p.

Woodward, F.I. 1987. Climate and plant distribution. Cambridge University Press, Cambridge UK. 174p.

Woodward, F.I. and Diament, A.D. 1991. Functional approaches to predicting the ecological effects of global change. Functional Ecology **5**, 202-212.

Evaluating the Terrestrial Water Balance from the Historical Climate Record

David R. Legates
Department of Geography
College of Geosciences
University of Oklahoma
Norman, Oklahoma 73019
USA

Cort J. Willmott
Center for Climatic Research
Department of Geography
University of Delaware
Newark, Delaware 19716
USA

Introduction

Water is necessary for life on earth. Spatial patterns of both vegetation and human activities are dictated largely by the availability of water. While seasonal or monthly averaged values of precipitation, evapotranspiration, soil moisture, and runoff are useful in describing a region's climate, the intra-annual variability of these variables also plays a key role in determining the characteristics of a given climate. To assess both the mean and intra-annual variability for terrestrial regions of the globe, the historical climate record and parameterizations of the land surface must be used. Although limitations exist in both climate data and land surface parameterizations, they can be used effectively to provide assessments of the terrestrial water cycle at both global and regional scales.

As modified from Thornthwaite (*cf.*, Willmott *et al.* 1985a; Mintz and Serafini 1992; Mintz and Walker 1992), the water balance in its most basic form is

$$\partial w / \partial t = P_r + M - ET - R \tag{1}$$

where $\partial w/\partial t$ is the rate of change in soil water content over time, P_r is the rainfall rate, M is the rate of snowmelt (in liquid water equivalent), ET is the rate of water lost to the atmosphere through the combined processes of evaporation from the soil and transpiration from plants (*i.e.*, evapotranspiration), and R is the rate at which runoff is

NATO ASI Series, Vol. I 31
The Role of Water and the Hydrological Cycle
in Global Change
Edited by Howard R. Oliver and Sylvia A. Oliver

produced and includes overland flow, lateral subsurface flow, and percolation losses to groundwater. Since snow provides a storage of water on the surface which is released to the soil through melt, water stored in the snowpack is governed by

$$\partial w_s / \partial t \quad = \quad P_s \quad - \quad M \tag{2}$$

where $\partial w_s/\partial t$ is the rate of change of water stored in the snowpack over time and P_s is the snowfall rate, both given as a depth of liquid water equivalent. Snowfall can be determined if the mean shelter-height air temperature falls below a threshold temperature (*cf.*, Willmott *et al.* 1985a) when specific information on the snowfall proportion of precipitation is not available. To evaluate these fluxes of moisture using the water balance, each of these terms must be estimated or measured at the desired spatial and temporal resolution. For regional to global scale representations of the water balance, spatial resolutions may range from 4° of latitude by 5° of longitude (*e.g.*, Mintz and Serafini 1992) to 0.5° of latitude by 0.5° of longitude (*e.g.*, Legates and Mather 1992). A one month time-step if frequently used although a pseudo-daily approximation may be employed (*cf.*, Willmott *et al.* 1985). This paper will focus on the approaches adopted by Willmott *et al.* (1985), Legates and Mather (1992), Mintz and Serafini (1992), and Mintz and Walker (1993) to produce regional-scale estimates of the terrestrial water balance.

Precipitation

While many assume that precipitation is the most accurately known term of the water balance, several attempts have been made to estimate the mean terrestrial precipitation over the past forty-five years. These estimates, however, exhibit considerable differences on both annual and seasonal time scales (*cf.*, Legates 1995). For example, estimates of mean annual terrestrial precipitation vary from 665 mm to 972 mm (Table 1). Three important problems associated with gage-based measurements of precipitation are responsible for these disparate results — the estimation of areal averages from point measurements, biases associated with the gage measurement process, and homogeneity of precipitation time series. All three of these problems are critical to obtaining accurate estimates of areally-averaged precipitation.

Researcher(s)	Year		Researcher(s)	Year	
Möller	1951	665 mm	L'vovitch	1969	732 mm
Reichel	1952	671 mm	Mather	1969	712 mm
Wüst	1954	671 mm	Baumgartner/Reichel	1970, 1972	671 mm
Budydo	1956	671 mm	Budyko	1970	719 mm
Mosby	1957	686 mm	Marcinek	1973, 1976	670 mm
Albrecht	1960	671 mm	Baumgartner/Reichel	1975	746 mm
Knoch	1961	671 mm	Manabe/Holloway	1975	972 mm
Budyko	1963	719 mm	Jaeger	1976, 1983	756 mm
Strahler	1965	679 mm	UNESCO	1978	†800 mm
Kessler	1968	671 mm	Legates/Willmott	1990	754 mm
Nace	1968	671 mm	Legates/Willmott	1990	†820 mm

Table 1: Estimates of mean annual terrestrial precipitation by various researchers (from Legates 1987). †Estimates adjusted for gage measurement biases.

Estimation of Areal Averages

The areal average problem arises from the fact that precipitation is measured only at selected points (gage sites) in the region but an areally-averaged estimate of precipitation for an area exceeding 2,000 km^2 (for 0.5° of latitude by 0.5° of longitude resolution at 45° latitude) is usually required. At this scale, the density of precipitation gages is quite sparse and despite the fact that precipitation has been sampled probably more extensively than any other climatic variable, its spatial and temporal variability is relatively large. Consider, for example, the continental United States where gage densities are among the largest found anywhere in the world. At present, there are about 13,000 gages spread over 7.8 million km^2 which provides an average of one gage per 600 km^2 (Mather 1974). Moreover, these gages are not uniformly distributed over the land surface since far more gages are located along the coastlines and near urban areas. Gage densities even in the United States are quite sparse over the Rocky Mountains (where spatial precipitation variability is greatest) and in the desert southwest. This paucity of gage measurements as well as the bias in the location of gages makes accurate estimates of areally-averaged precipitation difficult to compile. Note also that gage densities in Africa, South America, and most of Asia are not nearly as dense as the United States (*cf.* Legates 1987) and such problems are greatly accentuated in these regions.

To assess the impact of varying network densities on terrestrial precipitation averages, Willmott and Legates (1991) examined differences in long-term mean precipitation when the station densities were varied. They sampled randomly from the 24,635 stations provided by the Legates and Willmott high-resolution climatology (Legates 1987; Legates and Willmott 1990) to produce sub-networks consisting of sample sizes of 200, 500, 1000, 1500, and 2000 stations (Greenland and Antarctica were excluded). Fifty random samples of each network size were taken to provide an estimate of the mean and dispersion of the areally-averaged terrestrial precipitation estimate for each sub-network size. Mean precipitation from the Legates and Willmott climatology was taken as the "best" estimate of the true mean precipitation since it is the population from which the samples were drawn. Results indicate that smaller network densities tend to overestimate mean precipitation considerably while the bias decreases as the sample size increases (Figure 1A). Willmott and Legates (1991) then computed the long-term mean terrestrial precipitation from the Legates and Willmott climatology using the yearly station distribution given by Spangler and Jenne's (1984) World Monthly Surface Station Climatology for 1900 through 1984. As the number of

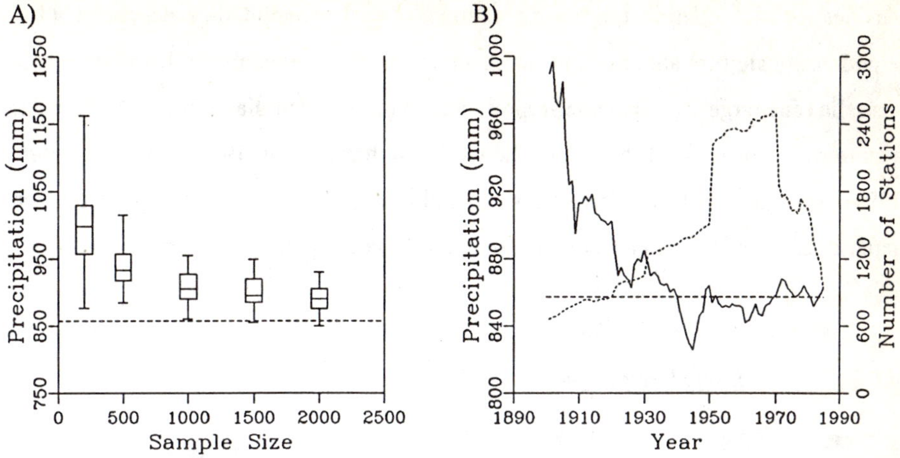

Figure 1: (A) Box plots of areally-averaged, long-term mean terrestrial precipitation obtained using gage networks of varying size. (B) Annual time series of areally-averaged, bias-adjusted terrestrial precipitation (solid line) estimated from the Legates and Willmott long-term mean precipitation climatology using the station availability for the given year (dashed line) from Spangler and Jenne (1984). The horizontal dashed line in both figures indicates the long-term mean from the Legates and Willmott climatology — the population from which the samples were drawn. (from Willmott and Legates 1991)

stations increased from 1900 to about 1945, the bias in the estimate of the long-term precipitation decreased (Figure 1B). Therefore, estimates of terrestrial precipitation may be considerably overestimated prior to the 1930s due to the global distribution of stations (Willmott and Legates 1991).

Willmott *et al.* (1994) further extended this analysis to continental precipitation averages and found that the findings of Willmott and Legates (1991) were valid for all continents except South America. South America was the lone exception where underestimates were associated with sparse gage networks. These conclusions result from the fact that with the exception of South America, gages tend to be co-located with urban development which is biased towards wetter regions. The Amazonian River Basin, however, is under-represented in most gage networks which leads to an underestimation of precipitation for South America. Using presently available gage densities for each continent, however, continental-scale precipitation averages tend to be underestimated by about 6% in South America, and overestimated by about 6% in Australia (due largely to the under-representation of the Australian Outback). The bias is quite small for most of the rest of the world (Willmott *et al.* 1994). Willmott *et al.* (1994) also found that sampling errors generally increase as mean annual precipitation increases and are larger in areas with strong precipitation and station density gradients.

Bias in Gage-Measured Precipitation

While the first source of error in precipitation measurement resulted from inadequate spatial sampling, the bias in precipitation gage measurements arises from the measuring process itself. It would seem that accurate measurements of precipitation could be easily obtained by simply placing a bucket on the ground. During the mid-nineteenth century, however, four researchers (Bache in 1838, Jevons in 1861, Symons in 1879, and Abbe in 1889) independently demonstrated that the gage catch decreases with increasing wind speed which, in turn, increases with height (Neff 1977). As a result of these observations, pit gages (raingages placed with their orifices at ground level) and gage shields (a device placed over the gage to reduce the wind flow across the orifice) were proposed to mitigate the deleterious effects of the wind. Nevertheless, pit gages are only used infrequently and gage shields, while drastically reducing the effect of the wind, do not eliminate this problem entirely (*cf.*, Goodison 1981; Peck 1993). Nearly all synoptic measurements of precipitation taken throughout the world are made with elevated gages, often without a shield, and consequently result in an underestimation of the actual precipitation.

Systematic biases in precipitation gage measurement can be attributed to seven sources — the effect of the wind, wetting losses, evaporation from the gages, splashing effects, blowing and drifting snow, the treatment of trace precipitation events, and the impact of automatic recording techniques. Random biases also exist and add an unsystematic component. The effect of the wind is the largest source of gage undercatch bias and accounts for a decrease in the annually-averaged, global precipitation of about eight percent (Legates 1987). This error tends to increase as latitude and elevation increase due to the greater impact of wind on snowfall. The reason for this wind-induced bias arises from the fact that a precipitation gage acts as an obstruction to the wind field which causes the air to flow around the sides and top of the gage. Near the gage orifice, air is forced to flow up and over the top of the gage producing a small updraft while the compression of the wind field resulting from the air flow over the top of the gage leads to an increase in the wind speed. Wind tunnel experiments have indicated that the wind speed increases by as much as twenty percent across the entire orifice (Sevruk 1988). Thus, the increase in wind speed coupled with the small updraft of wind lead to a decrease in the measured precipitation. Since national standard precipitation gages vary in size, shape, and design as well as in the elevation of their orifice above ground level, the effect of the wind is gage-dependent.

Wetting losses include moisture that adheres to the funnel during precipitation and to the collector during its emptying while evaporation from the gage may occur between the end of precipitation and its measurement by an observer. Evaporative losses, therefore, are potential problems only for gages that must be emptied and recorded manually and are largely significant only in the Tropics (Legates 1987). For the annually-averaged global precipitation, wetting and evaporative losses account for a decrease in the measured precipitation of approximately two and one percent, respectively (Legates 1987). Both effects are influenced by gage design, color, and size and therefore are gage-dependent.

Splashing losses include moisture splashed into the gage from surrounding surfaces or out of the gage from raindrops hitting the internal collecting surface. In general, however, most standard gage sitings place the gage orifice high enough above the surrounding ground so that splashing into the gage does not occur. Splashing into the gage may be problematic for pit gages although, as previously mentioned, pit gages are seldom used in standard practice (Sevruk 1979). Splashing out of the gage is rather small since nearly all gages employ a funnel which is deep enough to minimize this source of error. Consequently, splashing errors usually can be ignored except when pit gages are employed.

When a significant amount of dry snow lies on the ground and high wind speeds exist, some of the fallen snow may be lifted into the air. Under these circumstances, some gage shields act like a snow fence in trapping the blowing and drifting snow which results in an erroneous increase in the measured snowfall. This problem is significant for the Soviet Tretyakov gage in particular (Struzer and Bryazgin 1971) and may be significant in some instances for other gage shield designs as well. Nevertheless, this problem is mainly confined to high latitudes and elevations.

Mechanical procedures used in a variety of gage designs also can induce errors into the gage measurement process. Tipping bucket raingages, for example, employ a small, oscillating collector that produces an electrical signal every time a small amount of moisture falls into the collector (usually 0.01 mm or 0.01 in.). Some moisture, however, may be lost during the time it takes for the mechanism to "tip" and the other "bucket" to be positioned under the funnel, particularly during high intensity rains (*cf.*, Parsons 1941). Often the precipitation recorded by the tipping buckets can be compared with the amount of precipitation in the bottom of the collector. The discrepancy between these two measurements is the error induced by the tipping buckets and since the potential for this type of error increases with increasing rainfall intensity, the undercatch can be assigned to periods of high rainfall rates (Parsons 1941). During snowfall events, however, standard tipping-bucket gages cannot be used. To overcome this deficiency, some tipping bucket raingages are equipped with small, electrical heaters which melt the falling snow and allow its measurement by the tipping-bucket mechanism. This configuration, however, greatly enhances the evaporation of melted snow and sublimation of newly fallen snow which biases the precipitation measurement (Metcalfe and Goodison 1992).

Other gages also are adversely affected by the mechanics of the gage design. Frictional drag produced by the weighing mechanism of weighing gages, the floats of float gages, and the recording pen assemblage of most recording gages can result in a slight decrease in the measured precipitation. Some self emptying gage designs also inaccurately measure precipitation during the few moments it takes to empty the gage (Linsley *et al.* 1982). These mechanical biases generally lead to a decrease (albeit small) in the measured precipitation.

Random errors include those caused by the gage (*e.g.*, leakage or damage to the gage), the observer (*e.g.*, spilling the catch or inaccurate measuring or recording procedures), or by tampering by untrained individuals (Sevruk 1979). They are usually unsystematic biases since they sometimes result in an increase and sometimes in a decrease in the gage catch. Although some of these random errors can be significant, they usually can be minimized by frequent inspection, adequate observer training, and protecting the gage from tampering (Sevruk 1979). For long-term averages, the net effect of random errors is negligible since their expected value is zero.

Since (1) all biases except the effect of the wind are additive, (2) the effect of the wind influences the total gage catch including the wetting and evaporative losses as well as the bias induced by mechanical effects, and (3) the magnitude of these biases are different depending on whether the precipitation falls as rain or snow, Legates (1992) modified Sevruk's (1979) general model for precipitation correction to

$$\begin{aligned} P_c = \; & k_r(P_r + \Delta P_{wr} + \Delta P_{er} + \Delta P_{mr}) + \\ & k_s(P_s + \Delta P_{ws} + \Delta P_{es} + \Delta P_{ms}) - \Delta P_b \pm \Delta P_l \pm \Delta P_n \end{aligned} \tag{3}$$

where P_c is the bias-adjusted precipitation estimate, k is the wind speed deformation coefficient ($k \geq 1$) which increases as the wind speed increases, ΔP_w is the wetting loss, ΔP_e is the evaporative loss, ΔP_m is the adjustment for mechanical effects, ΔP_b is the adjustment for blowing snow, ΔP_l is the adjustment for splashing into or out of the gage, ΔP_n is the adjustment for random errors, and the additional subscripts r and s denote the adjustment for liquid (rain or drizzle) and solid (snow) events, respectively. To obtain an unbiased (or less biased) estimate of the actual precipitation for any time interval, each of the terms in this equation must be estimated although not all of them may be significant for all gages or for corrections on certain time-intervals (*cf.*, Legates 1992).

As an example of the magnitude of gage measurement biases, Legates and DeLiberty (1993a; 1993b) evaluated the biases in precipitation measurement for the continental United States. They reasoned that for the United States gage network as a whole, splashing biases, the effect of blowing snow, mechanical effects, and evaporative losses were negligible when compared to the impact of wind and wetting losses. For long-term mean values, the random bias could be neglected. On average, the bias in mean annual precipitation for the continental United States (Legates and DeLiberty (1993a) is 9% (Table 2) and a strong correlation exists with both latitude and elevation (Figure 2). In summer, the systematic undercatch in gage-measured precipitation is generally between 4% and 6% (Table 2) while slightly higher values were observed over the mountainous west. By contrast, biases in winter were much larger with the gage undercatch varying from less than 8% along the southern and western coastal areas to more than 30% in the northern Great Plains for an average bias of nearly 15% (Table 2). The strong latitudinal gradient observed by Legates and DeLiberty (1993b) is indicative of the increased effect of wind on snowfall. It should be further noted that these values are not always constant as some intra-annual variability in the bias is observed. Legates (1992) demonstrated that changes in wind speed and snowfall frequency that are easily observed in yearly fluctuations have considerable impact on the magnitude of the gage bias, particularly in winter. A more thorough discussion of the accuracy of precipitation data in the United States is given by Groisman and Legates (1994).

Homogeneity of Precipitation Time Series

The third source of bias in gage measurements is induced by inhomogeneities in the precipitation time series that may be caused by changes in instrumentation and

	Winter	Spring	Summer	Autumn	Annual
Gage-measured Precipitation (mm)	181.2	210.9	219.4	188.6	800.1
Bias-adjusted Precipitation (mm)	213.0	231.7	231.6	203.4	879.7
Gage-measurement Error (%)	14.9	9.0	5.3	7.3	9.0

Table 2: Areally-averaged precipitation estimates for the continental United States (from Legates and Deliberty 1993a,b).

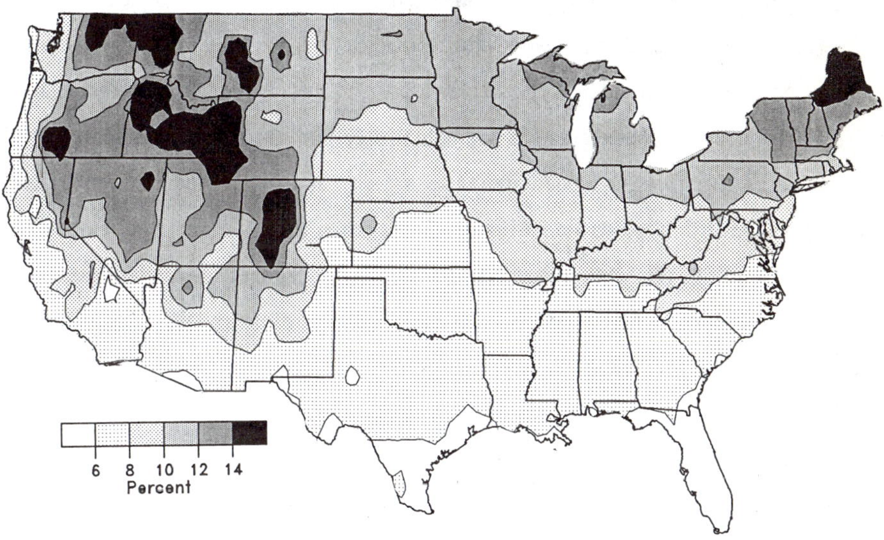

Figure 2: Biases in the gage-measured annual precipitation for the continental United States expressed as a percentage of the bias-adjusted precipitation (from Legates and DeLiberty 1993a).

recording practices, siting characteristics, and station location (Eischeid *et al.* 1991). As previously discussed, the magnitude of the gage measurement error varies with gage design and the use of a wind shield. As improvements in gage designs have been made, the adoption of new gages have introduced a marked discontinuity into the precipitation time series of various countries. For example, the Soviet Union changed from the Nipher-shielded to the Tretyakov-shielded gage between 1948 and 1953 while the United States adopted the use of Alter wind shields for some stations, particularly in the northwest, in the 1940s (Groisman 1991; Groisman *et al.* 1991). Canada, Finland, Norway, Poland, Sweden, and Switzerland also have changed gage designs or shields in the past one hundred years. Most of these innovations have increased the efficiency of the gage which, in turn, have resulted in an increase in the gage catch. This leads to an apparent increase in the precipitation for these countries.

Precipitation homogeneity also is affected by changes in the standard height of the gage orifice. Since the 1940s, the elevation of the gage for the national precipitation gage network has been decreased in both The Netherlands and China. This leads to an apparent increase in the actual precipitation since the wind-induced error increases as the wind speed increases which, in turn, increases with height. These inhomogeneities are particularly significant when the gage is moved from or to the roofs of buildings (Groisman *et al.* 1991). Similarly, changes in standard recording practices, particularly with respect to the measurement of snowfall, also adversely affect the accuracy of precipitation measurements (Goodison 1981; Metcalfe and Goodison 1992).

Over time, the environment surrounding the gage may change considerably. Bushes and trees may grow or be cut down, buildings and fences near the gage may be constructed or removed, increased urbanization may occur, or the grass under the gage may be replaced with dirt or concrete, just to provide several examples. Since each of these changes will affect the structure of the wind field and, in particular, the wind flow around the gage, they will adversely influence the gage catch (*cf.*, Brown and Peck 1962; Eischeid *et al.* 1991). Sudden changes in the gage environment caused, for example, by the cutting down of trees can produce a discontinuity in the station record which may be detectable through simple visual inspection or through double-mass analysis with a nearby station. More gradual changes such as the growth of trees over time or urban development may introduce a time-varying bias which can be difficult to detect by simply examining sudden discontinuities in the record.

It is unavoidable that some stations will be relocated, even slightly, during the time period of record. Of particular importance has been the movement of downtown (urban) stations to airports and other rural areas (Eischeid *et al.* 1991; Groisman 1991; Groisman *et al.* 1991). Such relocations can introduce a discontinuity into the station record resulting from changes in the local environment. Therefore, the combined effects of all of these potential discontinuities in the station record — instrumental and observational changes, variations in siting characteristics, and station relocation — can yield a misleading picture of temporal climate changes in precipitation.

Evapotranspiration

Continental- and global-scale analyses of the terrestrial water balance frequently have utilized some form of the Thornthwaite potential evapotranspiration concept (*cf.*, Willmott *et al.* 1985a; Legates and Mather 1992; Mintz and Serafini 1992; Mintz and Walker 1993) owing to its simple input requirements (only air temperature and precipitation) and assumptions. While the potential evapotranspiration concept may be outdated since more sophisticated and physically-realistic evapotranspiration models have been developed, they generally are not appropriate for evapotranspiration estimates at the spatial and temporal scales of interest here. This is because high-resolution global fields of net radiation, wind speed, atmospheric humidity and other specific plant canopy information are not available. Evapotranspiration estimates at this scale, therefore, must be determined from available information. Consequently, shelter-height air temperature, a surrogate for net radiation, is frequently used as a basis for estimating evapotranspiration at this spatial and temporal scale.

Modifying the Thornthwaite water budget to account for the accumulation of snow on the surface, the evapotranspiration rate becomes

$$ET = \begin{cases} P_r + M + \beta\,(ET^0 - P_r - M) & \text{if } (P_r + M) < ET^0 \\ ET^0 & \text{if } (P_r + M) \geq ET^0 \end{cases} \tag{4}$$

where ET^0 is the potential evapotranspiration rate and β is a function that specifies the ability of plants to extract the additionally needed moisture (above what is supplied by P_r and M) from the soil and is a function of the soil water content (Willmott *et al.* 1985a). Monthly potential evapotranspiration, in millimeters per month, was specified by Thornthwaite (1948) as

$$ET^0 = \begin{cases} 0 & \text{if } T < 0°C \\ [1.6\theta h\,(10T/I)^a]\,/\,360. & \text{if } 0°C \leq T < 26.5°C \\ \theta h\,(-1.155 + 0.08956T - 0.0012T^2) & \text{if } T \geq 26.5°C \end{cases} \tag{5}$$

where θ is the length of the month in days (poleward of 50° latitude, the daylength of 50° latitude must be used), h is the duration of daylight in hours on the fifteenth of the month, T is the monthly shelter-height air temperature in °C, a is an empirical constant given by

$$a = 6.75 \times 10^{-7} I^3 - 7.71 \times 10^{-5} I^2 + 1.79 \times 10^{-2} I + 0.49 \ , \tag{6}$$

and I is the heat index given by

$$I = \sum_{i=1}^{12} \left(0.2\,\overline{T}_i\right)^{1.514} \tag{7}$$

where $\overline{T_i}$ is the long-term mean monthly shelter-height air temperature for month i and the summation holds over all twelve months of the year. Thornthwaite argued that although there was a better physical relationship between net radiation and ET^0, surface air temperature could be used as a surrogate owing to the paucity of radiation, wind speed, and atmospheric humidity measurements. His relationship was obtained through measured evapotranspiration from well-watered lysimeters in the eastern and central United States and the difference between measured river runoff and monthly precipitation for selected well-watered drainage basins in the eastern United States regressed against monthly surface air temperature (Mather and Ambrosiak 1986).

While some evaluations have found that Thornthwaite's potential evapotranspiration equation provides estimates that compare rather favorably with other methods such as Jensen-Haise, Priestley-Taylor, and Penman (Mather and Ambrosiak 1986; Mintz and Walker 1993), others have argued that Thornthwaite's equation systematically underestimates potential evapotranspiration when compared to lysimeter-derived estimates (*cf.*, Willmott 1984; Willmott *et al.* 1985a). A reason for this possible underestimate was proposed by Legates and Mather (1992) where they observed that Thornthwaite used gage-measured precipitation as the input to the analysis. Thus, in order for Thornthwaite to reproduce the correct streamflow and runoff values given the biased underestimates provided by the precipitation gages, values from Thornthwaite's potential evapotranspiration equation must necessarily be underestimated to compensate for the missing water input. Moreover, Willmott (1984) notes that when the systematic bias is removed through simple linear regression, Thornthwaite's potential evapotranspiration equation provides relatively precise and probably accurate estimates commensurate with values obtained from radiation-based equations.

One problem with the use of potential evapotranspiration estimates derived from measurements taken under conditions where there is no soil moisture stress (*e.g.*, Thornthwaite's equation) is that the "apparent" potential evapotranspiration under stress conditions differs from the potential evapotranspiration under non-stressed conditions (Dooge 1992; Mintz and Walker 1993). During periods of soil moisture stress, this difference results from (1) an increase in the sensible heat flux caused by the reduction of evapotranspirative heat transfer, (2) a decrease in cloud cover due to the lower humidity and lower precipitation which, in turn, leads to an increase in insolation at the ground, and (3) decreased vegetative cover resulting in a higher surface albedo which is usually associated with low soil water amounts. The first two conditions lead to an increase in the difference between the measured shelter-height air temperature and the surface temperature (temperature of air in contact with the ground) while the third decreases the difference. Dooge (1992) pointed out that the relationship between the potential evapotranspiration under potential conditions, ET^o, and the potential evapotranspiration under non-potential (dry soil) conditions, ET_*^o, is given by

$$ET^o = \frac{1}{2}(ET + ET_*^o) \tag{8}$$

which demonstrates that the actual evapotranspiration rate and the potential evapotranspiration rate under non-potential conditions vary in an opposite manner.

To incorporate this process into the water balance equation already presented, Mintz and Walker (1993) propose that the equivalent temperature for computing the "true" potential evapotranspiration under potential conditions (in °C), T', computed from

$$T' = (0.83 + 0.17\beta)\,T \tag{9}$$

should be used in place of T (in °C) in equation (5). This will make the computed value of potential evapotranspiration commensurate with the conditions associated with Thornthwaite's definition of potential evapotranspiration. This compensates for the fact that if evapotranspiration occurred at the potential rate in an arid climate, the air temperature would be lowered. Thus, when the soil is at field capacity, T equals T' and as soil water content decreases, T-T' increases. The largest difference between ET_*^o and ET^o is 1.1 mm day^{-1} and occurs in west-central Sahara where the use of equation (9) reduces the true potential evapotranspiration under potential conditions to

5.1 mm day^{-1} (ET^{o}) from a value of 6.2 mm day^{-1} (ET_{*}^{o}) computed using the shelter-height air temperature, T. Due to the dry soil that occurs when T-T' is greatest, the actual evapotranspiration is increased by a maximum of only 0.3 mm day^{-1} (Mintz and Walker 1993).

A further limitation to the Thornthwaite definition of potential evapotranspiration is that homogeneous, closed vegetative covers seldom exist in nature. Most biomes are characterized by considerable variations in the number of species and for others, such as desert or tundra biomes, a closed vegetative cover is seldom observed. While this definition appears to be somewhat problematic, results obtained compare favorably with radiation-based methods (Mintz and Walker 1993) and particularly in areas where the gage measurement bias is minimal (*cf.*, Legates and Mather 1992). Nevertheless, a more realistic representation of the terrestrial biosphere used in the computation of evapotranspiration appears to be warranted.

Soil Moisture and Land-Surface Parameterizations

Thornthwaite's (1948) initial selection of a soil moisture resistance function, β, that specifies the ability of the plant to extract needed water from the soil was simply

$$\beta = \begin{cases} 1 & \text{for} \quad w > 0 \\ 0 & \phantom{\text{for}} \quad w = 0 \end{cases} \tag{10}$$

which indicates that the plants receive all the water the need from the soil as long as there is water available (*cf.*, Curve A, Figure 3). This seemed unrealistic since a significant plant water deficit may occur long before the soil becomes completely dry. As a consequence, Thornthwaite (Thornthwaite and Mather 1955) later replaced this equation with

$$\beta = w / w_{\max} \qquad \text{for all } w \tag{11}$$

where w_{max} is the soil moisture storage capacity in the root zone of the soil. Thus, a decreasing ability of the plants to extract needed water from the soil as soil moisture decreases exists (*cf.*, Curve B, Figure 3). A more realistic approximation of the true plant/soil relationship was developed by Nappo (1975) from the field measurements of Davies and Allen (1973) and is given by

$$\beta = 1 - \exp[-6.8\,(w/w_{\max})] \qquad \textit{for all } w \quad . \tag{12}$$

Soil moisture is readily available to the plant until it reaches about 30% of the soil moisture storage capacity. The ability to extract moisture from the soil then declines rapidly and exponentially at lower soil moistures (*cf.*, Curve C, Figure 3). This soil moisture resistance function was used by Willmott *et al.* (1985a), Mintz and Serafini (1992), and Mintz and Walker (1993) in their global water balance calculations.

Legates and Mather (1992) employed a different soil moisture resistance function that is generally more restrictive than Nappo's (1975) equation. The rationale behind their use of the function, given by

$$\beta = \begin{cases} 1 & \textit{for} \quad w > 0.7w_{\max} \\ w/(0.7w_{\max}) & \phantom{\textit{for}} \quad w \le 0.7w_{\max} \end{cases} \tag{13}$$

(*cf.*, Curve D, Figure 3), is based on the unrealistic treatment of precipitation in the simple monthly water balance. Even in studies where a pseudo-daily representation of the water balance is used (*e.g.*, Willmott *et al.*, 1985a), precipitation is assumed to fall evenly for every day of the month. One ramification of this assumption is that the soil does not dry out as much as it should since it is wetted each day and thus evapotranspiration will be overestimated and runoff will be underestimated. To compensate for this effect, Legates and Mather (1992) chose a soil moisture resistance function that makes it more difficult to extract moisture from the soil (*i.e.*, β for curve D is generally less than β for curve C). A stochastic representation of pseudo-daily precipitation, however, may be a more appropriate solution.

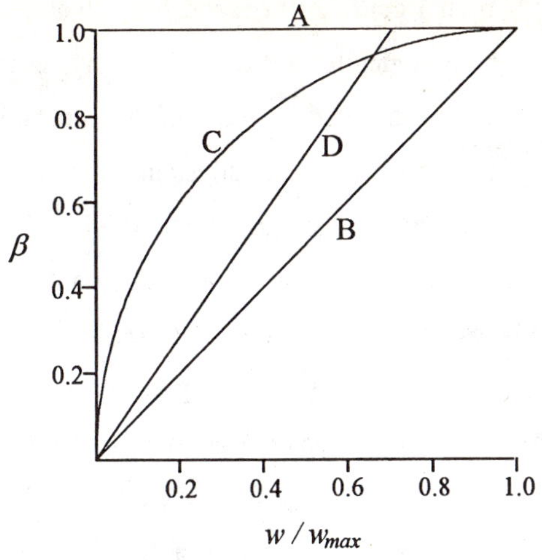

Figure 3: Soil moisture availability function, β, as a function of soil water content for four different parameterizations (after Mather 1974).

Only an additional specification of the soil moisture storage capacity in the root zone of the soil, w_{max}, is required for this simple model which often has been referred to as the Budyko or Manabe 'bucket'. In many analyses this figure is taken to be a constant of 15 cm for all locations of the globe (*cf.*, Willmott *et al.* 1985a; Legates and Mather 1992; Mintz and Serafini 1992; Mintz and Walker 1993). Milly and Dunne (1994) point out, however, that evapotranspiration and runoff estimates may be sensitive to the selection of w_{max} which is a function of the plant rooting depth and physical properties of the soil. Soil moisture storage capacity is not a constant globally since, for example, tropical rainforests have deeper rooting depths and are associated with different soils than grasslands. The selection of 15 cm probably originates with Thornthwaite's (1948) original specification but, according to Milly and Dunne (1994), may not be a bad first approximation, however. In an evaluation of the sensitivity of the global hydrologic cycle to variations in soil moisture capacity, Milly and Dunne examined a range of w_{max}'s from 1 cm to 240 cm and found that 15 cm is probably the most representative value hydrologically and, coincidentally, is the minimum value where the flux of water between the land surface and the atmosphere is maximized. This underscores the importance of plant water stress in determining the soil moisture holding capacity of the continents (Milly and Dunne 1994). Soil moisture storage capacity of the root zone, therefore, plays an important role if runoff and evapotranspiration are to be estimated accurately. Although little research has focused on properly specifying w_{max}, it is a crucial component to the investigation into hydrological and climatological interactions (Milly and Dunne 1994). Undoubtedly, future water balance analyses must treat w_{max} as a function of soil type and vegetation rather than simply specifying it as a global constant. Toward this end, a global assessment of the spatial distribution of soil moisture storage capacities is now available in digital form (Patterson 1990).

A further drawback to this simple approach is that the soil is treated as a uniform layer with no vertical stratification (*cf.*, Willmott *et al.* 1985a; Legates and Mather 1992; Mintz and Serafini 1992; Mintz and Walker 1993). Other model parameterizations, usually on much smaller spatial scales, may use two or more soil layers (*cf.*, Mahrt and Pan 1984; Kim *et al.* 1993). Evaporation can occur only from the upper soil layer and divisions between the soil layers may be made on the basis of rooting depths or soil horizons. For a global assessment, however, this information is difficult to obtain and small-scale variability exists to such a degree that the treatment of multiple soil layers globally may be impractical. As a consequence, the bucket model may be limited but it seems to be the only viable model that can be employed at this spatial scale from the available climatic data.

With respect to this spatial scale problem in hydrology, Becker and Nemec (1987) caution that there are no general laws of plant responses to environmental forcing and that both plant form and function are highly variable spatially. They cite biophysical models such as SiB and BATS that can be coupled to atmospheric general circulation models and raise the question that while they are indeed physically-based models, are they reasonable? Becker and Nemec argue that such models are useful for their conceptual value but they cannot describe the process. Considerable spatial variability in soils and vegetation exist within the resolution of global hydroclimatological analyses so that some form of spatial scaling process is needed. They propose a two-level approach: one component of the model accounts for the vertical fluxes of moisture and a second component accounts for the processes which describe horizontal flow. Their argument is that only by an explicit accounting of both vertical and flow processes and a coupling of them together can an accurate representation of soil moisture be accomplished. Nevertheless, Becker and Nemec (1987:436) conclude "from our own experience . . . models based on continuum mechanics and/or on existing knowledge of transpiration control of [the] vegetation canopy will hardly supply better results than the simplest models, such as the Budyko 'bucket' ". Bucket models are too simple but something simpler than SiB- or BATS-type models are warranted. This also is true particularly for models that simulate land-surface processes at the spatial and temporal scales discussed here.

Runoff

In the Thornthwaite water budget, runoff (or water surplus) is given simply by

$$R = \begin{cases} P_r + M - [ET + (w_{\max} - w)] & \text{for} \quad (P_r + M) > [ET + (w_{\max} - w)] \\ 0 & \phantom{\text{for}} \quad (P_r + M) \leq [ET + (w_{\max} - w)] \end{cases} \tag{14}$$

where runoff occurs only if there is excess moisture after evapotranspiration and soil moisture recharge have been satisfied (Willmott *et al.* 1985a). Runoff is usually the residual from the water budget and no distinction is made between percolation losses to groundwater, lateral subsurface flow, or overland flow. Routing of this water is not usually considered although Thornthwaite did develop a procedure to compare estimated runoff with streamflow by providing a delay in the time the water takes to reach the basin outlet (*cf.*, Mather 1978). Often, the model is validated by comparing modeled runoff with observations.

Problems associated with modeling runoff at the spatial and temporal scales of interest here arise from the fact that it is computed largely as a residual from the water balance (*cf.*, Mather 1978; Willmott *et al.* 1985a; Mintz and Serafini 1992; Mintz and Walker 1993). Inaccuracies in the measurement or estimate of any other variable (*e.g.*, precipitation, air temperature, evapotranspiration, soil moisture) or in the specification of any parameter (*e.g.*, land-surface processes, vegetation and soil characteristics) will usually result in errors in the estimate of runoff (*cf.*, McMahon *et al.* 1987; Russell and Miller 1990; Dümenil and Todini 1992; Marengo *et al.* 1993) unless these errors fortuitously offset one another. In fact, McMahon *et al.* (1987) found that the intercontinental variability of runoff is greatest in southern Africa and Australia because they have the highest variability of effective precipitation minus evapotranspiration (P_r+M–ET) and evaporative demand is high. Thus, inaccuracies in the measurement of runoff and parameter specifications may be most problematic in these regions.

Spatial resolution of a global or regional analysis, as with other hydroclimatic variables, also makes the estimation of runoff and its horizontal routing particularly problematic. For example, at a 2° of latitude by 2.5° of longitude resolution, three major rivers in the Himalayas pass through the same grid box (Russell and Miller 1990). One adaptation to account for these types of problems is to determine the proportion of runoff from a given grid box that will be routed to each of the adjacent grid boxes. This seems to be a better approach in that the simulated runoff can be more directly related to the observed streamflow. The bucket method is significantly limited in that the generated runoff cannot be related back to streamflow directly since it is assumed to disappear immediately. A limitation of all of these methods, however, including the approach of Russell and Miller (1990) and Marengo *et al.* (1993) is that no possibilities exist for subsequent evaporation while enroute to the ocean and no interaction on the water budget of downstream grid boxes is afforded.

Quality of regional and continental-scale estimates of runoff also are lacking (*cf.*, Probst and Tardy 1987; Russell and Miller 1990). An evaluation of estimates of continental-scale runoff from Baumgartner and Reichel (1975) and UNESCO (1978) indicate that considerable differences exist (Probst and Tardy 1987). For a global average, Baumgartner and Reichel estimate runoff to be 39,700 km^3 yr^{-1} while UNESCO estimates it to be 46,800 km^3 yr^{-1} — a difference of 7100 km^3 yr^{-1} (Table 3). Considerable differences between the two archives also exist for North America; 5840 and 8200 km^3 yr^{-1}, respectively. This analysis points out that our knowledge of runoff even at the continental-scale is considerably lacking.

	Budyko (1956)	Mather (1969)	Zubenok (1970)	Baumgartner and Reichel (1975)	UNESCO (1978)
Globe	244	274	310	266	294
Africa	160	129	260	114	139
Asia	220	272	290	276	293
Australia	61	92	50	27	---
Europe	241	266	250	282	319
N. America	269	247	340	242	287
S. America	490	647	930	618	583

Table 3: Estimates of runoff from selected investigations (from Legates and Mather 1992). Units are in millimeters depth.

Efforts of the Global Runoff Data Centre in Koblenz, Germany hopefully will provide an accurate and reliable database for analyses of global runoff (*cf.*, GRDC 1994). The purpose of this center, created under the auspices of the World Meteorological Organization through the World Climate Programme-Water, is to compile and provide a global database of streamflow time series. Although this center is not presently involved with research in runoff processes, the data and summary products available from this center should prove invaluable for a more proper assessment of the temporally-varying runoff from the land-surface at scales ranging from small basins to continents.

Given the inaccuracies in measuring precipitation, estimating evapotranspiration and soil moisture, and specifying land surface parameterizations, it is no wonder that runoff is so inaccurately known. Moreover, it seems that accurate estimates of regional-scale runoff will never be available until problems associated with these other variables and parameterizations are properly addressed. Dooge (1992) concludes that even though hydrologists have been aware of these problems for some time, it is unfortunate that there seems to be a significant lack of progress in accurately estimating large-scale runoff over the last decade.

Gridding Global Fields

Analysis and presentation of precipitation, evapotranspiration, soil moisture, and runoff at a global scale requires the interpolation of point values, usually surface air temperature and precipitation, to a regular lattice where the water budget then can be computed. Over the past forty years, a variety of methods have been developed to interpolate point values from irregularly-spaced data (*cf.*, Lam 1983; Bennett *et al.* 1984). These techniques include spectral filtering procedures, trend surface models, kriging, numerical approximation methods, and polyhedron techniques. While many advancements have been made in each of these methods over the years, recent developments have tended to focus more on "tuning" the methods to the problem at hand rather than making purely statistical modifications. As a consequence, the efficacy of many of these techniques has been increased greatly through the inclusion of physically-based dependencies in the interpolation algorithm.

There are many issues in associated with spatial interpolation. Topics to be discussed here include concerns associated with (1) spatial smoothing, (2) accounting for the sphericity of the earth, (3) including the spatial distribution of the observed data and extrapolating extremes, and (4) exploiting inherent dependencies with other observable data. While this is certainly not an exhaustive list of issues, these topics will serve as a introduction to some of the recent concerns that have been raised and should be considered when gridded climatological fields must be interpolated from existing irregularly-spaced data. Although the focus here is on hydroclimatological data, many of these concerns also apply to other geophysical fields which are spatially continuous and varying.

Spatial Smoothing

Most methods of spatial interpolation can be loosely grouped into two broad categories: spatial filtering techniques and local-search procedures. Most spatial filtering algorithms were developed for the interpolation of smooth fields where the low frequency spatial "signal" could be separated from the high frequency "noise". In meteorology, for example, these procedures have been used successfully to interpolate upper air fields such as 500 mb heights or wind velocities since these fields tend to be smoothly varying spatially and have little energy in the higher frequencies.

These procedures may not be very applicable for the interpolation of surface fields, however, due to the potentially large local spatial gradients that exist. Precipitation, for example, can vary considerably over small spatial scales (*e.g.*, in the Alps or the Rocky Mountains). In this case, characterization of the interpolation as the extraction of a low frequency signal from high frequency noise is inappropriate since the high frequency variations are important. Furthermore, aliasing of the higher frequencies into lower frequencies can greatly bias the results when the density of the observed data is greater than the sampling interval (*cf,* Holloway 1958; Burt 1986).

Interpolation of fields with large spatial gradients using these smoothing procedures is generally inappropriate even though they are often employed to produce areally-averaged estimates. It is more appropriate, however, to interpolate grid point estimates at a higher resolution and then average them to produce areal averages rather than assume that areal averages are commensurate with the lower frequencies.

Sphericity of the Earth

While it is obvious that the earth is a sphere, many interpolation algorithms implicitly assume that the interpolated surface is a flat plane. This assumption often is acceptable when small regions are considered and thus a cartesian approximation is reasonable. However, when large regions of the globe are to be interpolated, the "flat earth" assumption is inappropriate and considerable errors can be induced when cartesian-based interpolation procedures are employed.

An illustrative example of these potential errors in the "flat earth" assumption is given by Willmott *et al.* (1985b) where cartesian-based interpolation of mean annual surface air temperature for North America is accomplished using Shepard's (1968) nearest-neighbor algorithm for both conformal and equal-area cylindrical map projections (see their Figures 3 and 4). For these interpolations, the latitude and longitude of the observed values were projected into cartesian x- and y-coordinates using the equations for the selected map projection. Isotherms then were computed for the interpolated grid. In addition, Willmott *et al.* (1985b) modified the Shepard procedure to account for the curvature of the earth by computing distance, direction, and spatial gradients in spherical coordinates. These interpolated isotherms subsequently were projected and plotted together with the cartesian-based isotherms. One of the most striking features is that considerable differences exist between isotherms computed using the two cartesian-based projections and those computed from the spherically-based algorithm. Differences of more than 10°C can be found over Baffin Island, for example, for both map projections and these differences increase as the distance to the nearest observed value increases. A further problem is that in both cartesian-based interpolations, isotherms extend through the North Pole — the −10°C isotherm for Miller's and the −10°C, −5°C, and 0°C isotherms for Lambert's projection. Since the pole is, in reality, a single point on the earth's surface, isolines which pass through the pole, particularly when there are more than one and they have different values, are suspect. This is because the −10°C isoline, for example, delineates areas with air temperatures greater than −10°C from areas with temperatures less than −10°C. Multiple isolines passing through the pole indicates that the interpolation provides different values for the pole which obviously cannot be realistic.

Since the cartesian-based interpolation using Miller's projection differs considerably from the cartesian-based interpolation using Lambert's projection, the choice of a map projection may greatly affect the results (*cf.*, Willmott *et al.* 1985b). To account for the curvature of the earth, therefore, the interpolation must be computed in spherical coordinates <u>before</u> projecting the results since all map projections introduce significant distortion somewhere on the map. This issue is becoming widely recognized and several spherical adaptations of cartesian-based algorithms have been developed (*e.g.*, Wahba 1981; Renka 1984; Sardeshmukh and Hoskins 1984; Willmott *et al.* 1985b).

<u>Incorporating the Spatial Distribution of Data and Extrapolation of Extremes</u>

Many interpolation algorithms only incorporate distances from the observed data to the interpolate point but do not adequately account for the spatial distribution of the observed data. It is imperative, however, that the interpolation procedure account for the spatial structure of the observed data since, as Shepard (1968) demonstrated, serious biases can be induced if it is ignored. Consider, for example, a point that is to be interpolated from ten observed points — a small cluster of nine points located to the east and a single, lone point located at the same distance to the west. If distance is the sole factor in determining the value of the interpolate point, each observed point will be given approximately the same weight since their distances from the interpolate point are approximately the same. However, the lone point represents a measurement of the field to the west of the interpolate point while the cluster of nine points represents measurements to the east. Since spatial correlation exists, which is the impetus behind spatial interpolation, each point in the cluster to the east are not independent measurements. Thus, an equal weight to all ten points is inappropriate and the interpolation algorithm must account for the spatial scatter of the observed data.

A similar concern extends to the estimation of local maxima and minima (*i.e.*, the tops of peaks and the bottom of valleys) of the interpolate field by simple inverse distance weighted interpolation procedures since observations seldom exist at these locations. If only distance is considered, estimates of the extrema are limited to the maximum or minimum of the observed data. It is important, however, to consider not just the spatial distribution of the observed locations, but also the spatial gradients in the neighborhood of these observed data. Building on these considerations, Shepard (1968) modified the simple inverse-square distance weighted approach to include both the angular distribution of the observed data and the estimation of local spatial gradients. These modifications have resulted in an interpolation procedure that has proven to be very robust and has been widely used.

Comparisons of various techniques for interpolating surface fields indicate that local search procedures such as Shepard's (1968) method compare favorably with or even exceed the ability of other techniques. Bussières and Hogg (1989), for example, determined that the error associated with Shepard's method was less than that of other techniques for interpolating daily rainfall and was commensurate with a statistical procedure similar to Gandin's optimal interpolation (Table 4). Weber and Englund (1992) also observed that inverse-square distance weighting (similar to Shepard's method but without the correction for the spatial distribution of the observed data and

	Shepard	Barnes	Cressman	Statistical
Mean Bias (mm)	0.4	0	0.4	0
Root Mean Square Error (mm)	3.2	3.4	3.7	3
Mean Correlation	0.7	0.67	0.5	0.71

Table 4: Comparison of various interpolation algorithms to interpolate daily rainfall (from Bussières and Hogg 1989).

the extrapolation of extremes) was superior to other techniques, including kriging. Hamilton *et al.* (1988) similarly found that a method of distance weighting was preferable to other widely-used techniques to interpolate air temperature, precipitation, and solar radiation. Thus, Shepard's (1968) method appears to be a reasonable procedure to interpolate surface fields. Adaptations to this method, such as the optimization of the exponent of the distance weighting function to estimate surface air temperature and precipitation by Legates (1987), greatly increases the efficacy of this technique.

Exploiting Dependencies with Other Variables

Interpolated estimates should not be determined solely on the basis of mathematical and statistical considerations but other dependencies, both climatic and non-climatic, should be incorporated in the interpolation process where applicable. Surface air temperature, atmospheric pressure, and precipitation, for example, exhibit strong dependencies with both latitude and topography. Evaporation and transpiration also are tightly coupled to the surface air temperature, net radiation, land use, and vegetation cover. As examples of how these relationships may be included in the interpolation procedure, Isaaks and Srivastava (1989) have exploited these types of dependencies through cokriging which utilizes the cross-correlations between the interpolate variable and another correlated variable. Doswell (1977) modified the Barnes interpolation procedure to take advantage of the temporal as well as the spatial correlation structure present in meteorological fields. Similarly, Willmott and Robeson (1994) recognized that the spatial patterns are relatively consistent for a given field but the mean value may change over time. To exploit this fact, they developed an interpolation procedure that uses a long-term mean climatology at a relatively high spatial resolution to spatially interpolate monthly time series of air temperature from a lower resolution climatology. Inclusion of these inherent dependencies into the interpolation process will most certainly enhance the reliability of the interpolated field.

In addition to these explicit dependencies, some statistical parameters associated with the interpolation also may be field-dependent or vary spatially. Inclusion of these influences has been accomplished, for example, by Stephens and Stitt (1970), Ross (1986), Tronci *et al.* (1986), and Legates (1987). In these applications, the optimum radius of influence (or correlation distance) for several different interpolation procedures was derived from the observed data rather than based on some à priori assumption. A significant improvement in the interpolated values was observed in each of these studies. These results indicate, therefore, that the radius of influence and other statistical parameters should be determined by the data rather than based on purely mathematical or statistical considerations.

With the increased use of computers and geographic information system (GIS) software packages, the temptation to treat the interpolation of hydroclimatological data as a "black box" process has increased. This is unfortunate since the accuracy of the interpolated estimates can be greatly enhanced by tailoring the interpolant to exploit some of the information about the physical characteristics of the field. As a result, no interpolation procedure possibly ever will exist that is applicable for all fields, but rather, an appropriate interpolant will exploit as much available information about the field as possible. Statistical methodologies then should be used only as a vehicle to produce the best interpolated estimates. Consequently, interpolation on purely statistical grounds is no longer sufficient; the researcher must play an active role in tailoring the interpolation process to the problem at hand.

Recommendations

Over the years, many evaluations of the global water balance have been made. High spatial variability in precipitation, evapotranspiration, soil moisture, and runoff as well as soil and vegetative characteristics have required gross simplifications. Nevertheless, a representative picture of the global water balance has emerged despite the limitations in observational measurements, estimation techniques, sub-scale variability, and spatial interpolation. It is important, however, to recognize these limitations and to further refine our methods to improve our understanding of the global water balance.

Future advances in estimating the global water balance from observational data most certainly will include better representations and parameterizations in nearly all areas. Inaccuracies in the measurement of precipitation must be properly taken into account, particularly in areas where snowfall is significant. Moreover, a stochastic approach to the representation of precipitation in pseudo-daily water balance models is preferable to assuming precipitation falls equally every day of the month. Better evapotranspiration models should be developed for use with observational data at the spatial scales of interest here, particularly since the Thornthwaite potential evapotranspiration has indirectly incorporated the bias in precipitation measurement. Although many other models exist, they are limited in that they require measurements of variables, such as wind speed, atmospheric humidity, net radiation, and detailed soil and canopy parameters which are not available at high resolutions like air temperature. Spatially variable water holding capacities also will be incorporated now that global maps are available. More sophisticated routing methods for modeling runoff should be easily implemented through the emerging Geographic Information System (GIS) technology. Biases and inaccuracies introduced by the gridding process should be more properly addressed now that techniques for ameliorating their adverse impact have been developed. Moreover, estimates of the temporal and spatial (sub-grid scale) variability as well as the reliability of the estimated fields are now required in addition to the mean values. Hopefully, the observation of Dooge (1992) that a significant lack of progress has been observed over the last ten years will not be applicable to the coming decade.

References

Baumgartner A, Reichel E (1975) *World Water Balance: Mean annual global, continental, and maritime precipitation, evaporation, and runoff,* Elsevier, Amsterdam.

Becker A, Nemec J (1987) Macroscale hydrologic models in support to climate research. *The Influence of Climate Change and Climatic Variability on the Hydrologic Regime and Water Resources,* IAHS Publication No. 168, 431–445.

Bennett RJ, Haining RP, Griffith DA (1984) The problem of missing data on spatial surfaces. *Annals of the Association of American Geographers*, **74**:138–156.

Brown MJ, Peck EL (1962) Reliability of precipitation measurements as related to exposure. *Journal of Applied Meteorology,* **1**:203–207.

Budyko MI (1956) *Heat Balance of the Earth's Surface* (translated in 1958 by NA Stepanova). United States Weather Bureau, Office of Climatology, Washington, DC.

Burt JE (1986) Time averages, climatic change, and predictability. *Geographical Analysis,* **18**:279–294.

Bussières N, Hogg W (1989) The objective analysis of daily rainfall by distance weighting schemes on a mesoscale grid. *Atmosphere Ocean,* **27**:521–541.

Davies JA, Allen CD (1973) Equilibrium, potential and actual evaporation from cropped surfaces in southern Ontario. *Journal of Applied Meteorology,* **12**:649–657.

Dooge JCI (1992) Hydrologic models and climate change. *Journal of Geophysical Research,* **97**:2677–2686.

Doswell CA (1977) Obtaining meteorologically significant surface divergence fields through the filtering property of objective analysis. *Monthly Weather Review,* **105**:885–892.

Dümenil L, Todini E (1992) A rainfall-runoff scheme for use in the Hamburg climate model. *Advances in Theoretical Hydrology, A Tribute to James Dooge,* O'Kane JP (ed.), European Geophysical Society Series on Hydrological Sciences 1, Elsevier.

Eischeid JK, Diaz HF, Bradley RS, Jones PD (1991) *A Comprehensive Precipitation Data Set for Global Land Areas.* United States Department of Energy Monograph TR051, DOE/ER-69017T-H1, Washington, DC.

Global Runoff Data Centre (1994) *Global Runoff Data Centre Status Report No. 4*, Global Runoff Data Centre, Federal Institute of Hydrology, Koblenz, Germany.

Goodison BE (1981) Compatibility of Canadian snowfall and snow cover data. *Water Resources Research,* **17**:893–900.

Groisman PYa (1991) Unbiased estimates of precipitation change in the Northern Hemisphere extratropics. *Fifth Conference on Climate Variations,* American Meteorological Society, Denver, Colorado.

Groisman PYa, Legates DR (1994) Accuracy of historical United States precipitation data. *Bulletin of the American Meteorological Society,* **75**:215–227.

Groisman PYa, Mason EH, DelGreco S (1991) Metadata for construction of homogeneous long-term precipitation time series. *Proceedings, Seventh Conference on Applied Climatology,* American Meteorological Society, Salt Lake City, Utah, 119–122.

Hamilton JEM, Lennon P, O'Donnell B (1988) Objective analysis of monthly climatological fields of temperature, sunshine, rainfall percentage and rainfall amount. *Journal of Climatology,* **8**:109–124.

Holloway JL Jr (1958) Smoothing and filtering of time series and space fields. *Advances in Geophysics,* **4**:351–389.

Isaaks EH, Srivastava RM (1989) *An Introduction to Applied Geostatistics.* Oxford University Press, Oxford.

Kim J, Ek M, Lee RL (1993) A long-term simulation of surface fluxes and soil moisture. *Sixth Conference on Climate Variations,* American Meteorological Society, 407–410.

Lam NS-N (1983) Spatial interpolation methods: A review. *The American Cartographer,* **10**:129–149.

Legates DR (1987) A climatology of global precipitation. *Publications in Climatology,* **40**(1).

Legates DR (1992) The need for removing biases from rain and snowgage measurements. *Proceedings, Snow Watch '92,* Canadian Climate Centre and World Meteorological Organization, Niagara-on-the-Lake, Ontario, 144–151.

Legates DR (1995) Global and terrestrial precipitation: A comparative assessment of existing climatologies. *International Journal of Climatology,* forthcoming.

Legates DR, DeLiberty TL (1993a). Estimation of biases in precipitation gage measurements: An example using the United States raingage network. *Proceedings, Eighth Symposium on Methods of Instrumentation* and *Proceedings, Eighth Conference on Applied Climatology,* American Meteorological Society, Anaheim, California, J48–J51.

Legates DR, DeLiberty TL (1993b) Precipitation measurement biases in the United States. *Water Resources Bulletin*, **29**:855–861.

Legates DR, Mather JR (1992) An evaluation of the average annual global water balance. *Geographical Review*, **82**:253–267.

Legates DR, Willmott CJ (1990) Mean seasonal and spatial variability in gauge-corrected, global precipitation. *International Journal of Climatology*, **10**:111–127.

Linsley RK Jr, Kohler MA, Paulhus JLH (1982) *Hydrology for Engineers*, McGraw-Hill, New York.

Mahrt L, Pan H-L (1984) A two-layer model of soil hydrology. *Boundary Layer Meteorology*, **29**:1–20.

Marengo J, Russell G, Rosenzweig C, Abramopoulos F, Miller J (1993) Calculations of river-runoff in the GISS GCM: Modeling the hydrology of the Amazon River. *Proceedings, Sixth Conference on Climate Variations*, American Meteorological Society, 395–398.

Mather JR (1969) Average annual water balance of the world. *Proceedings, Symposium on Water Balance in North America*, American Water Resources Association, Banff, Alberta, 29–40.

Mather JR (1974) *Climatology: Fundamentals and Applications*. McGraw-Hill, Inc., New York.

Mather JR (1978) *The Climatic Water Budget in Environmental Analysis*. Lexington Books, DC Heath & Company, Lexington, Massachusetts.

Mather JR, Ambroziak RA (1986) A search for understanding potential evapotranspiration. *Geographical Review*, **76**:355–370.

McMahon TA, Finlayson BL, Haines A, Srikanthan R (1987) Runoff variability: A global perspective. *The Influence of Climatic Change and Climatic Variabilty on the Hydrologic Regime and Water Resources* Proceedings of the Vancouver Symposium, IAHS Publication Number 168.

Metcalfe JR, Goodison BE (1992) Automation of winter precipitation measurements: The Canadian experience. *Proceedings, WMO Technical Conference on Instruments and Methods of Observation*, WMO/TD No. 462, 81–85.

Milly PCD, Dunne KA (1994) Sensitivity of the global water cycle to the water-holding capacity of land. *Journal of Climate*, **7**:506–526.

Mintz Y, Serafini YV (1992) A global monthly climatology of soil moisture and water balance. *Climate Dynamics*, **8**:13–27.

Mintz Y, Walker GK (1993) Global fields of soil moisture and land surface evapotranspiration derived from observed precipitation and surface air temperature. *Journal of Applied Meteorology*, **32**:1305–1334.

Nappo CJ (1975) Parameterization of surface moisture and evaporation rate in a planetary boundary layer model. *Journal of Applied Meteorology*, **14**:289–296.

Neff EL (1977) How much rain does a rain gage gage? *Journal of Hydrology*, **35**:213–220.

Parsons DA (1941) Calibration of a Weather Bureau tipping-bucket gage. *Monthly Weather Review*, **69**:205.

Patterson, KA (1990) *Global Distributions of Total and Total-Available Soil Water-Holding Capacities.* MS Thesis, University of Delaware, Department of Geography.

Peck EL (1993) Biases in precipitation measurements: An American experience. *Proceedings, Eighth Symposium on Methods of Instrumentation,* American Meteorological Society, Anaheim, California.

Probst JL, Tardy Y (1987) Long range streamflow and world continental runoff fluctuations since the beginning of this century. *Journal of Hydrology,* **94**:289–311.

Renka RJ (1984) Interpolation of data on the surface of a sphere. *ACM Transactions on Mathematical Software,* **10**:417–436.

Ross SG (1986) Interpolation for contouring: An experiment in weighting functions. *Geo-Processing,* **3**:179–194.

Russell GL, Miller JR (1990) Global river runoff calculated from a global atmospheric general circulation model. *Journal of Hydrology,* **117**:241–254.

Sardeshmukh PD, Hoskins BJ (1984) Spatial smoothing on the sphere. *Monthly Weather Review,* **112**:2524–2529.

Sevruk B (1979) Correction of point precipitation measurement. *Versuchsanstalt für Wasserbau, Hydrologie und Glaziologie an der ETH Zürich,* **41**:267–279.

Sevruk B (1988) Towards the universal precipitation gauge of the future. *Vaisala News,* **113-114**:12–14.

Shepard D (1968) A two-dimensional interpolation function for computer mapping of irregularly-spaced data. *Proceedings, Twenty-Third ACM Conference,* 517–524.

Spangler WML, Jenne RL (1984) World monthly surface station climatology. National Center for Atmospheric Research, Boulder, Colorado.

Stephens JJ, Stitt JM (1970) Optimum influence radii for interpolation with the method of successive corrections. *Monthly Weather Review* **98**:680–687.

Struzer LR, Bryazgin NN (1971) Method of computing corrections to atmospheric precipitation measurements in polar regions. *Soviet Hydrology, Selected Papers,* **1**:38–47, translated from *Transactions of the Voyeykov Main Geophysical Observatory (Trudy GGO),* **260**:61–76.

Thornthwaite CW (1948) An approach toward a rational classification of climate. *Geographical Review,* **38**:55–94.

Thornthwaite CW, Mather JR (1955) The water balance. *Publications in Climatology,* **8**(1).

Tronci N, Molteni F, Bozzini M (1986) A comparison of local approximation methods for the analysis of meteorological data. *Archives for Meteorology, Geophysics, and Bioclimatology, Series B,* **36**:189–211.

UNESCO (1978) *World Water Balance and Water Resources of the Earth.* UNESCO Series Studies and Reports in Hydrology No. 25, Leningrad.

Wahba G (1981) Spline interpolation and smoothing on the sphere. *SIAM Journal of Scientific and Statistical Computing,* **2**:5–16 and Erratum, **3**:385–386.

Weber D, Englund E (1992) Evaluation and comparison of spatial interpolators. *Mathematical Geology,* **24**:381–389.

Willmott CJ (1984) On the evaluation of model performance in physical geography. *Spatial Statistics and Models*, Gaile GL and Willmott CJ eds., D. Reidel, Dordrecht, The Netherlands.

Willmott CJ, Legates DR (1991) Rising estimates of terrestrial and global precipitation. *Climate Research,* **1**:179–186.

Willmott CJ, Robeson SM (1994) Climatologically aided interpolation (CAI) of terrestrial air temperature. *International Journal of Climatology,* submitted.

Willmott CJ, Robeson SM, Feddema JJ (1994) Estimating continental and terrestrial precipitation averages from rain-gauge networks. *International Journal of Climatology,* **14,** forthcoming.

Willmott CJ, Rowe CM, Mintz Y (1985a) Climatology of the terrestrial seasonal water cycle. *Journal of Climatology,* **5**:589–606.

Willmott CJ, Rowe CM, Philpot WD (1985b) Small-scale climate maps: A sensitivity analysis of some common assumptions associated with grid point interpolation and contouring. *The American Cartographer,* **12**:5–12.

Zubenok LI (1970) Refined water balance of continents. *Soviet Hydrology: Selected Papers*, 6:516–517.

Hydrology in Climate Models and Effects on Climate

Peter R. Rowntree
Hadley Centre for Climate Prediction and Research, Meteorological Office, Bracknell, England

Lydia Dümenil
Max-Planck-Intitut für Meteorologie, Hamburg, Germany

1. Introduction

The main focus of this paper will be the representation of the relevant processes in climate models so far used in the prediction of global change. The topic has a large and steadily expanding literature and this paper can do no more than indicate general approaches, with any detail restricted to one or two models.

The ocean, the main source of water for the rest of the system, is regarded as outside the scope of this Workshop, so it is natural to start with the atmosphere which is responsible for transport of water from the ocean to the other parts of the surface. This is particularly appropriate because of the role of the atmosphere in modulating not only the precipitation input but also the radiative forcing at the land surface through the radiative effects of clouds, trace gases including water vapour, and aerosols.

At the land surface (Fig. 1), the vegetation and soil further modify the hydrological and radiative inputs and control the partitioning of both energy and water between turbulent fluxes back to the atmosphere and transfers into the ground. The near-surface layers, as well as storing water and influencing the surface exchanges through root extraction of water and nutrients, transfer water to rivers; these transfer part of their load back to the oceans though with some loss en route through evaporation, especially from lakes and swamps. The existence of water in frozen form on land as seasonal snow cover, ice sheets and permafrost is particularly important in the context of global change because warming would generally be expected to reduce it.

Some of these topics are as yet only represented in a rudimentary fashion in climate models. This includes snow, soil freezing and the representation of spatial heterogeneity in turbulent fluxes, percolation and runoff; ice sheets are simply

NATO ASI Series, Vol. I 31
The Role of Water and the Hydrological Cycle in Global Change
Edited by Howard R.Oliver and Sylvia A.Oliver

prescribed. The performance of climate models in the simulation of aspects of the hydrology will be assessed, though in many areas validation data are scanty.

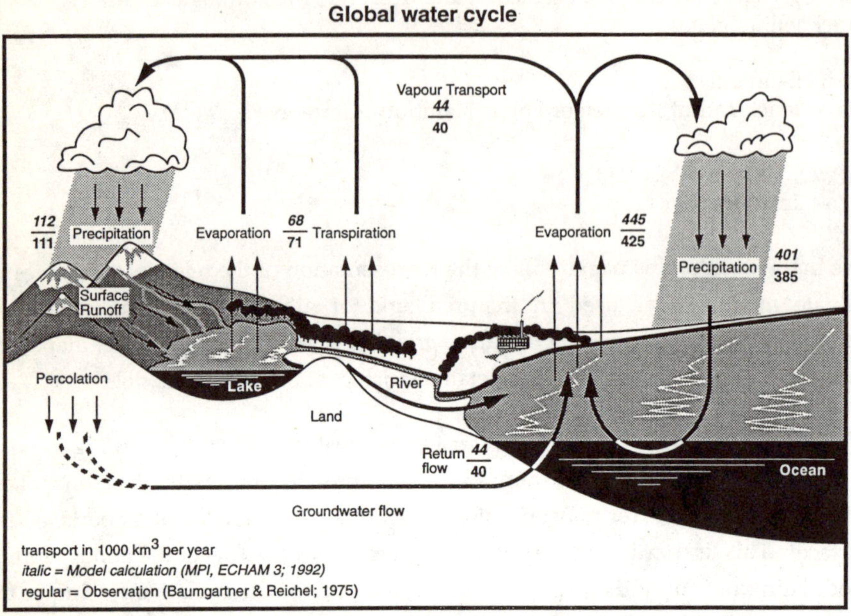

Fig. 1 The global water cycle. Comparison between observations (Baumgartner and Reichel, 1975) and a model simulation using the MPI model.

Climate prediction requires consideration of modelling of changes in the boundary conditions. The prediction of vegetation through inclusion of models of carbon assimilation is just one example of this. Atmospheric water vapour concentration and soil water content have long been regarded as model variables; vegetation is now beginning to be transferred from the list of parameters to that of variables, and soil characteristics are a future candidate. Prediction of sea level changes will require inclusion of models of ice sheets in the climate models.

The need for improving parametrizations and data sets of land surface characteristics and for new submodels can be assessed using sensitivity studies, made either with simple (land surface only or a single column) models or with full climate models. These can also provide indications of the possible causes of observed climate change such as the drought in the African Sahel or the potential climatic effects of changes in land use which are anticipated and already partly in progress such as tropical deforestation. These studies can also provide valuable insights into the role of the hydrological cycle in global change and are thus of particular relevance in this volume.

2. Water in the atmosphere

2.1 Atmospheric moisture

We start with water in the atmosphere so picking up the hydrological cycle en route from ocean to land. Water is advected in the atmosphere by the horizontal and vertical air currents, and this is represented in the models by the dynamical equations. These can be in spectral or finite difference form whereas the rest of the model - the physics - is performed on a finite difference grid. There are various problems which are encountered - concentrations of water in its various forms - vapour, cloud water and cloud ice, precipitation - whilst always remaining positive, vary over a wide range of magnitudes, so that care is needed to avoid negative values. Whilst this can be guaranteed by special approaches (predicting log q or using positive-definite advection schemes), it is more usual to use simple advection schemes and remove any negative values that may arise, whilst making adjustments (preferably locally) to maintain conservation of total moisture.

2.2 Clouds and precipitation

The model must represent the phase changes of water in the atmosphere which generate cloud and precipitation. This is one of the most difficult parametrizations to get right - we shall present for the Meteorological Office Hadley Centre (MOHC) model just the bare bones of the probably most important case, the formation of layer clouds of water droplets and their conversion to rainfall - thus the important (and considerable) complications due to the ice phase are completely omitted.

The model assumes a distribution of water vapour about its mean value which is triangular in shape with a spread about the mean of

$$\pm (1 - RH_c)\, a_L\, q_{sat}\, (T_L, p) \tag{1}$$

where $$a_L = \frac{1}{1 + \frac{L}{c_p} \cdot \frac{dq_{sat}}{dT}}$$

and RH_C is a critical humidity below which no cloud forms, typically set to about 0.95 in the lowest layers and 0.85 otherwise. From this can be calculated the fraction of the water vapour which is supersaturated and so is assumed to be in liquid form.

The precipitation processes then deplete this at a rate which should depend on the spectrum of drop size. This is not measured routinely but is normally expected to be biassed more to the small drops where there are large numbers of cloud condensation nuclei; the present assumption is that these are more plentiful over land than over sea; the rainout is therefore assumed to be more rapid over sea where droplets are fewer and larger on average. The equation used (Smith, 1990) is:

$$R_C = c_T \left\{ 1 - \exp\left[-\left(\frac{\rho q_L}{C \cdot c_w} \right)^2 \right] \right\} + \frac{P}{c_A} \quad (2)$$

where $c_T = 10^{-4} s^{-1}$, $c_w = 8 * 10^{-4} kg\ m^{-3}$ over land, $2 * 10^{-4} kg\ m^{-3}$ over sea and $c_A = 1\ kg\ m^{-2}$. P is rainfall entering from above, C is cloud amount and q_L is cloud liquid water.

The rain then falls towards the surface but is subject to evaporation as it falls. The equation governing this is of the form (Gregory, 1994):

$$\frac{dM}{dt} = \beta(\rho_v - \rho_{vs}(T)) \quad (3)$$

where ρ_V is the vapour density of water (with subscript s indicating saturation) and β is calculated assuming some dropsize distribution and fall-speeds related to drop size.

Having reached the land surface, the rainfall is then subject to evaporation on the surface of the vegetation canopy or soil. We postpone discussion of this as it depends on the energy balance at the surface, which is considered later. Part of the rainwater penetrates the vegetation canopy if present and reaches the surface as throughfall.

2.3 Movement of water in the soil and runoff

The movement of water in the soil is considered in the MOHC and some other land surface schemes including BATS (Biosphere Atmosphere Transfer Scheme) (Dickinson et al., 1986) and SiB (Simple Biosphere Model) (Sellers et al., 1986) in terms of some form of Darcy's equation. As an example, we will describe the new MOHC 4-layer scheme now in use in some experiments. The fractional volume water content θ is predicted by the flux convergence equation:

$$\frac{d\theta}{dt} = -\frac{df}{dz} - dE \tag{4}$$

where dE is the evaporative extraction per unit depth - considered later - and z is depth (positive downward). The downward flux

$$f = \frac{d}{dz}\left(K\left(\frac{d\psi}{dz} - 1\right)\right) \tag{5}$$

where K is the hydraulic conductivity, and ψ is the soil water suction or matrix pressure head.

The boundary condition at the surface is :

$$f = P_T - Y_G \tag{6}$$

where P_T is throughfall and Y_G is the surface runoff - we shall discuss this shortly. The bottom boundary condition is of free drainage or

$$f = -K \tag{7}$$

The surface runoff in such a scheme should occur when the top layer becomes saturated, or $\theta = \theta_{sat}$. However, near the surface, the saturated conductivity (K_s) can vary strongly over a shallow depth; for example, observations on pasture in the Brazilian Amazon show much smaller values within a few cm of the surface than is typical of the main body of the soil. In the MOHC scheme, therefore, a different value is used at the surface and the surface runoff is calculated separately as the excess above maximum surface infiltration.

The main problem area with this parametrization is defining the forms for the hydraulic conductivity K and the soil suction ψ. A commonly used approach is that proposed by Clapp and Hornberger (1978), which expresses them in terms of a normalised soil moisture $s=\theta/\theta_{sat}$

$$K = K_s\, s^c \tag{8}$$

$$\psi = \psi_s\, s^b \tag{9}$$

In the MOHC scheme we use a slightly more complicated scheme, that due to Mualem and van Genuchten (van Genuchten et al., 1991), which can give very similar values except near saturation (Fig. 2). A point worthy of note is that the exponents b and c in these equations are typically very large, of order 10 or more especially for clayey soils. Consequently, the drainage, though large - perhaps over 1 m/day - near the saturated value, decreases rapidly with decreasing s.

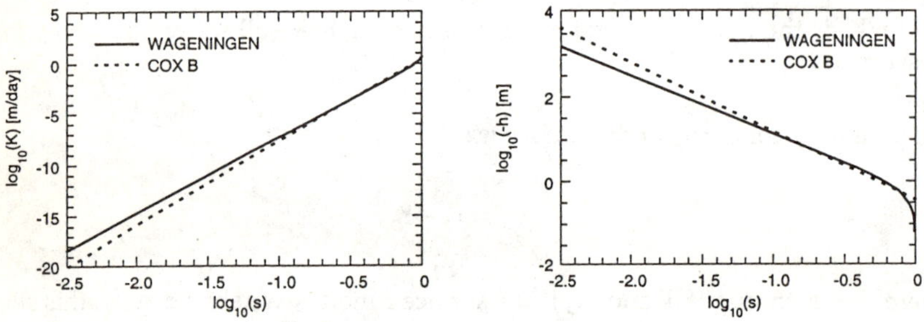

Fig. 2 Comparison, for (a) hydraulic conductivity and (b) soil suction, of the Clapp and Hornberger (1978) exponential fit (equ. 8, 9) (labelled "Cox B") to the more complicated van Genuchten et al. (1991) relation (labelled "Wageningen") over the range from s=0.003 to s=1 where s is soil moisture as a fraction of its saturation value.

An example of a simpler model is that of a single layer soil, still used in most experiments with the MOHC model (Warrilow et al, 1986), in which the equations reduce to the boundary conditions and ψ does not appear. The depth of the soil is limited to the rootzone so neglecting upward movement of water against gravity into the rootzone. Vertical transport in vapour form is ignored in most models; this can be important in winter when freezing prevents movement of water.

2.4 Representation of heterogeneity in rainfall and runoff

Fifteen years ago, hydrologists were very sceptical about the prospects of representing hydrology on the scales required in climate models; the detail they could see on the scale of a few metres seemed impossible to represent in a model with mesh size of hundreds of kilometres. (Some hydrologists are just as sceptical today!) Not only are soils very variable, but there are many other sources of heterogeneity - for example, the distribution of rainfall and snowmelt both due to random variations (showers) and forced (orographic), the effects of orography in redistributing the water and leading to great variations in soil moisture - for example between a ridge and a valley bottom, as well as differences in soil moisture due to evaporation contrasts due to cloudiness and slope induced variations in the radiative energy available for evaporation.

To meet these very real concerns, the more advanced models do attempt to represent aspects of this heterogeneity. One approach is the Arno or Nanjing model which assumes a distribution of soil moisture over the gridbox or catchment from small values up to saturation so that any precipitation produces runoff over part of the area and infiltrates over the rest, bringing an additional fraction to saturation. This has been used in the Hamburg GCM by Dümenil and Todini (1992) and in single column runs with the MOHC model (Rowntree and Lean, 1994). In a recent intercomparison of hydrology schemes funded by the CEC, Polcher et al (1995) showed that major differences in the runoff between the Hamburg and the LMD scheme were attributable to the inclusion of the Nanjing scheme in the Hamburg model (Fig. 3).

In the MOHC scheme Warrilow et al. (1986) made some allowance for the heterogeneity of rainfall by assuming it to have an exponential frequency distribution over a fraction of the gridbox, with no rainfall over the rest of the box. This allows rainfall over some fraction of the box to exceed the maximum infiltration rate and so generate surface runoff. It was assumed that rainfall at so high a rate would saturate the surface layer sufficiently quickly that there was no need to include a dependence on soil moisture. Even with this assumption, surface runoff is generally very small in the model except locally on clayey soils.

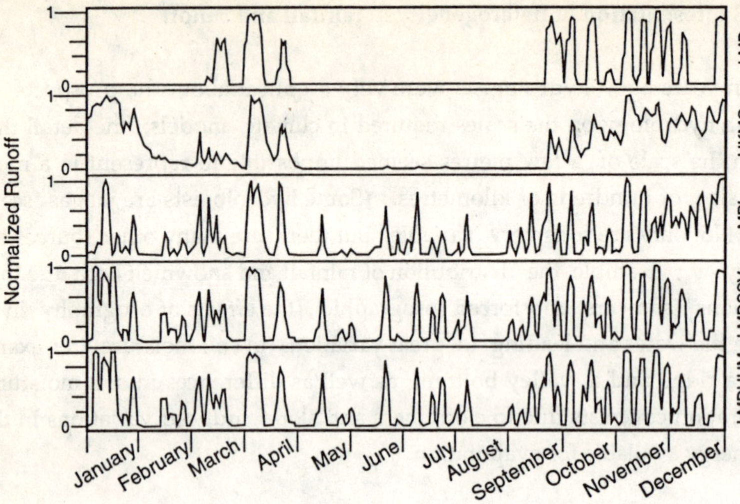

Fig. 3 Annual cycle of infiltration excess runoff as simulated by five models (Polcher et al, 1995)

a) LMD model

b) UKMO model

c) MPI model

d) LMD incl. Arno model of MPI

e) LMD incl. a different bucket size

Entekhabi and Eagleson (1989) allowed for heterogeneity both in the precipitation rate and in the soil moisture in the MIT model tested in the CEC intercomparisons mentioned above.

2.5 Lateral flow and river discharge

Once water has drained out of the surface layers, it is commonly ignored in climate models, although it is deposited in the ocean in the GFDL coupled model (Manabe et al., 1991) and in the latest version of the MOHC coupled model. Sausen et al.'s (1994) modelling of the river discharge in the Hamburg model is an exception. They include an advection of water between gridboxes using a flowspeed

$$u = \frac{c}{\Delta x}\left(\frac{dh}{dx}\right)^{a} \tag{10}$$

where dh/dx is the orographic gradient in the x-direction; calibration led to values of 4.2 m/s for c and 0.18 for a for a mesh width of 5.625°. This scheme is being developed further to use values of c and a derived for each major catchment.

3. Modelling the role of energy in hydrology

3.1 Water vapour and clouds

The energy balance is essential to the land surface hydrology because of its role in evaporation. It also drives the atmospheric motions which transport the water vapour around the globe and Fig. 4 , an approximate summary of the global heat balance, shows that the latent heat released when forming clouds and precipitation is on average the larger of two major terms (latent heat release (90 W/m^2) and solar atmospheric heating (68 W/m^2) approximately balancing the longwave radiative cooling of the atmosphere (174 W/m^2). Moreover, this radiative cooling is largely associated with two atmospheric trace gases, one of which is water vapour - the other is carbon dioxide. Water vapour accounts for much of the longwave cooling and 95% of the shortwave heating of the atmosphere (Fouquart et al., 1991). Possibly even more important than these, clouds considerably modify the radiative balance of the Earth, being responsible for about half of the reflected solar radiation of 105 W/m^2 (e.g. Slingo and Slingo, 1988) and reducing the outgoing longwave flux by about 30 W/m^2 (Slingo and Slingo, 1988). The longwave cooling effect is obtained by raising the level of emission from near the surface to cloudtop, and correspondingly weakening the emission (related to σT^4 where σ is Stefan Boltzmann's constant) as it comes from air with a lower temperature (T).

A full discussion of radiation schemes is beyond the scope of this paper. However it is worth noting, in the context of global change, that atmospheric pollutants can modify the surface radiative balance directly. For example, doubling of CO_2 reduces solar surface flux by about 0.5 W/m^2, and the increased water vapour associated with a warming of about 3K would decrease it further by about 2 W/m^2 (Mitchell et al., 1987). Changes in net surface longwave radiation approximately balance these. Increasing sulphate aerosols have been estimated to reduce surface solar flux by around 1 W/m^2 in the global mean and by up to about 4 W/m^2 locally.

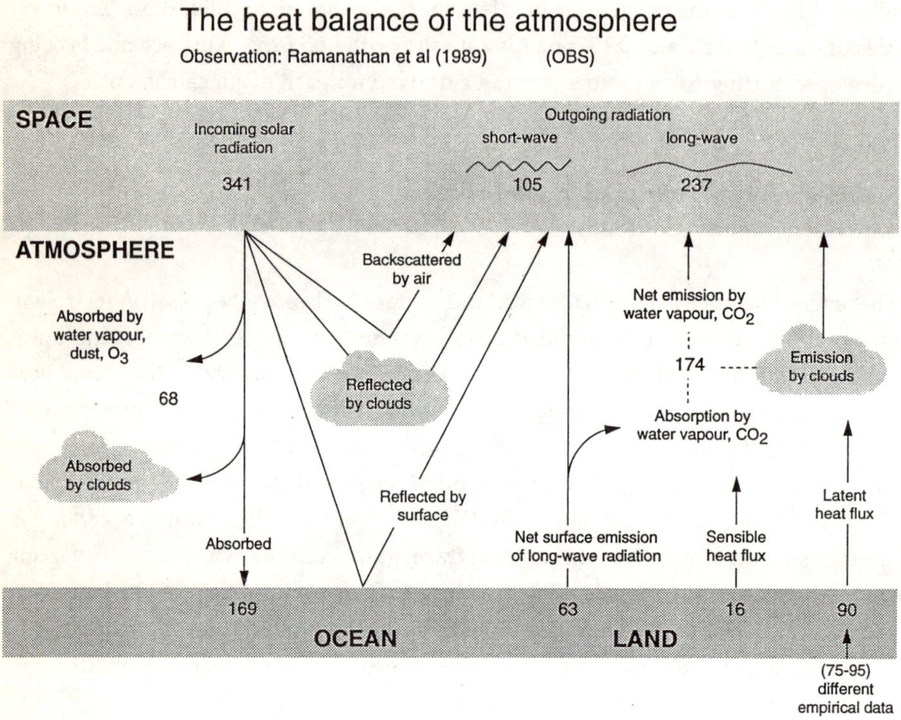

Fig. 4 The heat balance of the atmosphere (after Ramanathan et al. (1989))

The net radiation (R_N) at the surface can be represented as

$$R_N = (1-\alpha_s)R_{SD} + \varepsilon_D R_{LD} - \varepsilon_U \sigma T_*^4 \tag{11}$$

- where α_S is the albedo for solar radiation, the ε are the emissivities for upward and downward longwave radiation, σ is the Stefan Boltzmann constant and T_* is the surface temperature. R_{SD} and R_{LD} are the downward solar and longwave fluxes respectively.

3.2 Surface solar radiation

The surface albedo for solar radiation (α_S) has commonly been prescribed on the basis of observational studies. We will consider the possibility of using satellite data later. Table 1 shows data as used in the BATS , Dickinson et al (1986). Note the large contrast between the albedos for radiation of wave lengths above and below 0.7μm - typically the latter (near infrared) are around 5 times larger than the visible albedos, though for deserts the ratio is only about 2; the main exception is for snow for which the albedos are larger in the visible (see also Kondratyev et al. (1982)).

SELECTED BATS ALBEDOS

(Dickinson et al., 1986)

	< 0.7 μm	> 0.7 μm
DRY SOIL	0.10-0.23	0.20-.046
WET SOIL	0.05-0.12	0.10-0.24
SHORT GRASS	0.10	0.30
EVERGREEN NEEDLELEAF	0.05	0.23
EVERGREEN BROADLEAF	0.04	0.20
DESERT	0.20	0.40
TUNDRA	0.09	0.26
ICE CAP	0.80	0.60

Table 1

The spectral distribution of the radiation reaching the surface also varies, principally due to the preferential absorption and reflection by clouds; clouds typically absorb mainly in the infrared (above about 0.9μm) but reflect more in the visible. A much greater fraction of the diffuse beam is in the visible than for direct radiation. For example, for cloud with effective droplet radius of 7μm and a mean transmissivity of about 35% (typical of altostratus cloud), the transmissivity in the visible and near infrared (wavelength less than 1.19μm) is about 38%, while for longer wavelengths it is about 20% (Fig. 5). This means that the albedo cannot be regarded as a constant even for the same surface with the same solar elevation. In particular, the albedo of vegetated surfaces will be lower for diffuse radiation while for snow it will be higher. An alternative to prescribing the albedo for vegetation is to calculate it.

This is done in the SiB scheme (Sellers et al., 1986), so allowing account to be taken of varying solar zenith angles and seasonal variations associated with leaf loss and changes in leaf characteristics. It will also be necessary in models where the vegetation is predicted. The calculation requires knowledge of the leaf albedo characteristics and the typical plant structure inclu-ding the number of leaf layers and the orientation of the leaves.

3.3 Surface terrestrial radiation

Most models assume the surface emissivity for terrestrial or longwave radiation to be unity. The error in doing this is the spectral integral of the product of (1-ε) and the difference between the incoming radiation and the blackbody surface radiation. A first approximation to the error can be obtained as the product of the error in the blackbody emissivity (1-ε) and the net surface longwave radiation (70 W/m^2).

The resulting estimate would generally be an upper bound for the following reason. The integrated absorptivity ε_D for the downward radiation from the atmosphere (better thought of as the complement of the albedo (1-ε_D)) is not the same as the integrated emissivity ε_U which enters into the expression for the emission from the surface, even though, by Kirchhoff's law, the emissivity and absorptivity are the same at any given wavelength. This is because the spectrum of the downward radiation is well below the black-body spectrum of emission at the wavelengths (about 8 to 13μm, the "water vapour window") where water vapour emits only weakly. Temperature decreases with height through the troposphere so that in the window the incoming radiation comes from more remote and so colder emitters. Thus it is the emissivity at these wavelengths which dominates the net surface longwave radiation in Eq. (11). The above error estimate is thus generally too high because emissivity is relatively high in the water vapour window. Kondratyev et al (1982) quote typical emissivities in the window of 0.95 to 0.985. If a typical value is 0.97, the error is about 2 W/m^2, a small error, but one of consistent sign.

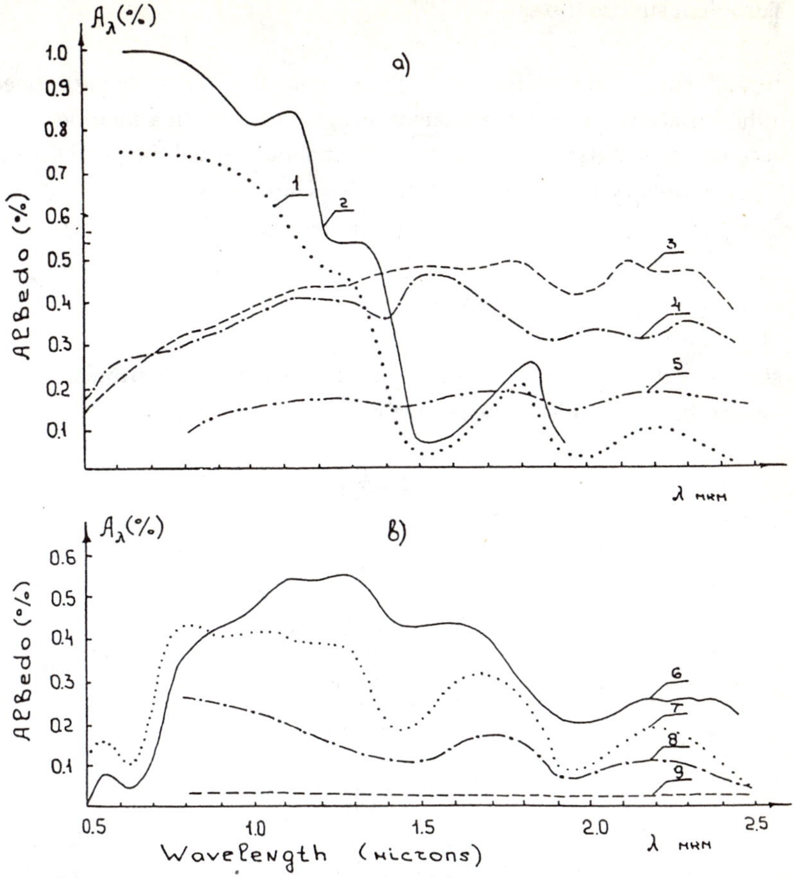

Fig. 5 The wavelength dependence of the spectral albedo for some natural formations (Kondratyev et al., 1982)

1 - new snow, h_o = 20-25°
2 - snow cover
3 - sandy loam, h_o = 25-30°
4 - dry sand, h_o = 40-45°
5 - chernozem, sandy loam
6 - blossoming meadow, h = 40-45°
7 - green leaf
8 - green coniferous forest
9 - water surface

4. Turbulent surface fluxes

The typical surface net radiation (R_N) of about 100 W/m^2 is partitioned between the surface sensible (H) and latent (LE) heat fluxes. In a long period mean, the flux into the soil (G*) is very near zero; it should equal the geothermal flux which is generally much less than 1 W/m^2. However, at any one time it also contributes significantly to the surface heat balance which is thus

$$R_N = H + LE + G. \tag{12}$$

The simpler schemes used in climate models express the turbulent fluxes of momentum, heat and moisture respectively as

$$\tau = \rho C_M V|\underline{V}| \tag{13}$$

$$H = \rho c_p C_H V \delta T \tag{14}$$

$$E = \rho C_E V \delta q \tag{15}$$

where $C_X = k^2/[\ln(z_r/z_{oM})\ \ln(z_r/z_{oX})]$ with k = 0.4. Here

$$\delta T = T_* - T_r \left(\frac{p_x}{p_r} \right)^{0.286} \tag{16}$$

is the potential temperature difference between the reference height and the surface, and

$$\delta q = q_s(T_o) - q_r \tag{17}$$

is the specific humidity difference between the reference height and the leaf interior.

The equations for the turbulent heat fluxes can also be written in resistance form as

$$H = \frac{\rho c_p \delta T}{r_{AH}} \tag{18}$$

$$E = \frac{\rho \delta q}{(r_{AE} + r_s)} \tag{19}$$

where r_{AH} and r_{AE} are aerodynamic resistances between the air at a reference height z_r and the air very near the surface, and r_S an additional stomatal resistance between the interior of the leaf and the leaf space. Schemes such as BATS and SiB which represent an explicit canopy include separate aerodynamic resistances beween the foliage and the canopy space, between the understorey and the canopy space and between the canopy space and the reference height, with appropriate additional canopy space variables. Models employed for climate studies have, as discussed by Milly (1992), used a parametrization for evaporation

$$E = \frac{\beta \rho \delta q}{(r_{AE} + r_s)} \tag{20}$$

with

$$\delta q = q_A - q_{sat}(T_s) \tag{21}$$

where T_S is the modelled surface temperature and the factor

$$\beta = \frac{w}{w_{crit}} \leq 1 \tag{22}$$

with w as soil moisture and w_{crit} a critical value of w, is used to represent the limitation by soil moisture. Milly points out that this formulation is inconsistent because the form of β is based on observed data analysed (Priestley and Taylor, 1972) in terms of a δq in which T_W, the temperature of a wet evaporating surface, appears rather than T_S, the surface temperature modelled for the (only partly wet) actual model surface. He concludes that this inconsistency initially gives excessive evaporation; this tends to dry out the soil too quickly, so providing a negative feedback which limits the subsequent evaporation. He also suggested a simple approximate means of correcting for the inconsistency and concluded that the inconsistency should not have grossly distorted the results of GCM experiments.

4.1 The aerodynamic resistance (r_A)

The resistances to the turbulent fluxes are ($1/C_X V$) where the C_X are the transfer coefficients for momentum, sensible heat and moisture in the absence of a stomatal resistance. According to Monin-Oboukhov theory, in the neutral (zero vertical density gradient) case these can, as noted earlier, be written

$$C_x = \frac{k^2}{\left[\ln\left(\frac{z_r}{z_{om}}\right)\ln\left(\frac{z_r}{z_{ox}}\right)\right]} \tag{23}$$

Thus the aerodynamic resistance for heat transfer can be expressed as $1/C_H V$. The magnitude of the wind speed V used in calculating r_A depends on the z_o for momentum. For a given geostrophic wind V_G, increasing z_o tends to decrease V but increase C_X. For moisture, the relation of r_{AE} to z_o can be calculated using Rossby similarity theory (Rowntree, 1991). Table 2 shows a selection of values.

	z_{oE} (m)				
	10^{-4}	10^{-3}	10^{-2}	10^{-1}	1
V_G = 10	141	103	70.9	44.6	24.2
V_G = 2.5	525	388	257	160	86.0

Table 2: Variations in r_A (s/m) for selected z_{oE} (m) and geostrophic wind (V_G, m/s) for Coriolis parameter $f = 10^{-4} s^{-1}$ (assumes $\ln(z_{oM}/z_{oE}) = 2$)

It shows that approximately r_{AE} varies inversely with V, while r_{AE} decreases roughly linearly as $\ln(z_o)$ increases. z_o for heat and moisture have been shown to be smaller than that for momentum, particularly in regions of rough orography. For most natural homogeneous surfaces, there is a body of literature suggesting that the ratio of the z_o value for momentum transfer to that for heat and moisture is of order 10 (Garratt et al., 1993) although much larger ratios have been suggested; Table 2 was calculated assuming a ratio of 7.3. Similar calculations with the roughness lengths equal give values of r_A smaller by from 15% with z_{oE} of 10^{-4} m to 45% with 1m.

For non-neutral conditions, the transfer coefficients increase slowly for unstable profiles, and decrease rapidly with increasing stability. In the Met Office model, stability is expressed in terms of the bulk Richardson number

$$Ri_B = \frac{\left(\frac{d\theta_v}{dz}\right)}{\left(\frac{dV}{dz}\right)^2} \tag{24}$$

where the differentials with height z are taken from the surface to the lowest model level, and θ_V is the virtual potential temperature (this allows for effects of moisture as well as temperature on the density) and V is wind speed.

There are problems with the concept of the roughness length for heat transfer associated with the different (molecular) character of the transfer very close to the surface where the temperature gradients are steepest. Koster and Suarez (1992a) used an additional resistance for heat transfer between the surface and the near-surface air similar to that used for moisture as discussed below.

4.2 Surface resistance for moisture transfer (r_S)

The surface resistance r_S is the stomatal resistance for dry, freely transpiring vegetation. The stomata control the transfer not only of water vapour but also of carbon dioxide for use in photosynthesis. The version of the MOHC model in current use specifies stomatal resistances for each vegetation type together with a dependence on soil moisture through a β factor. However, there is ample evidence in support of a formulation of r_S taking account of atmospheric stresses on the plant, and so introducing a dependence on radiation, temperature and vapour pressure deficit. To achieve this, the formulation used in BATS (Dickinson et al. 1986) includes dependences on visible solar radiation, temperature and soil moisture available to the roots . More recently similar equations, also including vapour pressure deficit, have been fitted to data for Brazilian tropical forest (Dolman et al., 1991) and post-deforestation pasture (Wright et al., 1994); these equations have been used in recent experiments with the MOHC model to estimate the effects of deforestation on climate.

These dependences appear to represent the external effects of the plant physics; an alternative approach to the problem is to represent this physics explicitly; this would have the advantage of also providing the dependence on CO_2. An assumption currently being tested for possible use in the Hadley Centre vegetation model is that the stomatal resistance is such as to optimise net photosynthesis (photosynthesis minus dark respiration) through the stomatal control of plant water potential and internal carbon dioxide concentration. Photosynthesis is favoured by increasing both the former - achieved by closing the stomata - and the latter - achieved by opening the stomata. The calculated carbon budget can be used to estimate the leaf area index which in turn can modify the albedo, roughness and interception capacity. Preliminary results with the scheme in the MOHC single column version showed that the response to doubling of CO_2 was considerably modified; instead of the intensification of the hydrological cycle obtained when stomatal resistance was held constant, a substantial decrease in precipitation and evaporation was obtained (Friend and Cox, 1995) (Table 3).

4.3 The role of land surface characteristics - evidence from simple models

Climate models have demonstrated substantial sensitivity to land surface characteristics as will be discussed in a later section. To appreciate the mechanisms involved, it is instructive to consider the basic equations controlling the surface fluxes and the use of simple (single column) models. The equations for the surface turbulent fluxes and surface energy balance can be combined to give the Penman-Monteith equation for the evaporative fraction (Monteith, 1973; for detailed derivation, see for example Rowntree (1991)):

$$\frac{LE}{(R_N - G_*)} = \frac{1 + \rho c_p \dfrac{\delta' q}{[(R_N - G_*) r_A \Delta]}}{1 + \left(\dfrac{c_p}{L\Delta}\right)\left(1 + \dfrac{r_S}{r_A}\right)} \qquad (25)$$

Here, Δ is the gradient of the saturation specific humidity q_S with temperature and $\delta' q$ is the vapour pressure deficit $(q_S(T_A) - q_A)$ at the atmospheric lowest or reference level.

RESULTS OF SINGLE COLUMN TESTS

a) CONSTANT STOMATAL RESISTANCE

		$1*CO_2$	$2*CO_2$	Change
Surface temperature	(K)	292.9	295.4	+2.5
Rain (= Evaporation)	(mm/day)	4.1	4.4	+7%
Transpiration	(mm/day)	3.1	3.3	+6%
GPP	gC/m^2/day)	8.2	11.0	+34%
LAI		6.2	7.5	+21%

b) SIMPLIFIED PGEN

		$1*CO_2$	$2*CO_2$	Change
Surface temperature	(K)	293.1	295.1	+2.0
Rain (= Evaporation)	(mm/day)	4.1	3.2	-22%
Transpiration	(mm/day)	3.1	2.3	-25%
GPP	(gC/m^2/day)	8.2	9.8	+19%
LAI		6.0	6.6	+10%

Table 3

Using this equation it is possible to demonstrate how the fraction of the available energy used for evaporation (or evaporative fraction) ($LE/(R_N-G)$) varies with changes in the aerodynamic and stomatal resistances (Rowntree, 1991). An interesting conclusion of the analysis illustrated by Fig. 6 and 7 is that there exists a temperature-dependent asymptotic value of the evaporative fraction which is approached (a) if $r_S=0$ and vapour pressure deficit tends to zero , (b) as r_S approaches a critical value for given temperature and $\delta q/(R_N-G)$ or (c) as r_A becomes large. The critical r_S is well within the range of observed r_S so care is needed in interpreting the response of evaporation to changes in roughness - opposite signs of variation will be found for evaporation with large r_S (e.g. near the wilting point) to those obtained for a wet canopy - while near the critical value a misleading impression can be obtained of small dependence on z_O. Evaporation was found to be much more sensitive to changes in stomatal resistance when r_A is small - i.e. for strong winds or rough surfaces.

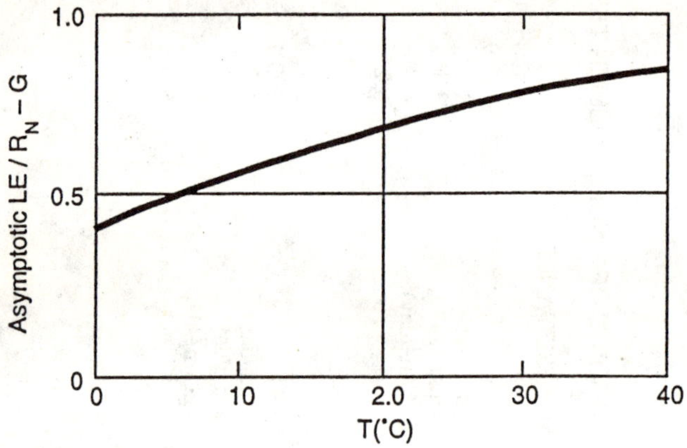

Fig. 6 Dependence of the asymptotic value of the evaporative fraction ($LE/(R_N-G)$) on temperature (Rowntree, 1991).

The Penman-Monteith equation can also be used to assess the effect of albedo changes on the evaporative fraction. However, a weakness of the approach is that atmospheric feedbacks are excluded. Thus the vapour pressure deficit (vpd) has to be prescribed whereas one would expect it to increase if, say, evaporation was increased leading to moistening of the boundary layer.

It was thus not possible to have confidence in the magnitudes of the changes in evaporation which were found. Experiments with the MOHC single column model for summer conditions in southern England have now been used to investigate the importance of these atmospheric feedbacks. Both daily mean and early afternoon model data showed that, with changes in albedo from 0.1 to 0.3, vapour pressure deficit $\delta'q$ tended to decrease only slightly less quickly than (R_N-$G*$). Consequently, neglect of variations in $\delta'q$ in particular, and also in Δ, is not justifiable. The rate of change of LE with (R_N-$G*$) from these experiments for southern England in summer was about 0.7 in the absence of limitation of evaporation by soil moisture deficits - only slightly smaller than the value of LE/(R_N-$G*$) itself. The estimate with the appropriate value of r_S/r_A (=2) from Rowntree (1991) taking account only of the change in (R_N-$G*$) was below 0.5.

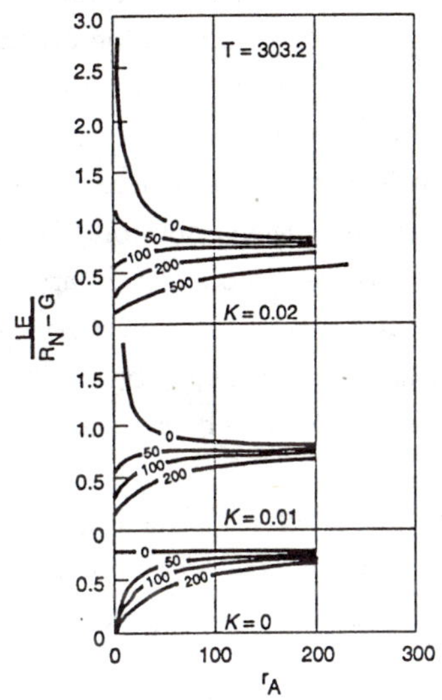

Fig. 7 The evaporative fraction (LE/(R_N-G)) as a function of the atmospheric resistance r_A for various values of $K = \frac{\delta' q}{(R_N - G)}$ and stomatal resistance r_S for a near-surface temperature of 303.2K or 30°C (Rowntree, 1991).

4.4 Representation of heterogeneity in evaporation

The typical climate model gridbox covers a region of scale 300 km or more. Within such an area there can exist a wide range of surface types with different characteristics. In some cases, particularly if surface characteristics do not differ strongly, surface parameters can be aggregated (Claussen, 1990; Blyth et al., 1993). For instance albedo can simply be linearly averaged, whereas for roughness length a more complex aggregation is necessary (Taylor, 1987; Mason, 1988; Claussen, 1990).

Generally, the calculation of regional surface fluxes is really difficult because fluxes and driving mean gradients are nonlinearly dependent. For example, where parts of the area are snow-covered with surface temperatures held at the freezing point, the mean vertical temperature structure over the area may imply a downward heat flux but, because the transfer coefficients are larger in the snowfree, statically unstable part of the area, the actual mean heat flux can be upwarded. Moreover, if surface types strongly differ 'parameter aggregation', as defined above, becomes unfeasible. For example, definition of an aggregated soil temperature diffusivity does not make sense if parts of the area consist of bare soil, where heat conduction is diffusive, and of open water, where heat can be advected horizontally as well as vertically by thermohaline circulation. In these, more complex landscapes, the method of 'flux aggregation' is recommended (Avissar and Pielke, 1989; Claussen, 1991). Flux aggregation implies computation of surface fluxes for each type in a gridbox separately. Consequently, a regional surface flux is obtained by a linear average.

The methods of parameter aggregation and flux aggregation are identical for roughness length and momentum in the case of a neutrally stratified boundary layer (Claussen, 1994a). Generally, a combination of parameter and flux aggregation is the most promising method (Claussen, 1994a).

Mahrt (1987) suggested use of an area-averaged exchange coefficient to relate the gridbox-mean fluxes to the gridbox-mean gradients, based on observed spatial distributions of Richardson number. In particular, his work implies that heterogeneity leads to some turbulence being present even for surface layers which are very stable in the mean;. this is taken account of in many climate models or weather forecast models by retaining non-zero transfer coefficients for very stable profiles, where Monin-Oboukhov theory would indicate a zero flux. However, as mentioned above, this method is not the most general one.

There are two different approaches to parameter or flux aggregation. Firstly, Wieringa (1986) and Mason (1988) suggested computing surface fluxes at the so-called blending height where the surface-layer flow changes from equlibrium with the local surface to horizontal homogeneity. Claussen (1990, 1991), Wood and Mason (1991), Dolman (1992), and Blyth et al. (1993) extended the concept of blending height to the aggregation of surface conductance and to stratified boundary layers. Moreover, Claussen and Klaassen (1992) and Claussen (1994b) included the effect of bluff roughness elements on regional momentum and energy surface fluxes using the concept of blending height. This concept has been incorporated into a mesoscale model (Claussen, 1994a), the Hamburg climate model (Grötzner et al., 1994), and a coupled sea-ice mixed-layer ocean model (Stössel and Claussen, 1993).

Another approach to the problem of heterogeneity involves division of the gridbox into a number of parts, each with different surface characteristics. Examples are the "tile" or "mosaic" scheme used by Koster and Suarez (1992a) with several surface types in a gridbox, and Ducoudre et al. (1993), whose SECHIBA scheme allows up to eight. Koster and Suarez (1992b), Blyth (1994) (Fig. 8) and Klaassen and Claussen (1994) point out that this approach is more suited to large-scale heterogeneity, whereas the "mixture" approach adopted in SiB is appropriate for smaller scale surface variability. Koster and Suarez (1992b) used their model in a comparative analysis and concluded that evaporation was larger, but only slightly, with the mixture approach. Both approaches assume the gridbox air at the lowest model level to have the same properties - i.e. to be well-mixed - whereas the blending height theory indicates that the boundary layer is modified to an increasing depth as the horizontal scale of the heterogeneity increases. An example of the use of the mosaic approach is the analysis by Taylor (1994) of evaporation from a partially wet canopy. He found that use of a single effective surface temperature as in the simpler climate model schemes overestimates the evaporation from the wet part of the canopy by more than it underestimates that from the dry part, the overall overestimate being by 15-20%.

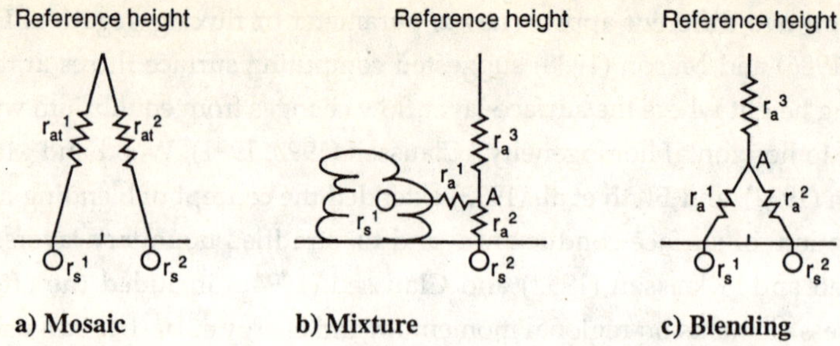

Fig. 8 Resistances to fluxes for (a) tile model, (b) sparse canopy model, (c) three resistance models.

A still unsolved problem concerns the effect of persistent subgrid-scale secondary circulations which can be generated at surface inhomogeneities at scales of the order of 10 km and larger. The development of such circulations is amply documented as in the case of boundaries between ice and icefree areas (Wefelmeier and Etling, 1991), between dry and wet areas (Schädler, 1990), and between grassland and forests (Pinty et al., 1989). Theoretical considerations (Schumann, 1992) and numerical experimentation (Claussen, 1994a) seem to indicate that secondary circulations strongly affect the development of the planetary boundary layer but only marginally influence area-averaged surface fluxes. This issue, however, deserves further attention.

5. Soil thermal processes

Heat storage in the ground has a significant role in the model surface energy balance on daily time scales with downward fluxes of tens of W/m^2 during the day and return heat flow at night. It can also be significant on annual time scales with fluxes exceeding 10 W/m^2 in middle and high latitudes. The flux is a function of the vertical temperature gradient in the soil

$$G = -\lambda dT/dz \tag{26}$$

where λ is the thermal conductivity of the soil. The temperature prediction equation is

$$C dT/dt = -dG/dz \tag{27}$$

with C as the thermal capacity. C and λ vary relatively little between soil types - from coarse (sandy) soils to fine (clayey) soils C varies from 1.9 to 2.9 $Jm^{-3}K^{-1}$ and λ from 0.5 to 0.7 $Jm^{-1}K^{-1}s^{-1}$ in the Met Office model. However, there is much greater variation with soil wetness. For example, for two soils measured during HAPEX MOBILHY, heat capacity varied linearly by a factor of about 2 and conductivity by a factor of from 5 (loam) to 10 (sand) from dry to saturated. Not all models include this moisture dependence - for example, the MOHC model still uses intermediate values. The inclusion of the dependence for conductivity appears particularly important.

The vertical resolution of the soil thermal model also varies between models. The MOHC model constructed by Warrilow et al. (1986) has 4 layers with layer thickness designed to represent variations on timescales from a few hours to several years. The actual thicknesses are expressed as a function of the thermal diffusivity. For a medium soil, the layer depths are about 3.5, 14, 49 and 154 cm, giving a total depth of about 220 cm. The Hamburg climate model uses a spacing of 6.5, 25.4, 91.3, 290.2, 570.0 cm.

Freezing of water in the soil influences the temperature changes in winter. many climate models make no allowance for the freezing of soil. It is represented in BATS, all but 15% of the water freezing progressively between 0 and -4°C. The Canadian Land Surface Scheme (CLASS, Verseghy, 1991) freezes soil water at 0°C; representation of freezing over a wide temperature range is not attempted because of the coarse vertical resolution and the uncertainty about actual behaviour.

6. Cryosphere

6.1 Snow

Having touched on the change of phase of water in the soil, we should now give some consideration to the representation of snow in climate models. Mostly this has been rather rudimentary. The separate existence of snow in the atmosphere, with different fall rates from that of rainfall and correspondingly different evaporation rates may be represented. Models vary in this. However, all models include snow on the ground: precipitation falling with air temperatures near or below freezing point accumulates as snow on the surface, and is melted when the temperature starts to rise above freezing point so that the surface temperature is restricted at or below zero until all the snow is melted. With favourable conditions - a near-surface humidity mixing ratio below the saturation mixing ratio at surface temperature - snow can sublimate while, with the opposite state, frost is deposited and adds to the mass of "snow". Snow may be allowed to influence the heat flux into the ground - this can also be quite crude as in the MOHC model where it reduces the thermal conductivity of the top "soil" layer. An example of a more realistic scheme is that in CLASS (Verseghy, 1991) with an explicit snow layer whose density increases exponentially with time from a new snow value of 100 kg/m^3 to 300 kg/m^3 with an e-folding time of 100 hours. New snow is mixed in at 100 kg/m^3. If the mean snow depth is less than 0.1m, the cover is assumed to consist of patches of snow of depth 0.1m.

The main influence of snow included in the models is its effect on land surface albedo. Marshall and Warren (1986) proposed a well-based parametrization of snow albedo which includes the dependence of snow albedo on solar zenith angle, snow mass, grain size and soot content; this has recently been implemented by Marshall and Oglesby (1994). Of particular interest are the effects of grain size and soot content, albedo decreasing strongly as these increase; the implied relation between albedo and snow age is allowed for in the BATS parametrization (Dickinson et al., 1986). The albedo of snow given by Marshall and Warren's parametrization is of course only appropriate as the apparent surface albedo when the snow is not masked by vegetation. This masking effect is roughly dependent on the vegetation height but also on its leaf area index or the total vegetation surface area per unit ground area. Thus evergreen trees tend to have a greater masking effect than deciduous trees. One of the most valuable uses of satellite data in estimating land surface characteristics has been in this context. Robinson and Kukla (1985) analysed the maximum surface albedo of seasonally snow-covered land in the northern hemisphere; this demonstated the large range of albedo appropriate for snowy conditions - from about 0.2 in forested areas to 0.8 in areas of low vegetation. This effect has been included in most models though sometimes only crudely. The GISS model (Hansen et al., 1983) was probably the first to approach the problem realistically, prescribing a deep snow albedo for each vegetation type which the actual albedo approaches asymptotically as snow mass increases. A similar method is still used in the MOHC model. BATS calculates the fraction of the grid square covered by snow as a function of vegetation height (about $20z_0$) and snow depth. Apart from the patchy snow treatment in CLASS noted above, little if any account is taken in the models of the large heterogeneity of snow cover generated by variable settling, drifting, and variations in melting due to inhomogeneities in cloudiness, aspect and slope.

6.2 Land ice

Three-dimensional climate models do not yet include representations of the glaciology needed to predict the evolution of ice sheets. Consequently, they cannot predict the evolution of sea level in a complete sense. For example, Gregory (1993) has analysed sea level changes in the transient response experiment run at the Hadley Centre but was able only to calculate sea level changes due to ocean thermal expansion, making estimates of those due to glacier melt and ice sheet changes. The accumulation of snow on ice sheets can be calculated - and typically is positive in simulations with the Hadley Centre and MPI models. However, the possible increase in calving, which would be expected at least partly to offset this in the long-term, cannot be predicted. As with snow, the surface temperature is limited to 0°C.

7. Land surface data specification for models

There is a wide range of land surface data which needs to be prescribed in current models (Table 4). A start has been made on use of vegetation models in predicting the vegetation and its characteristics such as leaf area index; these could be used to deduce not only the stomatal resistance as discussed earlier, but also the albedo, snow masking depth, roughness length, canopy capacity, and subsurface characteristics such as root depth. However, in current models these all have to be prescribed; generally this has been done on the basis of atlas-derived data on vegetation types and ground-based estimates of the surface characteristics appropriate to each vegetation type (e.g. Matthews, 1983; Wilson and Henderson-Sellers, 1985, Claussen, 1994c, Claussen et al., 1994). Soil characteristics similarly have been derived (e.g. by Wilson and Henderson-Sellers, 1985) from the FAO soil atlas which in turn is based on many soil pit observations.

It might have been expected that models would have used satellite-based estimates of albedo, as several compilations of this have been made over the last decade or more. However, except in the context of snow discussed earlier, the inconsistencies with ground-based estimates have generally been regarded as too large. For example, an estimate by Arino et al. (1991) for albedo over Africa gave values which look much more reasonable than many earlier ones with a realistic gradient between the high values over the Sahara and lower tropical forest values. However, there was a marked seasonal variation over the tropical region with lower values in the dry season; this is the opposite of what has been suggested by ground-based data (e.g. Oguntoyinbo, 1970; Culf et al., 1995) and by a recent estimate by Li and Garand (1994) which shows a seasonal variation in good agreement with that indicated by Culf et al. (1995) with wet season values lower by 0.02 to 0.04 (Fig. 9).

VEGETATION AND SOIL CHARACTERISTICS
FOR THREE VEGETATION TYPES

VEGETATION TYPE	CONIFEROUS FOREST	PASTURE	DESERT
Albedo: snowfree deep snow	.144 .255	.194 .65	.35 .80
Roughness length (m)	.76	.02	.003 *
Canopy resistance (s/m)	84,5	79	100
Canopy store (mm)	1.13	.63	.58
Root depth (m)	.84	.57	.10
Infiltration factor	5.5	1.68	.50
SOIL TYPE	CLAY	SAND	
x_s (saturation)	0.49	0.42	
x_c (critical)	0.25	0.087	
x_w (wilt)	0.18	0.055	
c (exponent)	11	5	
K_s (kg/m^2 /s)	.00036	.0081	
(saturated conduct.) (mm/hour)	1.3	29	

* z_0 decreases as snowdepth increases

N.B. Heat capacity and thermal conductivity and dependences on moisture content not in current MOHC scheme.

Table 4

As we shall see later, tropical circulation and precipitation are quite sensitive to albedo, and changes of this magnitude could, for example, improve the seasonal variation of rainfall in the Sahel, which exhibits a phase lead over reality in the Hadley Centre model.

8. Modelled hydrology

Data available for depicting the hydrological cycle in the models includes precipitation and evaporation, of which only the former can be readily validated against real data, and with reasonable accuracy only over the land; estimates do exist for ocean precipitation based on shipboard observations of frequency and, more recently, infrared and microwave satellite data, but it is not straightforward to convert these to quantitative values. It is also possible to compare modelled humidity with radio-sonde observations and some of the radiative effects of cloud can be validated against satellite data.

Figs. 10-12 show a comparison for the Hadley Centre model of the meridional fluxes of moisture averaged over the year against the estimates by Peixoto and Oort (1983). The observed profile of the total flux shows maximum poleward fluxes of around 20 $kgm^{-1}s^{-1}$ near 40° latitude in each hemisphere. This symmetry does not quite apply to the tropics where there are northward fluxes near the equator giving a peak of 15 $kgm^{-1}s^{-1}$ near 10°S. A southward flux of 15 kg $m^{-1}s^{-1}$ near 10°N is associated with the northern hemisphere trades. The Hadley Centre model simulates the middle latitude peaks though they are stronger than observed - by over 50% in the southern hemisphere. In the tropics, the northward equatorial flux is present, but the asymmetry is more marked with the 10°N peak being too weak.

Breakdown of the flux into its components shows that the middle latitude features are due to the transient eddies and those near the equator to the zonal mean flow or meridional circulation. Comparison of modelled and observed fluxes shows the shape for transient eddies to be correct but the modelled intensity to be higher than observed. The standing eddy fluxes near 25° latitude are correctly located but too large by around 100%. The Northern Hemisphere midlatitude peak near 60°N is well located but again too strong. Fluxes by the mean meridional circulation are quite close to the observed though the extra peak near the equator is absent from the observed analysis.

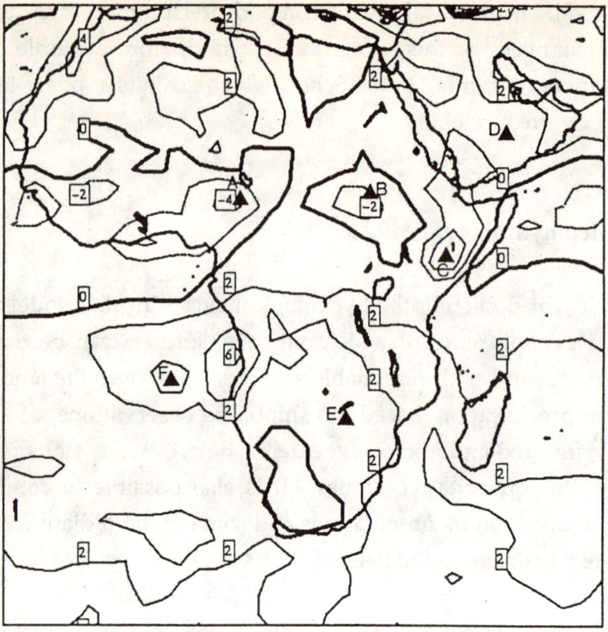

Fig. 9 Difference in surface albedo between July and January as estimated from satellite observations. Contour interval is 2% (Li and Garand, 1994) .

Comparison of Peixoto and Oort's flux divergence with the model's (P-E) (Fig. 13), the excess of precipitation over evaporation, shows that the model fluxes are generally more intense than observed, especially in the Southern Hemisphere; this could be due to errors in the model - e.g. the model being too moist, or the observed fluxes being too small due to sparsity of data (more especially in the Southern Hemisphere).

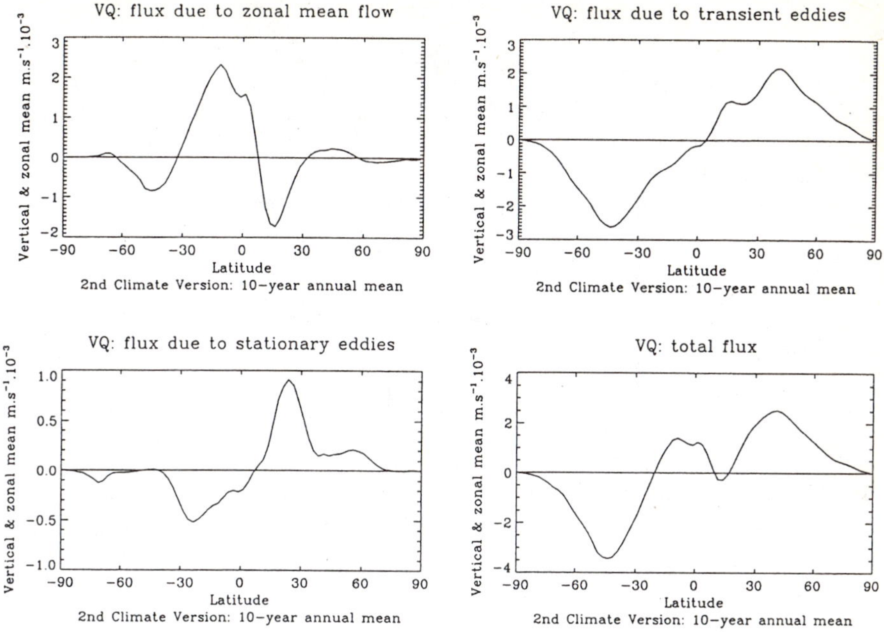

Fig. 10 Components of the poleward flux of moisture averaged over 10 years from the MOHC model (10^{-3} m/s in units of vq which approximately equals 10 kg m^{-1} s^{-1} for moisture flux).

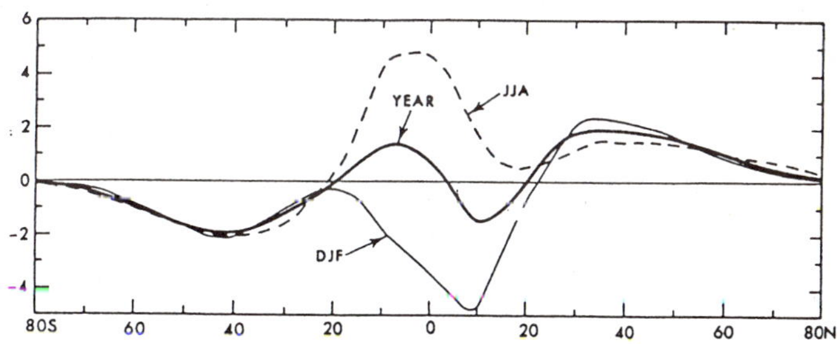

Fig. 11 Meridional profiles of the total meridional water vapour flux $[\overline{Q}_\phi](10\ kg\ m^{-1}s^{-1})$ across latitudinal walls in the atmosphere: —, year; __, DJF; ---, JJA. To convert to units of 10^8 kg s^{-1}, multiply by 4 x cosϕ.

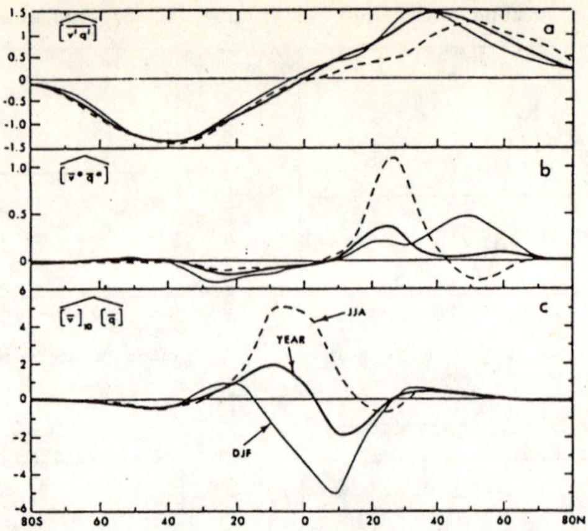

Fig. 12 Meridional profiles of the zonally- and vertically-averaged values of northward water vapour transport (g kg^{-1} m s^{-1}) by: a, transient eddies $[\overline{q'v'}]$; b, standing eddies $[\bar{q}*\bar{v}*]$; c, mean meridional circulations $[\bar{q}][\bar{v}]$.

Maps of the precipitation distribution from models developed at the Hadley Centre, LMD and MPI as part of the collaboration associated with the ECSN (European Climate Support Network, Arpe et al., 1993) showed overall similarity to the estimated observed patterns. The MPI and HC simulations were run with three different resolutions (T21, 42, 106 (the number following T is proportional to the number of waves represented east-west in the spectral model) for MPI, 5*7.5, 2.5*3.75 and 0.83*1.25° latitude by longitude for HC). For the HC runs the patterns were similar for all three, with mostly minor changes in magnitudes of features. MPI showed a little more dependence on resolution though not all features improved with increasing resolution (e.g. Indian summer monsoon).

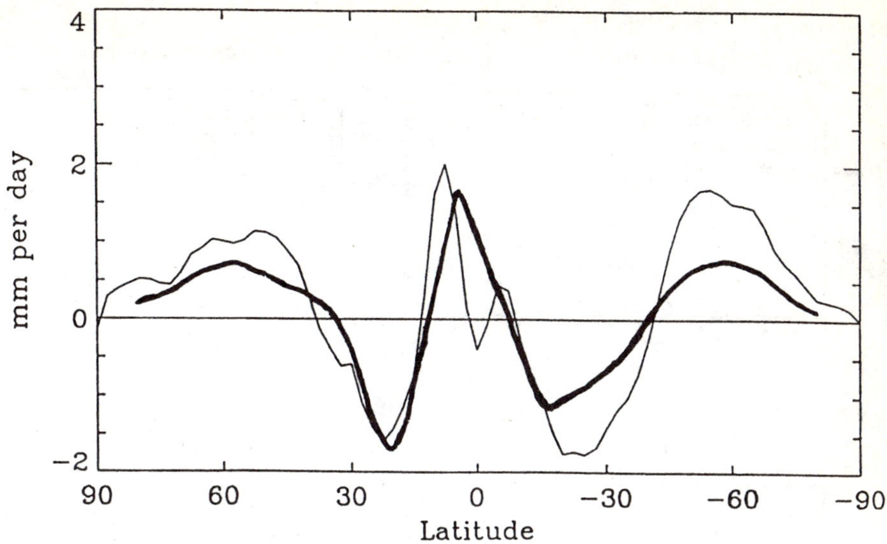

Fig. 13 Meridional profiles of the zonally-averaged convergence of the water vapour transport $\left[-div\,\vec{Q}\right]\left(mm\ day^{-1}\right)$: from observations (Peixoto and Oort, 1983) (thick line) and the Hadley Centre model (thin line).

The runoff simulations by the MPI model have been compared with observations in Sausen et al (1994) (Figs. 14a+b). There are major differences between observed and simulated river discharge at the lowest model resolution. The flow of the rivers is too fast in the model but also the modelled rainfall which is used as input to the system shows large systematic errors. In some catchments the representation is improved with increasing resolution because the horizontal distribution of the seasonal cycle of precipitation improves while in others it is the finer grid representation of the routing scheme at higher resolution which improves the simulation. Some modelled river discharge improves when the new model physics (e.g. ECHAM4) improves the representation of the precipitation (e.g. Indian monsoon).

The simulation of the Amazon river for example becomes worse when resolution is increased from T42 to T106 and improves only a little when input from the ECHAM4 model version is used instead of ECHAM3. At the same time the river discharge for the Ganges/Indus region deteriorates because it is overestimated by the model version ECHAM4, although precipitation climatologies suggest that the monsoon was underestimated by ECHAM3. Here, the step from T42 to T106 improved the representation.

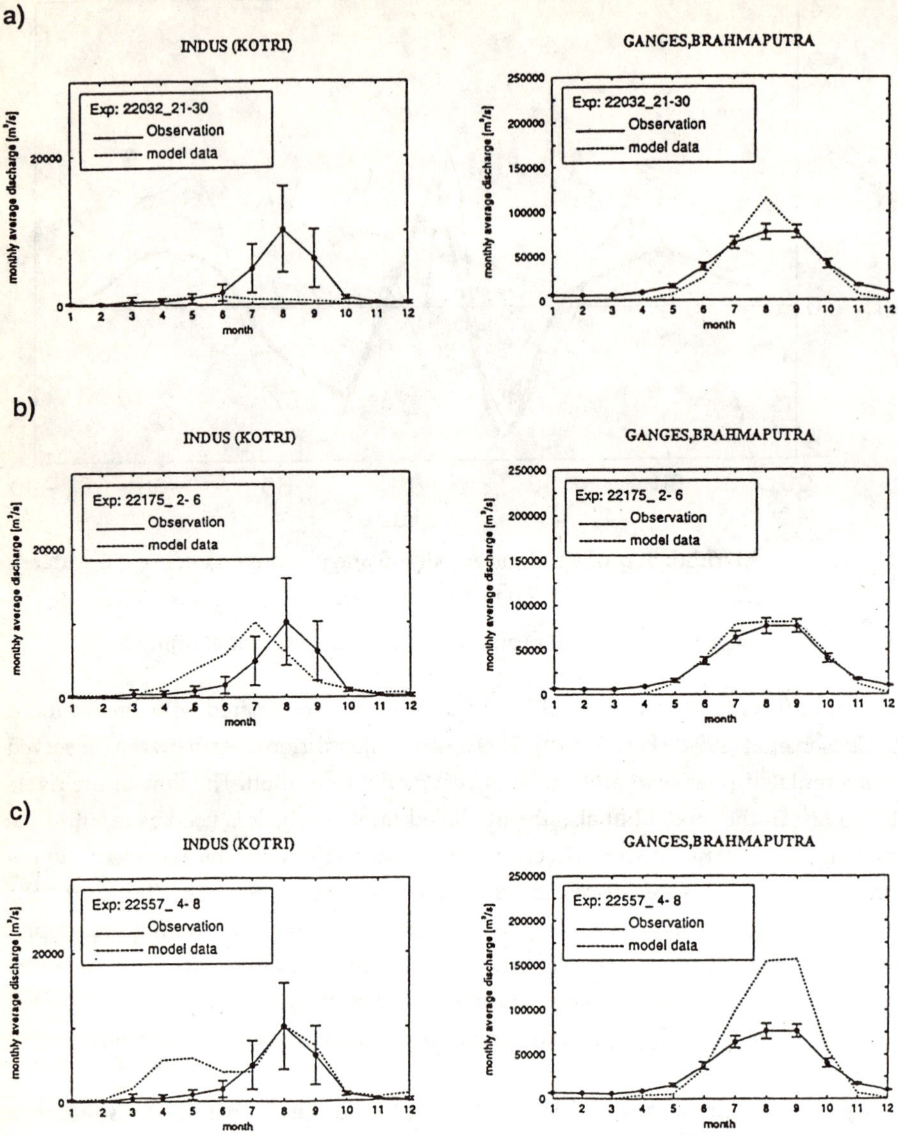

Fig. 14a Annual cycle of monthly averaged river discharge. Model data are courtesy of the Global Runoff Date Centre, Koblenz, Germany, averaged over all available years of data on record. Model data are from simulations using the MPI model

a) ECHAM3 at T42 resolution

b) ECHAM3 at T106 resolution

c) ECHAM4 at T42

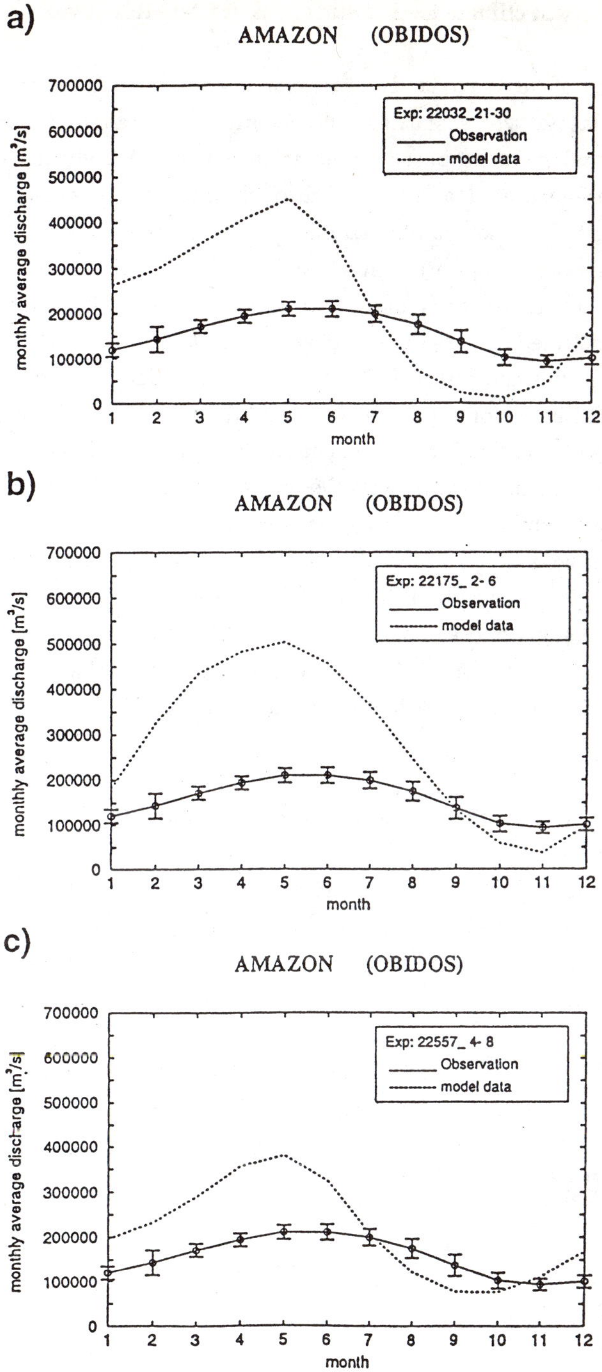

Fig. 14b Same as 14a but for the Amazon river.

9. Sensitivity of climate to land surface characteristics - evidence from GCMs

The sensitivity of climate to land surface changes has been the subject of many papers over the last 20 years starting with Charney (1975)'s paper on the "Dynamics of deserts" in which he proposed a feedback mechanism by which a decrease in vegetation cover generated a heat sink through increased albedos which reduced ascent and rainfall and so maintained the sparse vegetation cover. Subsequent papers have given broad support to this idea. Models also displayed an independent sensitivity to soil moisture (Walker and Rowntree, 1977). This was particularly clearly demonstrated by experiments in which the land was made all dry or all wet initially (e.g. Shukla and Mintz, 1982). The rainfall over land obtained by Shukla and Mintz (Fig. 15a) and more recently with the MPI model (Fig. 15b) was strikingly greater with wet soil than dry, and the modelled temperatures and surface pressures also showed major changes, with the dry land case having high temperatures and low pressure similar to those observed over the Sahara. Cunnington and Rowntree (1986) showed that such contrasts, triggered over North Africa either by differences in initial atmospheric humidity or soil moisture, could persist for over 50 days; they also suggested extending the Charney feedback mechanism to include moisture: an upper tropospheric heat sink generated by a lack of latent heat release could also require descent to maintain the heat balance.

9.1 Land surface change as a cause of the Sahel drought

The experiments discussed above clearly suggested the possibility that land surface changes produced by humans and their animals could cause reductions in rainfall in the Sahel region where a deficit of rainfall below previous normals which commenced around 1968 had led to persistent drought. The likelihood of such a link is also indicated by experiments (Joussaume et al., 1986) using tracer diagnostics to identify the evaporative source of the water vapour involved in precipitation. They found that for precipitation over the eastern Sahel, the evaporative source was Africa for 50% of the rainfall; for the western Sahel the figure was just over 30% (Fig. 16).

Model experiments with land surface characteristics changed in the Sahel region towards higher albedos and reduced soil water gave changes in rainfall similar in pattern and magnitude to those observed (e.g. Rowntree and Sangster, 1986; Xue and Shukla, 1993) (Fig. 17). However, there is only sparse evidence (Courel et al., 1984) of any actual changes in the surface characteristics of the Sahel over the last 30 years. Shukla and Doty (1994) have shown that there is a relation between a satellite-derived vegetation index and rainfall, which would suggest the likelihood of some change in albedo due to the drought. It would be particularly useful if the analysis of ERBE data by Li and Garand (1994) were extended to investigate interannual albedo variations.

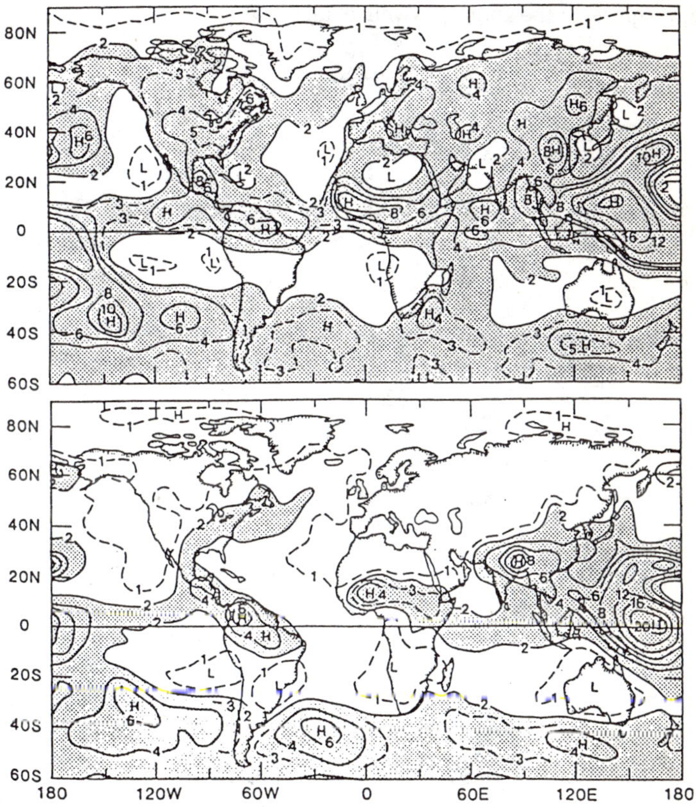

Fig. 15a Precipitation (mm/day) in wet-soil case (top) and dry-soil case (bottom), in experiment of Shukla & Mintz (1981). (Precipitation greater than 2 mm/day is shaded.)

An alternative explanation of the Sahel drought is that it is due to anomalies in sea surface temperatures, in particular a warming of the southern hemisphere oceans and Indian Ocean relative to the northern hemisphere oceans. This hypothesis was suggested by an analysis of the correlations between rainfall in the Sahel and global SSTs, which showed that for low frequencies (greater than 10 years) the 1949-90 correlations exceeded 0.7 in magnitude over substantial areas. It was given further support by numerical experiments in which the SSTs for particular dry and wet Sahel years were used as lower boundary conditions. Two series of experiments were run, each using SST data for April to October for ten years, but starting from initial atmospheric conditions for two different years. The correlations between modelled and observed Sahel mean July-September rainfall were 0.90 and 0.96, respectively. High correlations were also obtained over the Soudan and Guinea coast regions to the south (Rowell et al, 1995).

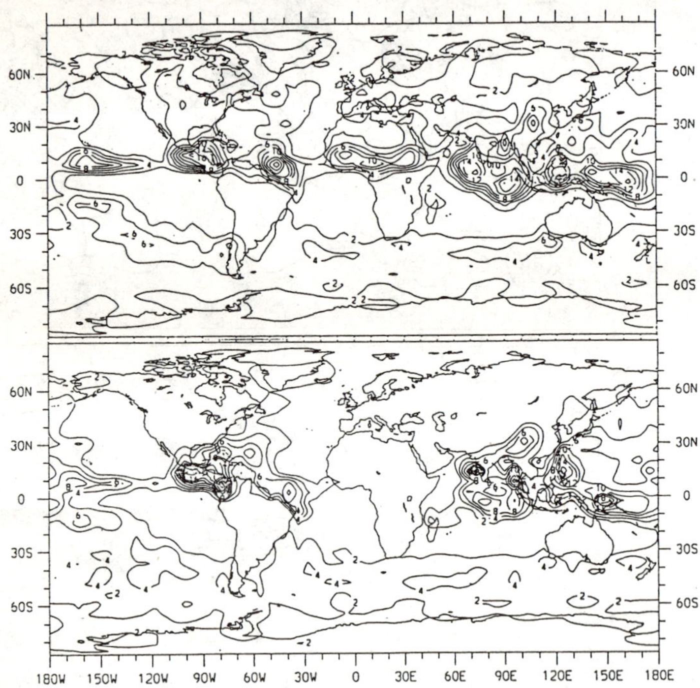

Fig. 15b Precipitation (mm/day) in wet-soil case (top) and dry-soil case (bottom) for a five year simulation using the MPI ECHAM model.

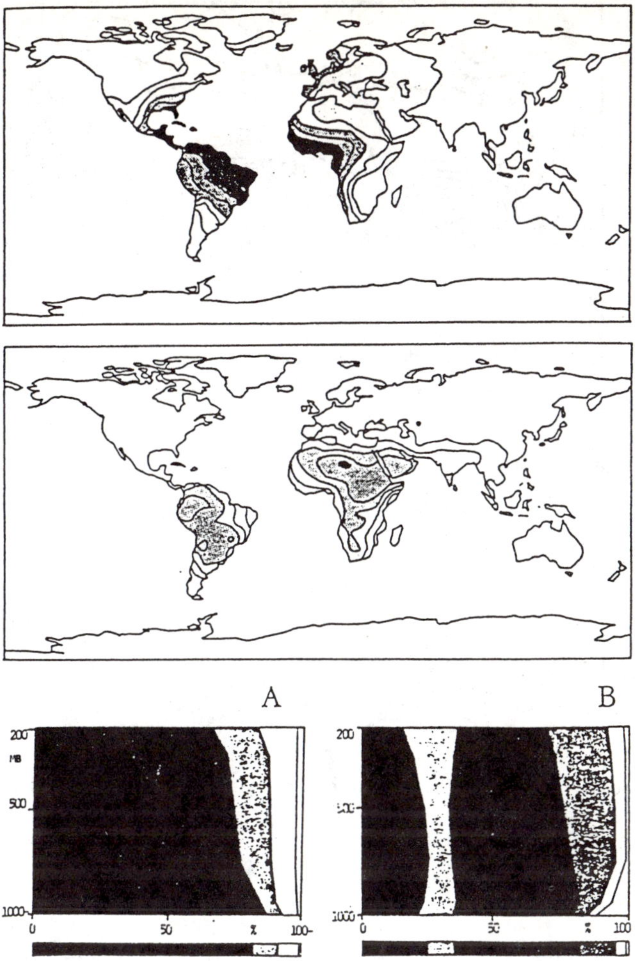

Fig. 16 Top: Precipitation ratio for Atlantic water. Isolines every 10%. Light shading: 10 to 30%; medium: 30 to 50%; heavy: 50 to 70%; black: more than 70% (after Joussaume et al., 1986). Middle: The same as top but for South-American and African water. Bottom: Distribution of water species in atmospheric water vapour (top) and precipitation (bottom) over the western (A) and eastern (B) Sahelian bands. White areas: respectively other origins and imbalance between total amount of tracers and water.

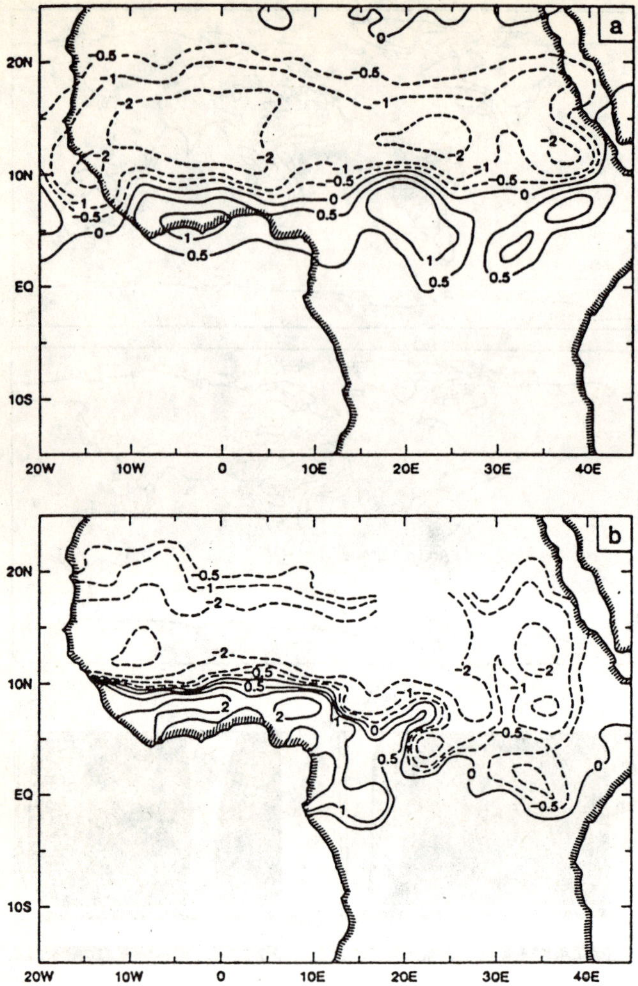

Fig. 17 Three-month mean precipitation. Contour intervals are 0, 0.5, 1, , 4 mm day^{-1}: (a) experiment D minus C and (b) observed rainfall differences between the two dry years (1983, 1984) and the two wet years (1950, 1958). From Xue and Shukla, (1993).

9.2 Tropical deforestation experiments

A major change in land surface characteristics has been imposed over the last few decades through tropical deforestation. Model experiments have been run to provide estimates of the potential effects of this on climate, especially for South America where the largest areas of forests existed and exist. The changes in the land surface from forest to pasture have been measured as part of the Anglo-Brazilian ABRACOS project; there is a shift towards higher albedos, lower roughness lengths, reduced rooting depths and decreased near-surface maximum hydraulic conductivity (Bastable et al., 1993; Wright, personal communication). The experiments exceeding three years (Lean and Warrilow, 1989, Nobre et al., 1991; Lean and Rowntree, 1993; Polcher and Laval, 1994) have mostly shown that rainfall and evaporation both decrease as would be expected with increases in albedo. The decreases in rainfall vary considerably however, with about 25% in Nobre et al., 20% in Lean and Warrilow, 14% in Lean and Rowntree and 12% in Polcher and Laval. The reduction between the second and third was attributed to a change in the parametrization of interception which was excessive in Lean and Warrilow's experiment. The importance of this follows from the very different effects on evaporation of a change in roughness length for canopy evaporation (r_S=0) and transpiration (r_S of about 100 s/m) discussed in the section on assessment of sensitivities. The results reported by Polcher and Laval (1994) refer to only a small part of the deforested area of South America; also, for the deforested regions of Africa and Indonesia they analysed, the reductions were only about 5%.

References

Arino O, Dedieu G, Deschamps P (1991) Accuracy of satellite land surface reflectance determination. Journal of Applied Meteorology 30: 960-972

Arpe K, Bengtsson L, Dümenil, L, Roeckner R (1994) The hydrological cycle in the ECHAM3 simulations of the atmospheric circulation, in: Global precipitation and climate change (eds M Desbois and F Deslmand) NATO ASI Series 126:361-377

Avissar R, Pielke RA (1989) A parameterization of heterogeneous land surfaces for atmospheric numerical models and its impact on regional meteorology. Mon Weather Rev 117: 2113-2136

Bastable HG, Shuttleworth WJ, Dallarosa RLG, Fisch G, Nobre CA (1993) Observations of climate, albedo and surface radiation over cleared and undisturbed Amazonian forest. Int J Climatology 13: 783-796

Baumgartner A and Reichel E (1975) The world water balance - mean annual global, continental and maritime precipitation, evaporation and runoff. Elsevier Amsterdam, 179pp

Blyth EM (1994) Using a simple SVAT scheme to describe the effect of scale on aggregation. Boundary Layer Meteorology, in press

Blyth EM, Dolman AJ, Wood N (1993) Effective resistance to sensible and latent-heat flux in heterogeneous terrain. Quart J R Met Soc 119: 423-442

Charney JG (1975) Dynamics of deserts and drought in the Sahel, Quart J R Met Soc 101:193-202

Clapp R B, Hornberger GM (1978) Empirical equations for some soil hydraulic properties. Water Resources Research 14:601-604

Claussen M (1990) Area-averaging of surface fluxes in a neutrally stratified, horizontally inhomogeneous atmospheric boundary layer. Atmos Environ 24a:1349-1360

Claussen M (1991) Local advection processes in the surface layer of th marginal ice zone. boundary layer. Boundary-Layer Meteorol 54:1-27

Claussen M (1994a) Flux aggregation at large scales: On the limits of validity of the concept of blending height. J. Hydrology, in press

Claussen M (1994b) Estimation of regional heat and moisture fluxes in homogeneous terrain with bluff roughness elements. J Hydrology, in print

Claussen M (1994c) On coupling global biome models with climate models. Climate Research, in print

Claussen M, Klaassen W (1992) On regional surface fluxes over partly forested areas. Beitr Phys Atmosph 65:243-248

Claussen M, Klaassen W (1994) Landscape variability and surface flux parameterization in climate models. Agric Forest Meteorol, in press

Claussen M, Lohmann U, Roeckner E, Schulzweida U (1994) A global data set of land-surface parameters.ISSN 0937-1060

Courel MF, Kandel RS, Rasool SI (1984) Surface albedo and Sahel drought. Nature 307:528-531

Cox PM (1993) A multi-layer soil hydrology scheme for inclusion in the unified model. Hadley Centre Internal Note 49 Meteorological Office Bracknell 14pp

Culf AD, Fisch G, Hodnett MG (1995) The albedo of Amazonian forest and ranchland. Accepted for publication in J. Climate.

Cunnington WM, Rowntree P R (1986) Simulations of the Saharan atmosphere - dependence on moisture and albedo. Quart J R Met Soc 112:971-999

Dickinson RE, Henderson-Sellers A, Kennedy PJ, Wilson MF (1986) Biosphere-Atmosphere Transfer Scheme (BATS) for the NCAR Community Climate Model. NCAR Technical Note NCAR/TN-275+STR

Dolman AJ (1992) A note on areally-averaged evaporation and the value of the effective surface conductance. J Hydrol 138:583-589

Dolman AJ, Gash JHC, Roberts J, Shuttleworth WJ (1991) Stomatal and surface conductance of tropical rainforest. Agric and Forest Meteorology 54: 303-318

Ducoudré NI, Laval K, Perrier A (1993) SECHIBA, a new set of parameterizations of the hydrologic exchanges at the land / atmosphere interface within the LMD atmospheric general circulation model. Submitted

Dümenil L, Todini E (1992) A rainfall-runoff scheme for use in the Hamburg climate model. In: J P O'Kane (ed) Advances in Theoretical hydrology, A Tribute To James Dooge. European Geophysical Society Series of Hydrological Sciences 1 Elsevier

Entekhabi D, Eagleson PS (1989) Land surface hydrology parametrization for atmospheric general circulation models including subgrid scale spatial variability. J Climate 2:816-831

Fouquart Y, Bonnel B, Ramaswamy V (1991) Intercomparing radiation codes for climate studies. J Geoph. Res 96:8955-8968

Friend AD, Cox PM (1995) Modelling the effects of atmospheric CO_2 on vegetation-atmosphere interactions. Accepted for publication in Agric and Forest Meteorology

Garratt JR, Hicks BB, Valigura RA (1993) Comments on the roughness length for heat and other vegetation parameters for a surface of short grass. J Applied Meteorol 32:1301-1303

Gregory D (1994) A consistent treatment of the evaporation of rain and snow for use in large-scale models, submitted for publication

Gregory JM (1993) Sea level changes under increasing atmospheric CO_2 in a transient coupled ocean-atmosphere GCM experiment. J Climate 6:2247-2262

Grötzner A, Sausen R, Claussen M (1994) On the impact of subgrid-scale sea-ice inhomogeneities on the performance of the atmospheric general circulation model ECHAM, submitted for publication (available as Report 143, Max-Planck-Institut für Meteorologie)

Hansen J and others (1983) Efficient three-dimensional global models for climate studies: Models I and II Mon Weath Rev 111:609-662

Joussaume S, Sadourny R, Vignal C (1986) Origin of precipitating water in a numerical simulation of the July climate. Ocean-Air Interactions 1:43-56

Kondratyev KYa, Korzov VI, Mukhenberg VV, Dyachenko LN (1982) The shortwave albedo and the surface emissivity. In: Land surface processes in general circulation models (ed PS Eagleson) 463-514

Koster RD, Suarez MJ (1992a) Modeling the land surface boundary in climate models as a composite of independent vegetation stands. J Geophys Res 97:2697-2715

Koster RD, Suarez MJ (1992b) A comparative analysis of wo land surface heterogeneity representations. J Climate 5:1379-1390

Lean J, Rowntree PR (1993) A GCM simulation of the impact of Amazonian deforestation on climate using an improved canopy representation. Quart J Roy Meteorol. Soc 119:509-530

Lean J, Warrilow DA (1989) Simulation of the regional impact of Amazon deforestation. Nature 342:411-413

Li Z, Garand L (1994) Estimation of surface albedo from space: a parametrization for global application. J Geophys Res 99:8335-8350

Mahrt L (1987) Grid-averaged surface fluxes. Mon Weath Rev 115:1550-1560

Manabe S, Stouffer RJ, Spelman MJ, Bryan K (1991) Transient responses of a coupled ocean-atmosphere model to gradual changes of atmospheric CO_2. Part I: Annual mean response. J Climate 4:785-818

Marshall S, Oglesby RJ (1994) An improved snow hydrology or GCMs. Part 1: snow cover fraction, albedo, grain size and age. Climate Dynamics 10:21-37

Marshall S, Warren SG (1986) Parametrization of snow albedo for climate models. In Snow Watch '85 (Kukla G, Barry RG, Hecht A, Wiesnet D eds). Glaciological Data Report GD-18, World Data Center A for Glaciology (Snow and Ice), Boulder, Colorado 215-223

Mason PJ (1988) The formation of areally-averaged roughness lengths. Quart J R Met Soc 114:399-420

Matthews E (1983) Global vegetation and land use: new high resolution data bases for climate studies. J Clim Appl Met 22:474-487

Milly PCD (1992) Potential evaporation and soil moisture in general circulation models. J Climate 5:209-226

Mitchell JFB, Wilson CA, Cunnington WM (1987) On CO_2 climate sensitivity and model dependence of results. Q J R Meteorol Soc 113:293-322

Monteith, JL (1973) Principles of environmental physics. Arnold, London

Nobre CA, Sellers PJ, Shukla J (1991) Amazonian deforestation and regional climate change. J Climate 10:957-988

Oguntoyinbo JS (1970) Reflection coefficient of natural vegetation, crops and urban surfaces in Nigeria. Quart J R Met Soc 96:430-441

Peixoto JP, Oort AH (1983) The atmospheric branch of the hydrological cycle and climate. In: Variations in the global water budget (Street-Perrott A, Beran M, Ratcliffe R eds). Reidel Dordrecht 5-65

Pinty JP, Mascart P, Richard E, Rosset R (1989) An investigation of mesoscale flows induced by vegetation inhomogeneities using an evaporation model calibrated against HAPEX-MOBILHY data. J Applied Met 28:976-992

Polcher J, Laval K (1994) A statistical study of the regional impact of deforestation on climate in the LMD GCM. Climate Dynamics, in press

Polcher J, Laval K, Dümenil L, Lean J, Rowntree PR (1995) Comparing three land surface schemes used in GCMs, submitted for publication

Priestley CHB, Taylor RJ (1972) On the assessment of surface heat flux and evaporation using large scale parameters. Mon Weath Rev 100:81-92

Ramanathan V, Barkstom BR and Harrison EF (1989) Climate and the Earth's radiation budget. Physics Today American Institute of Physics New York 5:22-32

Robinson DA, Kukla G (1985) Maximum surface albedo of seasonally snow-covered lands. J Climatol Appl Meteorol 24:402-411

Rowell DP, Folland CK, Maskell K, Ward MN (1995) Seasonal variability over tropical North Africa during July to September, Part I: Rainfall, submitted for publication

Rowntree PR (1991) Atmospheric parametrization schemes for evaporation over land: basic concepts and climate modelling aspects. In: Land surface evaporation: measurement and parametrization (Schmugge TJ, Andre JC ed), Springer-Verlag New York 5-29

Rowntree PR, Lean J (1994) Validation of hydrological schemes of climate models against catchment data. J. Hydrology 155:301-323

Rowntree PR, Sangster AB (1986) Remote sensing needs dentified in climate model experiments with hydrological and albedo changes in the Sahel. Proc ISLSCP Conference, Rome, European Space Agency ESA SP-248, 175-183

Sausen R, Schubert S, Dümenil L (1994) A model of river runoff for use in coupled atmosphere-ocean models. J. Hydrology 155:337-352

Schädler G (1990) Triggering of atmospheric circulations by moisture inhomogeneities of the earth's surface. Boundary-Layer Meteorol 51:1-29

Schumann U (1992) A simple model of the convective boundary layer over wavy terrain with variable heat flux. Beitr Phys Atmosph 64:169-184

Sellers PJ, Mintz Y, Sud YC, Dalcher A (1986) A simple biosphere model (SiB) for use within general circulation models. J Atmos Sci 43:505-531

Shukla J, Doty B (1994) On the expansion and contraction of the Sahara Desert, submitted for publication

Shukla J, Mintz Y (1982) Influence of land surface evapotranspiration on the earth's climate. Science, 215:1498-1501

Slingo A, Slingo JM (1988) The response of a general circulation model to cloud longwave radiative forcing. I. Introduction and initial experiments. Quart J R Met Soc 114:1027-1062

Smith RNB (1990) A scheme for predicting layer clouds and their water content in a general circulation model. Quart JR Met Soc 116:435-460

Stössel A, Claussen M (1993) On the momentum forcing of a large-scale sea-ice model. Climate Dyn 9:71-80

Taylor C (1994) Aggregation of wet and dry land surfaces in interception schemes for general circulation models. Accepted for publication in J Climate

Taylor PA (1987) Comments and further analysis on effective roughness lengths for use in numerical three-dimensional models. Boundary-Layer Meteorol 39:403-418

van Genuchten MTh, Leij FJ, Yates SR (1991) The RETC code for quantifying the hydraulic functions of unsaturated soils. U.S. Environmental Protection Agency Report EPA/600/2-91/065

Verseghy DL (1991) CLASS - a Canadian land surface scheme for GCMs. I. Soil model. Int J Climatology 11:111-133

Walker J, Rowntree PR (1977) The effect of soil moisture on circulation and rainfall in a tropical model. Quart J R Met Soc 103:29-46

Warrilow DA, Sangster AB, Slingo A (1986) Modelling of land surface processes and their influence on European climate. Meteorological Office. Met O 20 Tech Note DCTN 38

Wefelmeier C, Etling D (1991) The influence of sea ice distribution on the atmospheric boundary layer. Z Meteorol 41:333-342

Wieringa J (1986) Roughness-dependent geographical interpolation of surface wind speed averages. Quart J R Met Soc 112:867-889

Wilson MF, Henderson-Sellers A (1985) A global archive of and cover and soils data for use in general circulation climate models. J Climatology 5:119-143

Wood N, Mason PJ (1991) The influence of static stability on the effective roughness lengths for momentum and heat transfer. Quart J R Met Soc 117:1025-1056

Wright IR, Manzi A, da Rocha HR (1994) Canopy surface conductance of Amazonian C4 pasture grass. Agric and Forest Meteorology, in press

Xue Y, Shukla J (1993) The influence of land surface properties on Sahel climate. Part 1: Desertification. J Climate 6:2232-2245

Plant Control on Evapotranspiration: Models and Measurements

Willem Bouten
Landscape and Environmental Research Group
University of Amsterdam
Nieuwe Prinsengracht 130
1018 VZ Amsterdam
The Netherlands

SUMMARY

Plants play a dominating role in the evaporation of water from terrestrial systems. They act as a bridging mechanism in the transfer of water from soil to atmosphere. They thus lower the surface resistance to vapour exchange.

The Penman-Monteith combination equation is capable of describing the evaporation at the earth's surface on the basis of the available energy, the meteorological conditions at some reference height - air temperature and vapour deficit - and two resistance parameters. It is applicable to almost any surface, at landscape scale and at leaf scale as well. It is sometimes used in simplified forms, such as the Priestly-Taylor equation. It can also be used as part of comprehensive multi layer models which can describe stomatal behaviour as function of environmental factors, micro-meteorological exchange within plant canopies or vertical patterns of soil water uptake and feed back mechanisms upon drought. Different modelling aims and different temporal and spatial scales generate the need for different model complexity.

Three types of models with different complexity are evaluated: Single Big Leaf models, SPAC models (Soil-Plant-Atmosphere Continuum) and Multi Layer models.

NATO ASI Series, Vol. I 31
The Role of Water and the Hydrological Cycle
in Global Change
Edited by Howard R. Oliver and Sylvia A. Oliver

Some soil water uptake models are also presented. Finally, rainfall interception models are discussed as the interception evaporation can be as high as half the precipitation amount and more than the transpiration on a yearly basis.

Several techniques are discussed for measuring the evaporation at different spatial scales including meteorological techniques to determine vapour fluxes above the canopy, the measurement of sap flow in plants, and the measurement of components of the water balance. Special attention is paid to the usefulness of these techniques in parametrizing and validating the various models.

INTRODUCTION

Water supply is the most important factor determining the distribution, species composition and growth of plants. In the equatorial zone, for example, contrasts occur between semi-deserts and rain forests although mean temperatures and radiation characteristics are almost similar in both cases. Precipitation amount and biomass tend to be highly correlated. Net primary productions range from 3000 g m^{-2} y^{-1} for tropical rain forests, with a precipitation of 3000 mm y^{-1}, to productions of 25-100 g m^{-2} y^{-1} for arid regions with less than 200 mm of precipitation per year.

Water supply is also an important factor in the partitioning of the available energy at the earth's surface. The global incoming radiation is only partly reflected or returned as long wave radiation. Most of it is returned as latent or sensible heat flux of which the distribution is dependent on the availability of water at the soil surface. Due to the non-parallel temporal distribution of rainfall and evaporation, the availability of water depends on the water storage capacity of the soil and the system properties which determine the transport of water from a certain storage location to the evaporating surface.

For bare soils, water is transported due to hydraulic gradients according to the law of Darcy:

$$F = -K(\theta)\,(dh/dz + 1) \tag{1}$$

where F is the vertical water flux density (m day^{-1}), K is the conductivity (m day^{-1}) which is a function of soil water content (θ), z is depth (m) and h is the soil matric potential (m). h and θ are related through the water retention characteristic.

As the top layer of a bare soil dries in a few days without rain, the water content decreases at the surface. Due to the non-linear K(θ) relation, the conductivity strongly decreases with small reductions of water content. As a result, water can't be transported to the soil surface and the soil dries even more rapidly. This positive feed-back mechanism is strongest for sandy soils and causes very steep water content gradients which can range from dry to very wet within a few centimetres below the soil surface. As a result the characteristic total water supply to the surface is below 20 mm (Fig. 1A). This means that evaporation is strongly reduced or even ceases within a few days after a rainfall event. This is different for soils covered with plants. Roots can take up water from greater depth, thus enlarging the volume of soil which can supply water to plants. With a root zone of 40 cm, a characteristic total water supply for a sandy soil can amount to 60 mm (Fig. 1B). When the root zone dries during a prolonged dry period, plants can extend their root zone by growth, to increase the water availability. Plant covered soils thus loose much more water than bare surfaces, especially in case of sandy soils.

Apparently, plant growth is correlated with precipitation and evapotranspiration is correlated with plant growth, which means that evapotranspiration in correlated with precipitation. This of course is not amazing. The point is, to what extent plants can control these mechanisms.

Plants contain lots of water. Tropical hard wood species contain 50% but some vegetables such as spinach can contain 99% of water. High amounts of water enable a plant to increase its leaf area index very fast without the need of large amounts of assimilates.

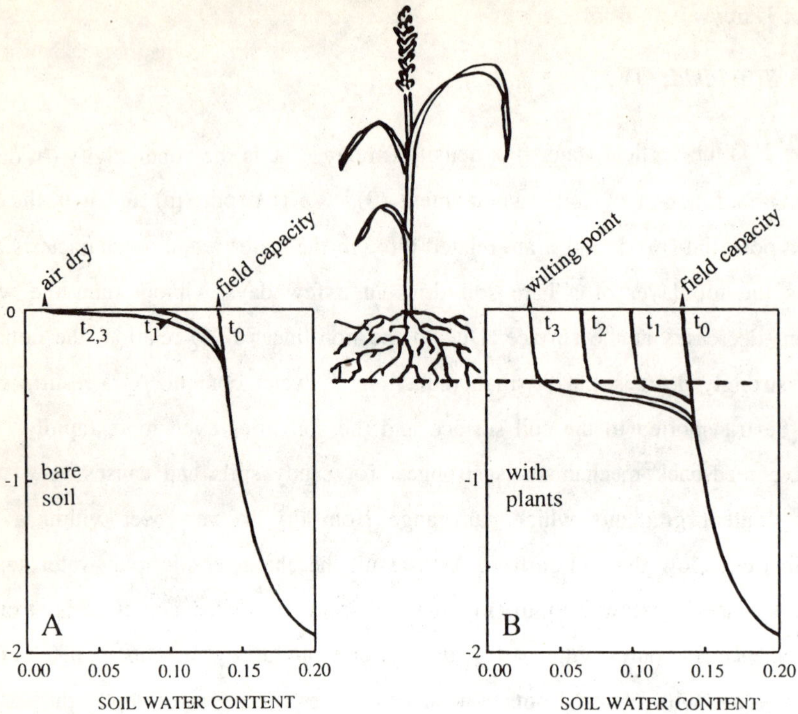

Figure 1. Typical water content profiles in sandy soils: (A) a bare soil evaporating at the surface and (B) a plant covered soil with water uptake for the root zone.

Plants also need enormous amounts of water, up to 1000 times as much as their biomass production. In order to understand the relation between plant growth and water use, we should try to answer the question of why plants developed in the evolutionary process in such a way that they waste so much water. A reason may be that they transpire to decrease the leaf temperature at high irradiation. However, plants are often not capable of supplying enough water and stomata close especially at high temperatures and associated high vapour pressure deficits when transpiration is needed most to keep the leaf temperature within acceptable limits. Another reason may be that plants transpire to transport nutrients from the roots to places where plant tissue is formed. On the other hand, plants are capable of reallocating sugars with water flow driven by osmotic potentials. In the same way plants could have developed mechanisms to force cation transport. It is likely that one of the main reasons why plants transpire is that it enables them to force convective water flow through the soil. Basically they manipulate their root environment by redistributing nutrients towards the roots where they can be taken up. A nutrient supply by diffusion only would never be sufficient for plant growth.

Plants must control their water use as the water storage in the soil and the water supply by the roots may be too limited to fulfil all transpiration demands. The capability to extend the root zone to greater depth to increase water uptake was already mentioned. Moreover, plants have the possibility to reduce their water loss by adjustment of their stomatal opening. Stomata tend to close at high evaporative demand (high vapour pressure deficit) and low water supply. Different opinions exist whether the true cause behind this mechanism is low relative humidity (Ball et al., 1987), high saturation deficit (Aphalo and Jarvis, 1991), high transpiration rate (Mott and Parkhurst, 1991), or low leaf water potential (Lynn and Carlson, 1991). Although the true causes of these mechanisms are poorly understood, many models have been made to describe the evapotranspiration dynamics of leaves, plants or entire plant covered surfaces.

MODELS

Before we start to discuss different types of models, we should give it a moment's thought why we want to model. Different models are needed to either *describe* the hydrologic behaviour and energy exchange at the earth's surface, or to *understand* or *evaluate* this behaviour, or to *explain* the mechanisms which cause this behaviour or to *predict* the effects of for example enhanced CO_2 concentrations on evapotranspiration. We also need to know what exactly we want to model. Is it the mean energy and water exchange at the earth's surface or do we want one model for energy and water exchange with differentiation for bare soil, vegetated surfaces and oceans, or do we want to model specific feed-back mechanisms of plants upon drought? We also need to think of the scale. Do we want a model on a time scale of seconds, hours, days or years? Do we want a model for the spatial scale of single stomata, plants, a field, a region or a GCM-grid cell. All these points make clear that there is no "best" model. Different focuses and aims generate the need for different model generalities and complexities. Moreover, (lack of) possibilities for parametrization also limit the useful complexity of models.

Some history

The most famous model for evaporation is the Penman equation (Penman, 1948):

$$\lambda E_0 = \frac{sR_n + \gamma\lambda E_a}{s+\gamma} \quad \textit{with} \quad \lambda E_a = (c_1 + c_2 u)(e_s - e_a) \qquad (2)$$

where λE_0 is the evaporation of a wet surface, s is the rate of change of the saturated vapour pressure with respect to temperature, R_n is the net radiation, γ is the psychrometric constant, u is the wind speed at 2 m above the surface and e_s and e_a are the saturated and actual vapour pressure. The coefficients c_1 and c_2 were originally proposed as 74 and 40 respectively. Many different values exist in the literature and it is probably best to use a local calibration.

It was the first physically sound description of evaporation. It is well known because i) it is physically realistic and thus promotes understanding, ii) it needs meteorological parameters at only one height, and iii) many other models were later based on the same concept. It is often referred to as the combination equation as it combines an energy balance term with a ventilation term in one single equation. The equation was derived for open water surfaces and has been used as a reference for vegetated surfaces by including a plant specific factor. Later, Rijtema (1965) and Monteith (1965) eliminated the empirical plant factor by explicit formulation of the aerodynamic and surface resistance:

$$\lambda E = \frac{s(R_n - G - M) + \rho c_p(e_s - e_a)/r_a}{s + \gamma(1 + r_s/r_a)} \tag{3}$$

where G is soil heat flux, M is the heat flux into biomass, ρ is the air density, C_p is the specific heat of air, r_a is the aerodynamic resistance and r_s is the surface resistance. This equation solves the energy balance depending on the resistances of the latent and sensible heat fluxes. It is applied for different scales ranging from leaf scale to regional scale. The difference for the various scales lies in the parametrization and in the relative importance of the energy term and the ventilation term (Fig. 2). This figure shows that the larger the scale, the more important the energy balance term as the aerodynamic resistance increases with the scale and the corresponding higher reference. Equation A (Fig. 2) shows the extreme for $r_a \rightarrow \infty$. Conversely, at small scale and low aerodynamic resistance, such as forests, the ventilation term is more important with the extreme of equation B (Fig. 2) for $r_a \rightarrow 0$. For decreasing surface resistance, such as wet surfaces, equations A and B change to equations C and D respectively. For example, equations D satisfies for the evaporation of a wet forest canopy, especially during a clouded night when the net radiant energy is close to zero.

Other equations, such as by Makkink (1957):

$$\lambda E_o = (1-r)\frac{s}{(s+\gamma)} c_1 K\downarrow + c_2 \tag{4}$$

when r is the reflection coefficient, $K\downarrow$ is the global radiation and c_1 and c_2 are empirical constants, or Priestley and Taylor (1972):

$$\lambda E = 1.26 \frac{s}{(s+\gamma)}(R_n - G) \tag{5}$$

or Thom and Oliver (1977):

$$\lambda E = \frac{s(R_n - G) + 2.5\,\gamma\, E_a}{(s + 2.4\,\gamma)} \tag{6}$$

are all simplifications of the Penman-Monteith equation, or re-evaluations with focus on the larger scale and often parameterized for specific situations or regions.

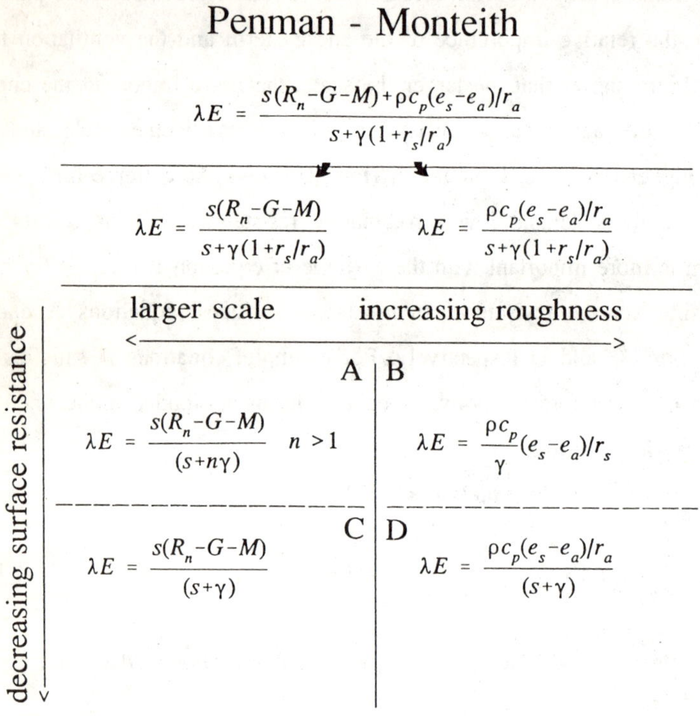

Figure 2. The Penman-Monteith equation, split into an energy term and a ventilation term. Differences in spatial scale and aerodynamic roughness or changing surface resistance control the significance of the various contributions to evaporation (cf. text).

Single Big Leaf models

The simplest model for the latent heat flux from a dry vegetated surface is the Single Big Leaf model (Jarvis, 1976; Stewart, 1988; Bosveld et al., 1992). In this concept the earth's surface is represented by a single structural element with the properties of a leaf. Tan and Black (1976) have shown that the surface conductance of a Douglas fir forest is to a large extend determined by the stomatal conductance of the needles. The Penman-Monteith equation is applied with a surface conductance which is modelled as a product of the maximum conductance ($g_{s,max}$) and a number of reduction functions representing effects of leaf area index (L), vapour pressure deficit (δq), radiation (K), temperature (T) and water availability in the soil (ψ):

$$g_s = g_{s,\mathrm{max}}\ L\ f_1(\delta q)\ f_2(K)\ f_3(T)\ f_4(\psi) \qquad (7)$$

The model describes the surface resistance and the instantaneous latent heat flux as a function of the actual environmental conditions, without any influence of preceding conditions. This kind of rather simple model has proven to be capable of fitting measured water fluxes for the dynamics within the day (Fig. 3a) as well as daily totals over the year (Fig. 3b).

Many different types of functions and parameterizations have been used to fit measured data as well as possible. Due to atmospheric feed back mechanisms and the resulting correlation between "independent" variables, different functional relations can produce equally good fits of measured fluxes. The models are not conclusive upon the true mechanisms of stomatal control.

SPAC models

Soil-Plant-Atmosphere-Continuum models represent the transfer of water from soil to atmosphere with a sap flow which is driven by the difference in hydraulic

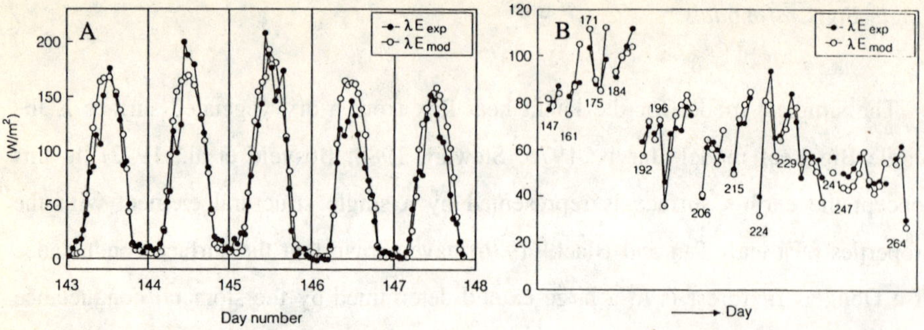

Figure 3. Transpiration dynamics, measured and modelled with an SBL-model; A: hourly values of eddy correlation; B: daily averaged sap flux measurements (reprinted from Bosveld et al., 1992)

potential between root (or soil) and leaves. The representation of the resistances to water flow mostly include resistances for water uptake and xylem sap flow. The concept for vapour exchange is to a large extent equal to the SBL models. The main difference is that the reduction function for soil water potential in eq. 7 is replaced by another reduction mechanism to incorporate the control by the leaf water potential. The leaf water potential is calculated from the water content of the plant which follows from water balance calculations. The stomatal conductance is no longer an instantaneous representation of environmental conditions but it also depends on the evolution of transpiration and water status of the plant during the day.

SPAC models simulate water sap flow as well as latent fluxes above the canopy with a time lag due to the water buffering capacity of the vegetation, which is often found when combining measurements of sap flow and atmospheric fluxes (Fig. 4). The models tend to be capable of fitting measurements of water and vapour flow as well as measured plant water potentials (Kowalik et al., 1988).

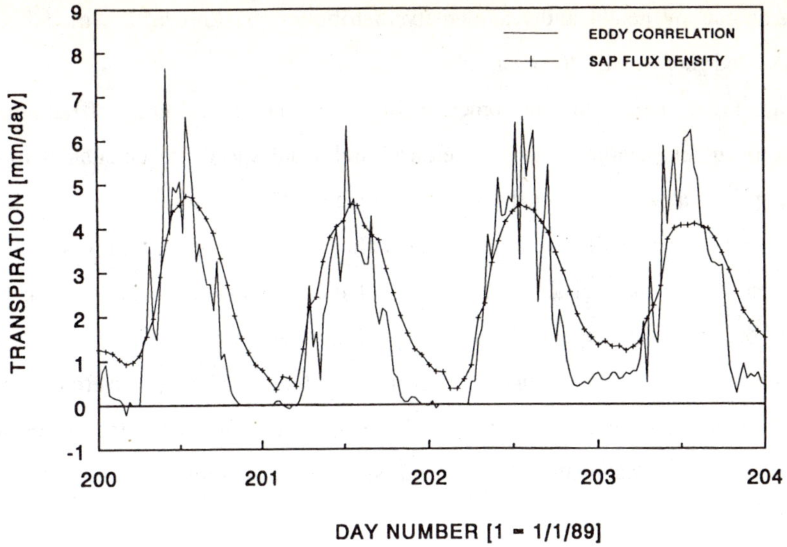

Figure 4. Water flow dynamics measured with eddy correlation and heat pulse velocity (repreinted from Bouten et al, 1992a)

Multi layer models

Multi layer models are more comprehensive models in which the interaction of the plant with its environment is described on a leaf level to reduce the empiricism as much as possible. For example, single layer models and multi layer models both use response curves for the sensitivity of photosynthesis or transpiration for photosynthetically active radiation. Due to the change in height of the sun, the number of leaves which contribute to the total transpiration changes during the day with a maximum around noon. As a strong correlation exists between the height of the sun and the irradiation, SBL models can suffice with an empirical response function with only one independent variable to include effects of radiation. The parameterization of this function not only depends on plant species properties but also on latitude. Multi layer models treat both effects separately. If system properties such as canopy height and leaf density distribution are known, the model can be extrapolated more easily to other sites.

In multi layer models, the micrometeorological conditions in terms of radiation, distribution, air temperature, vapour pressure, and wind speed are calculated from the meteorological conditions at some reference height above the vegetation. Basically, the models calculate the energy balance and the partitioning between sensible heat and latent heat fluxes by means of the Penman-Monteith equation for a number of vegetation layers. These energy fluxes are considered as sinks and sources in the energy exchange between layers and above the canopy. Aerodynamic resistances between vegetation layers and the resistance between the top layer and reference height are calculated from wind speed and stability to calculate the turbulent exchange.

Several models have been developed with different purpose. Goudriaan (1977) developed the MICROWEATHER model to describe the micrometeorology in a maize field in support of studies of the growth and development of crops and their pests and diseases. This model was extended to model rainfall interception, evaporation, and canopy wetness in forests (Bouten and Bosveld, 1992a) in support of studies of atmospheric deposition of acidifying compounds. Sellers and Lockwood (1981) developed a multi layer model to simulate the effects of different crop types on a large scale. The model was more recently applied to evaluate the effects of climatic change on evapotranspiration (Lockwood, 1992).

Root water uptake models

Concepts with different complexity also exist for calculating the distribution of water uptake by plant roots. Models range from simple buckets to the description of multi-dimensional water flow to clustered roots. Bucket models simulate the water balance of a bucket with a maximum amount of available water which equals rooting depth multiplied by the difference of water content at field capacity and wilting point. Transpiration equals potential transpiration as long as some water is available or the ratio between actual and potential transpiration decreases linearly from 1 to 0. In multi layer soil compartment models, the potential water uptake is distributed over depth on the basis of root density or root length. The combined influence of soil properties and root behaviour represented by empirical functions simulating transpiration reduction or compensation of reduced uptake by wetter parts within the root zone (Tiktak and Bouten, 1992). These models have sufficient number of parameters so that they can fit observations of soil water content or transpiration. Most complicated are the models which are derived from Gardner's single root approach (Gardner, 1960). Some include a clustered representation of the roots with a finite element grid for the soil around each root (Lafolie et al., 1991). None of the theories on effects of soil water properties and the distribution and activity of roots has been generally accepted. Problems with measurements of root length, root potential and soil water content and the temporal and spatial variability of processes impede the understanding.

Evaporation of intercepted rainfall

With respect to transpiration, we discussed that a plant can regulate its water loss. The evaporation of intercepted rain is a purely physical process not controlled by the plant. Nevertheless, the evaporation flux of intercepted rainfall is mentioned here as it can be important. Especially in forests with low aerodynamic resistance, evaporation rates can be very high compared to transpiration rates. Over longer time scales (year), the total interception evaporation amount can be as high or even higher than the total transpiration.

Numerous interception models have been published in the past. Most simple are regression models linking interception to rainfall with linear, quadratic or logarithmic functions without any physical significance (Jackson, 1975). All other models calculate the water balance for the canopy layer. Rainfall is input and meteorological conditions are used as boundary conditions to calculate the evaporation. Variation between models occurs due to differences in spatial (single or multi layer) and temporal scales, the simplifications implemented to reduce the number of system parameters needed, and the mathematical techniques to solve the equations (analytical or numerical techniques). Only few models are mentioned here, as the differences with other models are not very large.

The Rutter model (Rutter et al., 1971; Rutter et al., 1975) was the first physically based dynamic model. The model discriminates between leaves and stems in terms of rainfall interception, evaporation and drainage (Fig. 5).

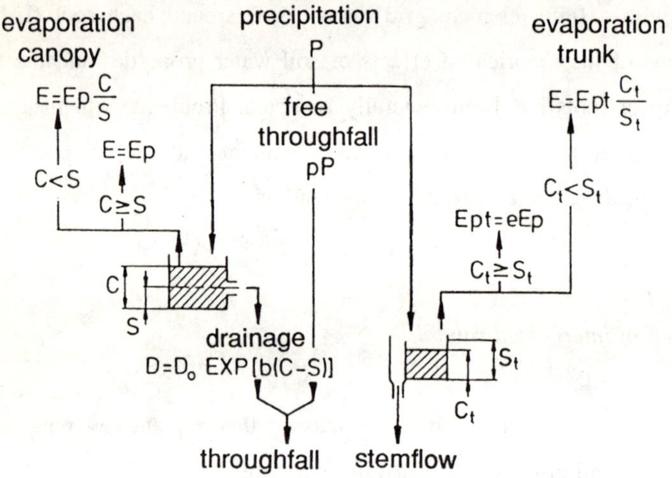

Figure 5. Schematic representation of the Rutter model (Rutter, 1971)

The model uses rainfall and meteorological conditions on an hourly basis and a number of system parameters such as the gap fraction and the water holding capacity. Drainage and evaporation rates are dependent on the actual water storage in the canopy. Other dynamic models use the same basic concept. Calder (1977) and Sellers and Lockwood (1981) extended the model to a multi layer approach. Gash (1979) developed an analytical model with daily input for rainfall amounts with mean values for rainfall intensity and evaporation rate. Hourly values were used to obtain these mean values which appeared to be rather constant over Great Britain (Gash et al., 1980). Mulder (1985) used thrice daily observations of air temperature and relative humidity, daily means of wind run and daily totals of precipitation and the number of rainy hours and of bright sunshine. These data are often available from standard meteorological stations. All models have been validated with varying success. No systematic study has been performed until now to assess the limitations to parameterization and model prediction reliability.

MEASUREMENTS

Many different kinds of measurements exist to quantify the evapotranspiration at the earth's surface. Here we distinguish the determination of vapour fluxes in the atmosphere above the canopy, measurements of sap flow within the plant, and soil water balance measurements.

Energy balance/Bowen ratio

The energy balance or Bowen ratio method evaluates the ratio between the sensible (H) and latent (λE) heat fluxes above the evaporating surface against the available energy which is measured independently:

$$\lambda E + H = A = R^{*} - G - M - D - \mu A \tag{8}$$

in which R^* is net radiant flux, G is soil heat flux, M is the energy storage change in biomass, D is the advective flow of energy, and μA is the energy consumption for photosynthesis. The Bowen ratio (β) is determined from vertical gradients of temperature and vapour pressure:

$$\beta=\frac{H}{\lambda E}=\frac{-K\rho c_p(\delta T/\delta z)}{-K\rho c_p(\delta e/\gamma\delta z)}=\gamma\frac{\delta T}{\delta e} \tag{9}$$

in which the turbulent transfer coefficient (K) is assumed equal for heat and water vapour. The latent heat flux can be calculated from measurements of the net radiation and measurements or estimates of the other contributions to the available energy and the calculated Bowen ratio:

$$\lambda E=\frac{A}{1+\beta} \tag{10}$$

Highly accurate measurements are required especially if the small gradients found over forests are to be measured adequately (Stewart and Thom, 1973). Sensors are sometimes mounted on vertically moving trolleys, to avoid problems with sensor specific offsets. The method is applicable to uniform surfaces and flux estimates can be obtained with a temporal resolution of about 15 minutes with an accuracy around 15%. The technique is not applicable at night when the available energy tends to zero. Latent and sensible fluxes then have the same value but opposite direction, β equals -1, and λE becomes indeterminate.

Eddy correlation

Eddy correlation has become a technique for routine measurements of sensible and latent heat fluxes above an evaporating surface. The technique relies on measurements of the fluctuating components of wind and of the associated fluctuations in temperature and humidity. Since water vapour is carried from the canopy to the atmosphere by turbulence (eddies), the flux density (F) is given by the correlation (Webb et al., 1980):

$$\lambda E = \frac{\lambda \rho}{\tau} \int_0^{\tau} w'(t).q'(t)dt \tag{11}$$

where w′ is the fluctuation of the vertical wind velocity around its mean value, q′ is the fluctuation of the specific humidity around its mean value and τ is the integration period.

The broad range of frequencies and scales of the transporting turbulent eddies poses serious demands on the temporal resolution of the instruments. Vertical wind fluctuations can be measured with a sonic anemometer. This instrument measures wind velocity by means of the travel time of sound pulses. Water vapour density fluctuations can be measured by infrared or UV attenuation. The emitted radiation is filtered to obtain a beam with a specific wavelength at which the radiation is absorbed by water vapour. Wind and vapour instruments are mounted in a tower above the evaporating surface in direct vicinity of each other to avoid spectral loss due to sensor separation.

The integration period in eq. 11 is a compromise between a statistically stable average of the stochastic dynamics of turbulence, and an average which is not contaminated by trends in meteorological conditions. An advantage of the eddy-correlation method is the high time resolution (eg 10 minutes). Drawbacks are problems due to fetch requirements (Businger, 1986). The instruments sense a mixture of the flux of the local stand and water vapour coming from the surrounding upwind area.

Branch chambers

Growth chambers for agricultural crops or branch chambers and leaf cuvettes for trees are used to study photosynthesis and transpiration in a manipulated or non-manipulated environment. In an open system, a flow of air passes through the chamber and the transpiration rate is calculated from the difference in water vapour content of the air entering and leaving the chamber. In a closed system, a humidity sensor monitors the rate of increase of humidity in the chamber. Compared with undisturbed references, the chambers show higher temperatures, lower light intensity, lower wind speeds and higher vapour pressure (Leuning and Foster, 1990). In open chambers, these environmental factors are monitored to control air conditioning equipment in order to keep the differences as small as possible. Nevertheless, transpiration rates may diverge from reality and effects on transpiration are difficult to assess. Moreover, high costs of artificial control limit the number of replications on the one hand, while large numbers are necessary to obtain reliable spatial averages on the other hand. Chambers are therefore most useful in comparative studies with a manipulated environment such as elevated CO_2 or air pollution (Leadley and Drake, 1993).

Sap flow

Xylem sap flow measurements are based on the flow of heat which is applied at a specific point along the vessel and which is partly dissipated by diffusion but also carried along with the flow of water through the vessel. Several techniques to measure the xylem flow have been proposed:

The *sap flow velocity* is determined by inserting a heating device and a thermocouple at a known vertical distance in the transporting area of a tree. The heat-pulse velocity is then determined from the time interval between a heat pulse applied by the heating device and the detection of a temperature increase at the sensor. Subsequently, the mean rate of water transport can be calculated if the water content of the conducting vessels is known. This technique is useful for determining transpiration dynamics. Without calibration, it has a limited applicability with respect to absolute values because of difficulties with estimating the effective area of xylem.

The *sap mass flow* is determined with a heating device which is wrapped around the stem or inserted into the trunk of trees (Fig. 6). The technique is applied either by permanent heating with a constant amount of energy (Sakuratani, 1981; Granier, 1985, 1987) or with a constant temperature difference between heating device and its surroundings (Cermak and Kucera, 1981; Schulze et al., 1985). From the energy supply to the heating device and the temperature difference between the device and its surrounding, the heat dissipation is calculated. By subtracting the zero-flow dissipation the convective component due to sap mass flow is calculated. The equipment is rather rugged and once installed it can monitor continuously for months without any problems. The main limitation is the accuracy of the measurement of the temperature difference and the zero-flow calibration in situations where the occurrence of zero-flow is questionable.

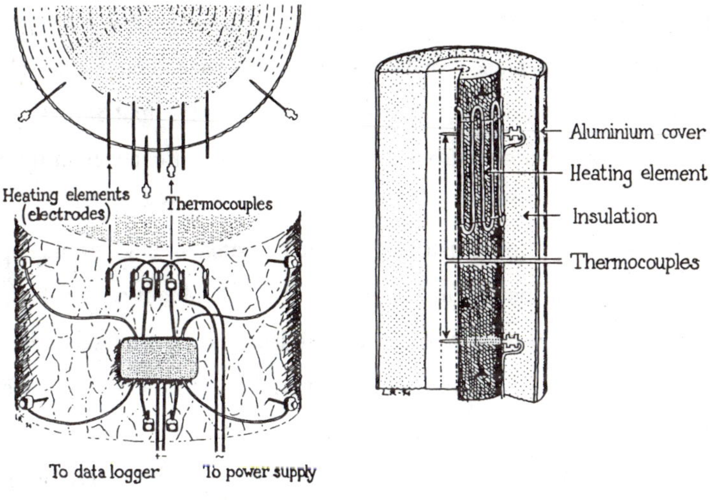

Figure 6. Schematic representation of the Heat Balance device as inserted into the trunk of a tree (left) or wrapped around the stem (right) (reprinted from Cienciala, 1994)

All water which is transpired has been taken up by roots before and has been transported to the leaves through xylem vessels. The transpiration thus can be assessed by measuring the sap flow of plants. Due to the buffering capacity of the plant, however, the temporal dynamics of the sap flow can differ from the transpiration dynamics. This makes the technique applicable for measuring daily totals but not for instantaneous transpiration rates.

Soil water balance

Evapotranspiration can be calculated from soil water budgets of clearly defined soil volumes in combination with input and output fluxes. Catchments, lysimeters, or exclusively one-dimensional systems can be used for this purpose. Rainfall or throughfall funnels with many replicates are used for measuring input fluxes. Weirs or drainage tubes are used to measure the output. In case of deeply drained systems calculations are restricted to periods with a clear transpiration surplus. During these periods, a soil dries as a result of soil evaporation and root water uptake. A zero-flux plane often develops just below the root zone or drainage can be neglected. The existence of a zero-flux plane can be checked with tensiometers (Cassell and Klute, 1986; Marthaler et al., 1983). With known input and output fluxes, evapotranspiration rates can be calculated from the soil water content changes in time.

Several techniques exist to measure soil water contents. Due to the spatial variability large numbers of measurements are necessary to obtain a reliable spatial average. Besides the destructive way of taking soil samples, neutron scattering was the only widely used technique until now. Recently, TDR (Time Domain Reflectometry) has become a better alternative.

A neutron probe, consisting of a source and detector, is lowered into an access tube. The Am-Be source emits high energy neutrons that collide with protons which are present in the soil. After a certain number of collisions the neutrons have lost so much energy that they form a static cloud of which the concentration is measured by a selective detector. As water is the main proton source whose quantity changes in time, a change of the concentration of neutrons can be attributed to a change of water content. Due to differences in soil properties, site specific calibrations are recommended to obtain reliable results (Halbertsma et al., 1987). The size of the measuring volume (0.01-0.1 m^3) is an advantage of the technique. The rather poor reproducibility and safety standards are disadvantages. The same holds for the dual probe gamma ray attenuation technique (Fritton, 1969; Rawitz et al., 1982).

Topp et al. (1980) introduced TDR as a dielectric measurement technique for the indirect measurement of soil water content. The method is based on the measurement of the velocity of electromagnetic waves along the sensor in the soil. The TDR probe, mostly three stainless steel parallel rods, is installed in the soil in such a manner that the soil forms the dielectric material of the transmission line. The apparent dielectric permittivity, as calculated from the velocity, can be related to the soil water content with an empirical relation which holds for most soils (Topp et al., 1980). Extremely high resolution and reproducibility (Heimovaara, 1994) and the possibility of automated measurements (Heimovaara and Bouten, 1990) are distinct advantages of TDR. The technique proved to be very valuable in determining both temporal (Fig. 7a) and spatial (Fig. 7b) dynamics of soil water content.

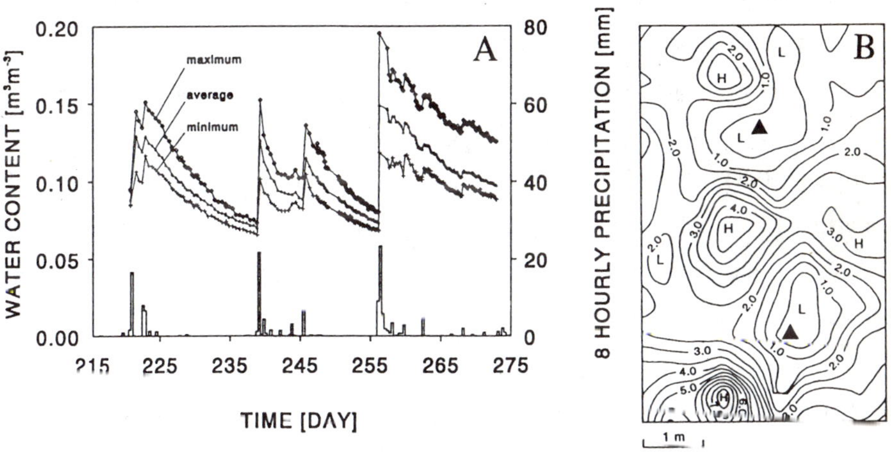

Figure 7. Temporal (A) and spatial (B) dynamics of water content of the top soil (0-50 cm) in a Douglas fir forest as measured with TDR (reprinted from Heimovaara and Bouten, 1990, and Bouten et al., 1992b)

Rainfall interception

Models of evaporation of intercepted rainfall were mostly validated on measurements of net rainfall. Due to the spatial variability of throughfall, large numbers of recording instruments are needed. Different attempts have been made to sample throughfall as accurately as possible. Ford and Deans (1978) used 104 gauges distributed around one tree, Aussenac (1970) randomly distributed fixed and periodically moved gauges, and Calder and Rosier (1976) have designed a large plastic sheet rain gauge. Although these techniques are useful to obtain net rainfall amounts, they are not very useful for validating the models that describe the evaporation and drainage as a function of canopy water storage and leaf wetness.

Many different techniques have been used to measure canopy water storage and periods of leaf wetness during and after rainfall events.

First of all, there are measurements based on the change of the electrical conductivity upon the wetting of a grid sensor (Shuttleworth 1976, Getz 1978). These measurements are very easy to perform, but can diverge strongly from reality because it is almost impossible to construct a sensor with the same stiffness, heat characteristics and water retention properties as natural leaves.

A second technique is based on the measurement of β-ray transmission and has been used for dew detection on soil surfaces (Jones 1973) as well as on leaves (Barthakur 1983). A Radium204 source and Geiger Müller detector were placed 4 cm apart at both sides of a leaf, tied in a metal frame. However, the method is not suitable for automated monitoring under field conditions because of several drawbacks, notably the problem of protecting the GM detector against rain and the limited distance between source and detector.

A more direct technique has been proposed by Hancock and Crowther (1979). They measured the deflection of water-laden branches. The method proved to give a good but only qualitative insight on the wetting and drying cycles of vertically distributed branches. However, measurements on single branches are hard to calibrate and to extrapolate to the entire forest canopy.

A far better integrating method was developed by Calder and Wright (1986). They measured the gamma ray attenuation over a 25 to 35 meter horizontal transmission line, scanning vertically, right across the canopy. They succeeded to calibrate their measuring system and obtained direct measurements of canopy mass and water storage. They used a 200 millicurie Caesium137 radioactive source, emitting a beam of 660 keV gamma-rays. Unfortunately, this equipment is not suitable for unattended automated monitoring due to safety standards. Moreover, statistics on radiation show that long counting times are necessary in order to obtain reproducible estimates of densities. This limits the possibilities of studying vertical distributions of water amounts during highly dynamic rainfall events.

Recently, transmission measurements were carried out by Bouten et al. (1991). They used the attenuation of microwaves (10.26 GHz) instead of gamma ray. Vertical scans were made every 5 minutes to measure the total canopy water storage and its vertical distribution. Measurements proved to be very reproducible with a typical standard deviation of only 0.035 mm of water storage for the canopy as a whole (Bouten et al., 1994).

The choice for either measuring throughfall or canopy water storage depends on the aims of the study.

CONCLUDING REMARKS

The Penman Monteith equation has proven to be applicable to describe the energy exchange of an evaporating surface. It has been successfully applied in a wide variety of cases with different meteorological conditions and at different scales, ranging from individual leaves to regional studies. Also for heterogeneous surfaces such as landscapes or partially wet canopies the basic equation can be further extended into various concepts (Fig. 8) with different complexity. The basic difference for application of the PM-equation to different scales lies in the (empirical) parametrization of aerodynamic and surface resistances.

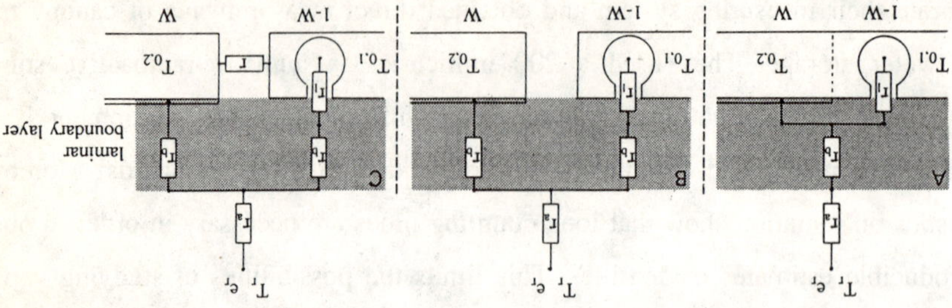

Figure 8. Various concepts for energy and water vapour exchange of a partially wet canopy. The same concepts are applicable to heterogeneous surfaces; w is wet fraction, T_0 is surface temperature, T_r and e_r are air temperature and vapour pressure at reference, r_a, r_b and r_l are aerodynamic, boundary layer and stomatal resistance (reprinted from Bouten and Bosveld, 1992b)

Many models have been built in attempts to decipher the lengthy chains of plant control on the surface resistance and the repercussions of complex feed-back mechanisms on evapotranspiration. These models have shown to be able to describe the measured dynamics of transpiration. However, they are not necessarily conclusive upon the true mechanisms. Bouten and Bosveld (1992b), for example, showed that measured transpiration dynamics were equally well fitted by a SBL model with instantaneous response to environmental conditions and a multi layer model which incorporated the course of the plant water potential over the day. This was explained by feed back mechanisms causing a high correlation between the modelled plant water potential with the associated stomatal conductance and the measured vapour pressure deficit (Fig 9). This shows that a validation tells you more about the model's capability to fit measurements than it tells you about the true mechanisms of control.

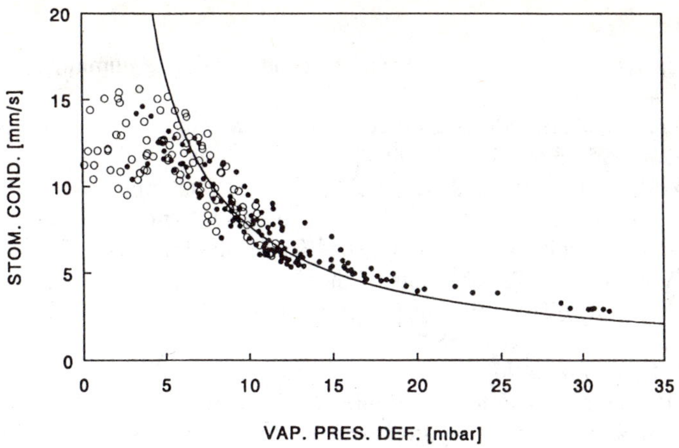

Figure 9. Stomatal conductance controlled by plant water potential as simulated with a multi layer model for a dry (•) and a wet (o) canopy. The line shows the input SBL-function for instantaneous control by vapour pressure deficit (reprinted from Bouten and Bosveld, 1992b)

It is questionable whether we need to know about the true mechanisms to be able to predict the consequences of climatic change on evapotranspiration at the earth's surface. Strong negative feed back mechanisms facilitate the modelling of the behaviour of complex systems with simple models (McNaughton and Jarvis, 1991). These empirical models can be used to run sensitivity analysis with respect to changed boundary conditions (McKenney and Rosenberg, 1993). The only prerequisite is that the behaviour is successfully simulated under changed boundary conditions. This can be tested by systematic application of a model to different locations as the variety of climatic conditions over the world is large enough. Only changing CO_2 concentration then is not incorporated although it can have large implications on evapotranspiration as discussed in another chapter of this book.

As a conclusion in this context, I could say that I agree with the content of the article by Raupach and Finnigan (1988), although their title 'Single layer models of evaporation from plant canopies are incorrect, but useful, whereas multi layer models are correct but useless' suggests a different opinion.

REFERENCES

Aphalo P.J., Jarvis P.G. (1991) Do stomata respond to relative humidity? Plant Cell Environ., 14: 127-132.

Aussenac, G. (1970) Action du couvert forestier sur la distribution au sol des precipitations. Ann. Sci. For. 27: 383-399.

Ball J.T., Woodrow I.E., Berry J.A. (1987) A model predicting stomatal conductance and its contribution to the control of photosynthesis under different environmental conditions. In: Biggens (Ed.), Progress in Photosynthesis research. Vol. IV.: 221-225, Marinus Nijhoff, Dordrecht.

Barthakur N.N. (1983) The beta-ray gauge as a leaf wetness detector. Int. J. Appl. Radiat. Isot., 34:1549-1552.

Bosveld F.C., Bouten W., Noppert F. (1992) Transpiration dynamics of a Douglas fir forest II: Parametrisation of a Single Big Leaf model. In: W.Bouten, Monitoring and modelling forest hydrological processes in support of acidification research. PhD-thesis: 163-180, University of Amsterdam.

Bouten W., Bosveld F.C. (1992a) Modelling rainfall interception and canopy wetness in a douglas fir stand. In: W.Bouten, Monitoring and modelling forest hydrological processes in support of acidification research. PhD-thesis: 65-88, University of Amsterdam.

Bouten W., Bosveld F.C. (1992b) Stomatal control in a partially wet Douglas fir canopy. In: W. Bouten, Monitoring and modelling forest hydrological processes in support of acidification research. PhD-thesis: 181-200, University of Amsterdam.

Bouten W., Bosveld F.C., Noppert F., Tiktak A. (1992a) Transpiration dynamics of a Douglas fir forest I: Evaluation of three measuring techniques. In: W. Bouten, Monitoring and modelling forest hydrological processes in support of acidification research. PhD-thesis: 149-161, University of Amsterdam.

Bouten W., Heimovaara T.J., Tiktak A. (1992b) Spatial patterns of soil water dynamics in a Douglas fir stand. Water Resources Research, vol.28, no. 12:3227-3233.

Bouten W., Swart P.J.F., de Water E. (1991) Microwave transmission, a new tool in forest hydrological research, J. Hydrol., 124, 119-130.

Bouten W., Schaap M.G., Aerts J.C.J.H., Vermetten A. (1994) Monitoring and modelling canopy water storage amounts in support of atmospheric deposition studies. J. Hydrol. (Submitted)

Businger J.A. (1986) Evaluation of the accuracy with which dry-deposition can be measured with current micro-meteorological techniques. Journal of Applied Meteorology (25):1100-1124.

Calder I.R. (1977) A model of transpiration and interception loss from a spruce forest in Plynlimon, central Wales. J. Hydrol. 33:247-275.

Calder I.R., Rosier P.T.W. (1976) The design of large plastic-sheet net rainfall gauges. J. Hydrol., 30: 403-405.

Calder I.R., Wright I.R. (1986) Gamma ray attenuation studies of interception from Sitka spruce: some evidence for an additional transport mechanism. Water

Resour. Res., 22: 409-417.

Cassell D.K., Klute A. (1986) Water potential: Tensiometry. In:A. Klute (Ed.): Methods of soil analysis Part I, Physical and Mineralogical Methods 2[e] ed.: 563-569.

Cermák J., Kucera J. (1981) The compensation of natural temperature gradients at the measuring point during sap flow rate determination in trees. Biol. Plant., 23: 469-471.

Cienciala E. (1994) Sap flow, transpiration and water use efficiency of spruce and willow in relation to climatic factors. PhD-thesis, Swe. Univ. Agric. Sci., ISSN 0348-422x.

Ford E.D., Deans J.D. (1978) The effect of canopy structure on stemflow, through-fall and interception loss in a young Sitka spruce plantation. J. Appl. Ecology, 15: 905-917.

Fritton D.D. (1969) Resolving time, mass absorption coefficient and water content with gamma-ray attenuation. Soil Science Society of America Proceedings, 33:651-655.

Gardner W.R. (1960) Dynamic aspects of water availability to plants. Soil Sci. 89: 63-73.

Gash J.H.C. (1979) An analytical model of rainfall interception by forests. Quart. J. R. Meteorol. Soc. 105: 43-55.

Gash J.H.C., Wright I.R., LLoyd C.R. (1980) Comparative estimates of interception loss from three coniferous forests in Great-Britain. J. Hydrol.. 48: 89-105.

Getz R.R. (1978) Dew monitoring network in the south east. Bull. Am. Meteor. Soc., 59: 1150-1154.

Goudriaan J. (1977) Crop micrometeorology: a simulation study, 249pp, Pudoc, Wageningen.

Granier A. (1985) Une nouvelle methode pour la mesure du flux seve brute dans le tronc des arbres. Ann. Sci. For., 42: 193-200.

Granier A. (1987) Mesure du flux de seve brute dans le tronc du Douglas par une nouvelle methode termique. Ann. Sci. For. , 44: 1-14.

Halbertsma J., Przybyla C., Jacobs A. (1987) Application and accuracy of a dielectric soil water content meter. In: Proc. Conf. on measurement of soil and plant water status, Logan USA, 11-15, July 1987.

Hancock N.H., Crowther J.M. (1979) A technique for the direct measurement of water storage on a forest canopy. J. Hydrol., 41: 105-122.

Heimovaara T.J. (1994) Design of Triple Wire Time Domain Reflectometry Probes in Practice and Theory. Soil Sci. Soc. Am. J. 58: 1410-1417

Heimovaara T.J., Bouten W. (1990) A computer controlled 36-channel Time Domain Reflectometry system for monitoring soil water contents. Water Resources Research, 26:2311-2316.

Jackson I.J. (1975) Relationships between rainfall parameters and interception by tropical plant forests. J. Hydrol. 24:215-238.

Jarvis P.G. (1976) The interpretation of the variations in leaf water potential and stomatal conductance found in canopies in the field. Phil. Trans. R. Soc. Lond. B. (273):593-610.

Jones H.G. (1973) Estimation of plant water status with the beta gauge. Agric. Meteorol., 11: 345-355.

Kowalik P.J., Borghetti M., Busoni E., Sanesi G. (1988) Measured and simulated water relations in a Douglas-fir forest during the development of drought in the Apennines, Central Italy. For. Ecol. Manage. (25):181-194.

Lafolie F., Bruckler L., Tardieu F. (1991) Modeling root water potential and soil-root water transport: I. Model presentation. Soil Sci. Soc. Am. J. 55: 1203-1212.

Leadley P.W., Drake B.G. (1993) Open top chambers for exposing plant canopies to elevated CO_2 concentrations and for measuring net gas exchange. Vegetatio 104/105: 3-15.

Leuning R., Foster L.J. (1990) Estimation of transpiration by single trees: Comparison of a ventilated chamber, leaf energy budgets and a combination equation. Agric. and For. Meteorol. 51:63-68.

Lockwood J.G. (1992) The sensitivity of the water balance of a wet multi layer model pine canopy to variations in micrometeorological input, Clim. Change, 20, 23-56.

Lynn B.H., Carlson T.N. (1991) Simulating transpiration plateaus: the importance of leaf water potential. Ecol. Modelling, 58: 199-208.

Makkink G.F. (1957) Testing the Penman formula by means of lysimeters. Journ. Int. of Water Eng. 11: 277-288.

Marthaler H.P., Vogelsanger W., Richard F., Wierenga P.J. (1983) A pressure transducer for field tensiometers. Soil Science Society of America Journal, 47:624-627.

McKenney M.S., Rosenberg N.J. (1993) Sensitivity of some potential evapotranspiration estimation methods to climate change. Agric. For. Meteorol., 64: 81-110.

McNaughton K.G., Jarvis P.G. (1991) Effects of spatial scale on stomatal control of transpiration. Agric. For. Meteorol., 54: 279-301.

Monteith J.L. (1965) Evaporation and environment. In: The state and movement of water in living organisms. 19th Symp. Soc. Exp. Biol. (G.E. Fogg, ed.), pp205-235. Cambridge University Press, London.

Mott K.A., Parkhurst D.F. (1991) Stomatal response to humidity in air and helox. Plant Cell Environm., 14, 509-515.

Mulder J.P.M. (1985) Simulating interception loss using standard meteorological data. In: Hutchison, B.A. and B.B. Hicks (eds), The forest-atmosphere interaction: 177-196, Reidel Publ. Comp.

Penman H.L. (1948) Natural evaporation from open water, bare soil and grass., Proc. Roy. Soc. London, Ser A, 193, pp120-146.

Priestley C.H.B., Taylor R.J. (1972) On the assessment of surface heat flux and evaporation using large scale parameters. Monthly Weather Review, 100, 2, pp 81-92.

Rawitz E., Etkin H., Hazan A. (1982) Calibration and field testing of a two-probe gamma-gauge. Soil Science Society of America Journal, 46:461-465.

Raupach M.R., Finnigan J.J. (1988) Single layer models of evaporation from plant canopies are incorrect, but useful, whereas multi layer models are correct but

useless: Discuss. Aust. J. Plant Physiol., 15: 705-716
Rijtema P.E. (1965) An analysis of actual evapotranspiration. Agric. Res. Rep.659, Pudoc, Wageningen, The Netherlands.
Rutter A.J., Kershaw K.A., Robins P.C. and Morton A.J. (1971) A predictive model of rainfall interception in forests. I. Deviation of the model from observations in a plantation of Corsican Pine. Agric. Meteorol. 9:367-384.
Rutter A.J., Morton A.J., Robins P.C. (1975) A predictive model of rainfall interception in forests, II. Generalization of the model and comparison with observations in some coniferous and hardwood stands. J. Appl. Ecol., 12: 367-380.
Sakuratani T. (1981) A heat balance method for measuring water flux in the stem of intact plants. J. Agric. Meteorol., 37: 9-17.
Sellers P.J., Lockwood J.G. (1981) A computer simulation of the effects of differing crop types on the water balance of small catchments over long time periods. Quart. J. R. Meteorol. Soc. 107:395-414.
Schulze E.D., Cermák J., Matyssek R., Penka M., Zimmerman R. (1985) Canopy transpiration and water fluxes in the xylem of the trunk of Larix and Picea trees - a comparison of xylem flow, porometer and cuvette measurements. Oecologia 66: 475-483.
Shuttleworth W.J. (1976) Experimental evidence for the failure of the Penman-Monteith equation in partly wet canopies. Boundary layer Meteorol., 10: 91-94.
Stewart J.B. (1988) Modelling surface conductance of pine forest. Agr. and Forest Met. (43):19-35.
Stewart J.B., Thom A.S. (1973) Energy budgets in a pine forest. Quart. J. Roy. Met. Soc., 103: 345-375.
Tan C.S., Black T.A. (1976) Factors affecting the canopy resistance of a Douglas-fir forest. Bound. Layer Met. (10):475-488.
Thom A.S., Oliver H.R. (1977) On Penman's equation for estimating regional evaporation. Q.J.R. Meteorol. Soc., 103:345-357.
Tiktak A., Bouten W. (1992) Modelling soil water dynamics in a forested ecosystem III: model description and evaluation of discretization. Hydrol. Proc. 6: 455-465.
Topp G.C., Davis J.L., Annan A.P. (1980) Electromagnetic determination of soil water content: Measurements in coaxial transmission lines. Water Resources Research 16:574-582.
Wallin J.R. (1963) Dew, its significance and measurement in phytopathology. Phytopathology 53: 1210-1216.
Webb E.K., Pearman G.I., Leuning R. (1980) Corrections of flux measurements for density effects due to heat and water vapour transfer. Quart. J. Met. Soc. (106):85-100.

SOIL - VEGETATION - ATMOSPHERE RELATIONS: PROCESS AND PROSPECT

William James Shuttleworth
Department of Hydrology and Water Resources
The University of Arizona
Tucson, Arizona 85721
U.S.A.

1. Introduction

The driving mechanism of the climate system and the primary instrument of climate change is the capture of solar energy, and the way this energy is shared within the atmosphere. On average about half of the sun's energy reaches the earth's surface. From there it returns to the atmosphere in different forms - as reflected or emitted radiation, or as latent or sensible heat, depending on the nature of the underlying surface (Verstraete and Dickinson, 1986). The way in which such solar energy is shared is determined in substantial measure by the extent and availability of water at or near the surface.

Taking the global average, only about 40% of the precipitation measured as falling on continental surfaces is then measured to run off to the oceans (UNESCO, 1978) - which implies that there is "double counting" of precipitation, *i.e.* liquid or solid water falls, is re-evaporated, and falls again. The extent to which such recycling occurs is influenced by soil-vegetation-atmosphere processes, and by the structure and the level of activity of the biosphere. The size and shape of vegetation influences the amount of solar radiation, momentum and rain water captured at the ground, and the depth to which soil water is accessible to the atmosphere via roots. The level of activity of vegetation controls the rate of water loss directly via stomatal control on transpiration rate; it also has more subtle influence, by changing the hydrological properties of the soil to alter infiltration and bulk soil porosity.

The nature of the earth's surface - whether it is water or land, and what vegetation covers

NATO ASI Series, Vol. I 31
The Role of Water and the Hydrological Cycle
in Global Change
Edited by Howard R.Oliver and Sylvia A.Oliver

the land is known to influence the atmosphere, sometimes visibly by moderating overlying cloud (*e.g.* Shuttleworth, 1988). The extent of human intervention through land management is still expanding, but is already substantial: typically 30-50% of continental surfaces are now managed as cropland and permanent pasture (Shuttleworth 1991). Given the extent and rate of land cover change, and the importance of the associated soil-vegetation-atmosphere processes, it is clear that describing these processes with adequate realism is important for the prediction of global change. Indeed, providing such description is at the very center of human concern since it relates directly to the formulation of public policy. Since local, regional and global climate can be altered by human intervention, and since people live on continents, *it is the prediction of near-surface weather and climate over continents which matters most to human welfare.*

2. The Basic Elements in Soil-Vegetation-Atmosphere Processes

The exchange of energy over land surfaces is controlled by the nature of the vegetation and soil. The spectrum of natural radiant energy interacting with the earth falls into two almost distinct bands (*e.g.* Monteith and Unsworth, 1991), namely shortwave and longwave radiation. These bands are determined by Plank's Law, and are related to the temperature of the sun (about 6,000°K) and the temperature of the earth (about 300°K) respectively. When writing models of soil-vegetation-atmosphere processes it is therefore convenient to describe these two radiation streams separately. Specifying (or calculating) the reflectivity of continental surfaces for solar radiation (integrated over the wavelength range 0.15-4.0 μm) is therefore important. This is complicated by the fact that the reflectivity for green vegetation and soil differ above and below 0.7 μm because plants absorb visible light preferentially for photosynthesis. More complex models allow for this difference by using "two stream" models of radiation capture (*e.g.* Dickinson *et. al.*, 1986).

The description of longwave radiation exchange is made simply, even in complex models. The net emission of radiant energy in the "thermal" wave band (wavelength range 3-100 μm) is calculated from the Stefan-Boltzmann Law, given a value of average surface emissivity over this wavelength range. Current models of soil-vegetation-atmosphere processes in Global Models normally prescribe an average value of emissivity: most simply set the value to unity.

Once radiant energy exchange has been described, the next priority is to provide a description of the way in which the captured radiant energy is shared between latent and sensible heat, or "stored" temporarily in the soil or vegetation. Temporary storage of thermal energy in soil is in principle a comparatively simple physical process, especially for uniform soil. The equation of thermal conductivity and the divergence equation are the basic descriptive equations used, often in discrete form, to describe soil heat flow between a few (typically 2 or 3) layers (e.g. Dickinson *et. al.*, 1986; Sellers *et. al.*, 1986). The computational process is sometimes further simplified by introducing the so-called "force restore" method (Deardorff, 1978) which, in effect, assumes that the major features of soil heat flux can be adequately represented by two Fourier components, namely those corresponding to the daily and annual cycles.

The latent heat of vaporization of water , λ, provides a critical bridge between energy exchange and water exchange when describing soil-vegetation-atmosphere processes. The physical process of evaporation is a dual exchange: some water molecules gain sufficient energy to become vapor - the number being statistically linked to temperature, while some vapor molecules are captured at the water surface - the number being related to the vapor pressure of the air. A balance between these two rates occurs with a saturated atmosphere (see Shuttleworth, 1975), and the fact that saturated vapor pressure, e, and the rate of increase of saturated vapor pressure, Δ, are both well-defined functions of temperature is also extremely important in writing models of soil-vegetation-atmosphere processes. How such models describe the partition of "available energy" (equal to net radiant energy input less soil heat flow) into latent-heat, *i.e.* evaporation, and sensible heat depends on Δ, and on the degree of saturation of the atmosphere.

The other basic physical process which determines energy partition is atmospheric diffusion. All atmospheric entities are transferred by diffusion, and in particular sensible and latent heat transfer and momentum transfer depend both on molecular diffusion close to the surfaces of vegetation and soil, but also on turbulent diffusion, this being the major transfer mechanism further from the ground. Most models of soil-vegetation-atmosphere processes include description of transfers within the atmospheric "Surface

Layer" up to a few meters above the ground. Within this layer the turbulent diffusion mechanism is usually a mixture of forced convection, which is determined by wind speed and by the aerodynamic roughness of the underlying vegetation and soil; and free convection, which is influenced by the amount of energy leaving the ground as sensible heat. Models of surface energy exchange must necessarily include simultaneous description of momentum exchange, since this determines forced convection; and such models often include some iteration, *i.e.* they first-compute energy fluxes without free convection, then include this process as a correction using the "first guess" estimate of sensible heat.

Very commonly the movement of energy, momentum and other atmospheric entities (*e.g.* carbon dioxide) by diffusion processes is represented in soil-vegetation-atmosphere models by use of resistance analogies. In general, the diffusion equations are integrated so that fluxes of energy, for example, are proportional to differences in equivalent potentials (vapor pressure for latent heat; temperature for sensible heat), divided by a resistance (see Shuttleworth, 1991 for details). Some more complex models will include a resistance to represent exchange by molecular diffusion through the boundary layer adjacent to transpiring leaves and soil, but more commonly they will include a "surface resistance", which describes diffusion through the stomata from inside to outside leaves; and an "aerodynamic resistance", which describes diffusion from within the vegetation canopy to a prescribed height in the atmospheric surface layer above. Using the resistance analogue approach, it is possible to create quite complex, perhaps multilayer, models of canopies (Shuttleworth, 1991), but the most complex models of soil-vegetation-atmosphere processes used in models of global climate rarely represent more than one layer of vegetation. They usually do so using equations which assume the general behavior observed for the stomatal resistance of individual leaves, but they are "recalibrated" to describe the surface resistance of whole canopies (*e.g.* Sellers *et. al.*, 1989).

When rain falls on vegetation, the primary source of evaporating water is the free water on the canopy, not the moist cells within the leaves as is the case for transpiring vegetation. In effect, the stomatal resistance is "shorted out", and evaporation rates increase. The wetting by rainfall and subsequent drying of vegetation is usually represented by a canopy water balance model, which includes the description of rainfall capture, of water dripping from the canopy, and of evaporation from the surface of the leaves. The most common form of mass balance used is that first described by Rutter (Rutter *et. al.*, 1971; 1975), though most global models use a simplified version of this model.

Water movement within the soil is also described in models of soil-vegetation-atmosphere processes, albeit usually in simple terms, with the Darcy equation applied to compute vertical water flow between a few (usually 2 or 3) layers of soil. Furnishing values for the soil parameters required to enable a description based on the Darcy equation is arguably the most difficult problem in providing globally available, yet realistic models of soil-vegetation-atmosphere processes. Even at plot scale the variability in the value of these hydrological parameters from one place to the next is know to be very large indeed. The most common approach is to assign typical parameters by specifying the soil as belonging to one of (say) 10 classes. Very often the classification of Clapp and Hornberger (1978) is used, and small-scale and meso-scale heterogeneity in soil properties around these typical parameters is simply ignored.

In summary, models of soil-vegetation-atmosphere processes normally start by describing radiation exchange by treating this as two distinct streams of (solar and longwave) radiation. Having allowed for energy temporarily stored in the soil, the outstanding problem is to partition the remaining energy either as sensible heat, or for evaporation. The latent heat of vaporization is used to link the energy and mass balance equations, and diffusion models created to describe the movement of water vapor and sensible heat, these being implicitly or explicitly framed in terms of analogue resistances. Once energy and mass movement has been described for one time step in the model, the state variables (*i.e.* the stores of water and energy in the vegetation or soil) are updated, and these new values are then used to compute energy and mass flows at the subsequent time step.

Clearly the above general approach to describing soil-vegetation-atmosphere processes is capable of very diverse implementation, and there are numerous models of varying complexity currently available. Arguably the most important, unresolved issue is the definition of which of the several processes outlined above must be represented, and at what level of accuracy. Further, if a process is necessarily in the model, how can the parameters in the equations describing that process be specified globally?

3. Range of Complexity in Present Day Models

The broad range of complexity used in current models of soil-vegetation-atmosphere processes is in part generated by their diversity of purposes. Originally such models were mainly rather simple descriptions of evaporation for use in hydrological models. Over the last decade their use in atmospheric models has expanded rapidly, both in magnitude and importance. In the future such models will increasingly be needed to provide the vital coupling in models which include descriptions of both atmospheric and hydroecologic processes. Notwithstanding this diversity, as a basic minimum they must describe

- bulk momentum exchange
- exchange of radiation (both solar and longwave)
- partition of radiant energy between soil heat flux, and latent and sensible heat flux.

The number of models of soil-vegetation-atmosphere processes described in the literature is large and expanding - there are at least 60 distinct models, perhaps 100, or more - but many are locally relevant "research" models with process emphasis relevant to specific research issues. The range of models applied at the global scale (in General Circulation Models) is less extensive, and they are usually one dimensional, mainly with representation of only a few soil and canopy layers. They exist in three general forms, namely

(a) "Bucket" models, with no representation of a vegetation canopy;
(b) "Micrometeorological" models, with substantial representation of canopy features; and
(c) "Intermediate" models, with some canopy features included.

These three classes are overviewed in the following sections.

Bucket Models

The simplest models of evaporation were originally used in hydrological models and describe evaporation rate as potential evaporation moderated by soil moisture. Such models were also the first to be incorporated into atmospheric models (*e.g.* Manabe; 1969), with the addition of features and parameters relevant to energy and momentum, specifically with inclusion of

albedo (to describe solar radiation capture)
emissivity (to describe longwave radiation exchange)
roughness length (to describe momentum exchange)
soil thermal properties (to describe temporary storage of energy in the soil)

In such models energy partition is determined by calculating actual evaporation from potential evaporation, with sensible heat exchange then given as the residual in the energy balance. The moderating factor (in the range 0 - 1) used to scale actual to potential evaporation is set to be the fractional fill of a "bucket" of specified capacity, typically around 150 mm. This fractional fill is calculated from a running water balance between precipitation input and evaporation loss.

Micrometeorological Models

The most complex models currently used in General Circulation Models are typified by the Biosphere-Atmosphere Transfer Scheme (BATS) (Dickinson *et. al.*, 1986) and the Simple Biosphere (SiB) Model (Sellers *et. al.*, 1986). These are essentially micrometeorological models capable of representing surface exchanges reasonable realistically for a uniform plot of vegetation, but they are applied at a much bigger scale to areas up to (say) 10^5 km.

The exchange of solar radiation is represented in detail in BATS and SiB using the two-stream approximation (Ross and Monteith, 1975) mentioned earlier, while upward longwave radiation is calculated from the Stefan-Boltzmann law with a prescribed value of emissivity, and an effective radiometric surface temperature, this being an appropriately weighted combination of the surface temperatures of the vegetation and the soil surfaces in the model. Both models describe momentum transfer traditionally, in terms of aerodynamic roughness which is prescribed for each cover type in BATS, but calculated during a pre-processing phase from prescribed canopy morphology in the case of the SiB model.

Transpiration is calculated in BATS and the SiB model, and indeed all micrometeorological models, using a "big leaf" model, with whole-canopy surface resistance (effectively) represented with a Jarvis-Stewart model (*e.g.* Stewart, 1988) as a product series of soil and vegetation cover dependent stress functions. Most micrometeorological models adopt a derivation (usually a simplification) of the Rutter model of canopy water balance to describe wet canopy evaporation. The critical parameter in describing this last process is the storage capacity of the canopy: this is normally defined as proportional to the leaf area of the vegetation, and tests against field data (Sellers *et. al.*, 1989) suggest use of values in the range 0.1 to 0.25 mm per unit leaf area. Micrometeorological models include a simple, vertical soil hydrology, with water balance calculation typically for 2 or 3 soil layers.

In general, micrometeorological models require specification of many parameters for different canopy covers and soil types, and accomplish this by allowing a restricted number of classifications of each. Never the less, many land cover classes (BATS has 20; SiB has 12), and soil types (typically 10) are allowed. The primary problem with the use of micrometeorological models is that the values of these parameters largely remain speculative.

Intermediate Models

Intermediate models trade rigor in process description for simplicity. Commonly such models allow a significant range of vegetation covers and soil types, but simplify the physical representation of the exchange processes for each. The UK Meteorological Office model, for instance, prescribes fixed values not only for radiation and aerodynamic parameters for each land cover, but also for surface resistance (Lean and Warrilow, 1985). Soil moisture restriction is allowed for as a multiplicative function related to the moisture content within a rooting layer. Other models, such as that of the French Meteorological Service (Noilhan and Planton, 1989), are similar, but parameterize surface resistance in a way broadly equivalent to BATS.

An alternative approach is merely to simplify the computational processes involved in more complex models. "Simple SiB" (Xue *et. al.*, 1991), for example, essentially uses the same assumptions as the SiB model, but codes these more efficiently by using empirical parameterizations to replace some of the time consuming direct calculations used in the original version of SiB. At the other end of the spectrum, bucket models can be made significantly more realistic simply by including a typical value of surface resistance [which is usually about 70 cm^{-1}, see Shuttleworth, 1993], to give a so-called "super bucket" model.

The use of this class of models with intermediate complexity is increasing, primarily because they represent a practical but still plausible alternative to more complex models pending evidence of the need for the more complex micrometeorological models.

4. Intercomparison of Present-Day Models

The Project for Intercomparison of Land-Surface Parameterization Schemes (PILPS) is an ongoing international initiative to evaluate the performance of sub-models of soil-vegetation-atmosphere interaction processes in comparison with each other, and in due course in comparison with relevant field data (Henderson-Sellers and Dickinson, 1992). This project is still in an early stage of implementation, but has already provided interesting first results (Pitman *et. al.*, 1993) which are summarized below.

PILPS currently involves around 25 land-surface parameterization schemes with a wide ranging range of complexity, and includes examples of the three classes overviewed in the previous section. So far these several models have been used with two synthetic series of climatological data calculated using a General Circulation Model, namely the National Center for Atmospheric Research (NCAR) Community Climate Model (CCM). The two data series was selected to correspond to a temperate grassland site (at 52°N, 0°E), and a tropical forest site (3°N, 60°E).

(a)

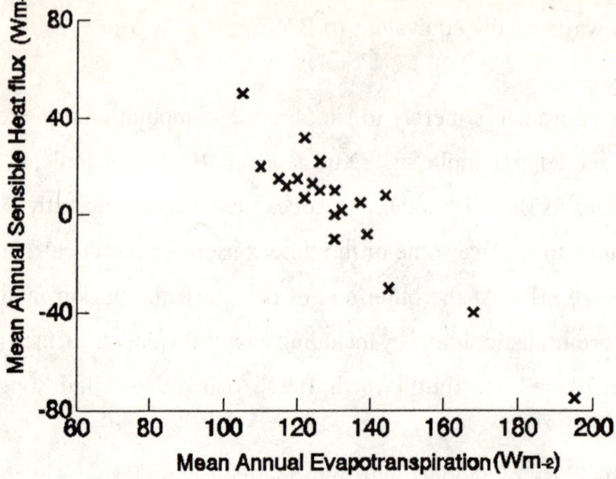
Mean Annual Sensible Heat flux (Wm-2)
Mean Annual Evapotranspiration (Wm-2)
80
40
0
-40
-80
60
80
100
120
140
160
180
200

(b)

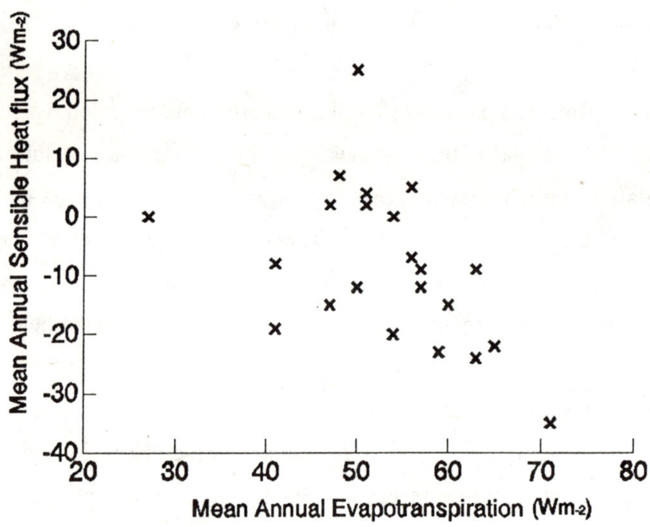
Mean Annual Sensible Heat flux (Wm-2)
Mean Annual Evapotranspiration (Wm-2)
30
20
10
0
-10
-20
-30
-40
20
30
40
50
60
70
80

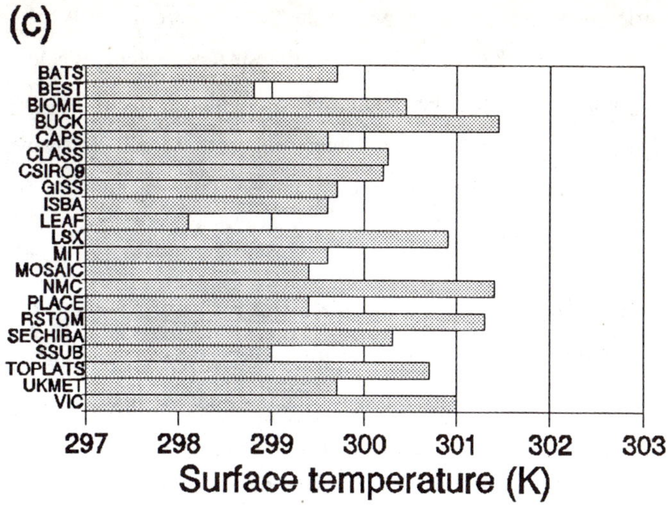

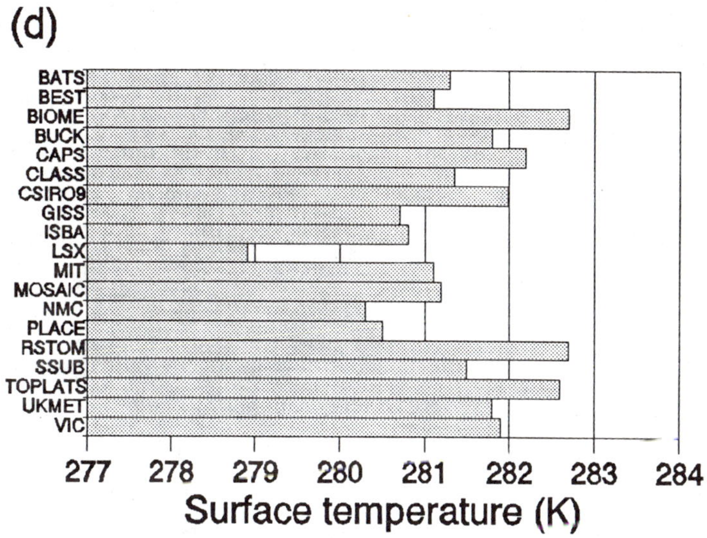

Figure 1. *Mean annual evapotranspiration and sensible heat given by models involved in PILPS calculated from near surface weather variables for (a) "tropical forest", and (b) "grassland" sites. Mean annual temperature for (c) "tropical forest" and (d) "grassland" sites.*

These data series were used as "driving variables" which were input to the several land-surface parameterization schemes, and the resulting surface exchanges calculated in each case. The outgoing fluxes from the ground are decoupled from the atmosphere above in these studies. That is, the driving variables remain those calculated by the NCAR-CCM, unaltered by the calculated surface response. This will tend to emphasize the distinction between calculated fluxes.

Figure 1(a) and 1(b) summarize the results of the PILPS comparison for the annual average values of latent and sensible heat for forest and grassland sites respectively. Figures 1(c) and 1(d) show similar results for mean annual temperature.

There are some very substantial outliers in these diagrams, and still substantial spread in the order ± 20 Wm^{-2} for both forest and grassland amongst the remaining group of models if we neglect these outliers. The essential ecoclimatological difference in energy partition between forest and grassland is in fact resolved, but only just. The intercomparison of annual temperatures is not good - but neither is it terrible. Neglecting outliers again, the agreement is to within about 1 - 2 °K. However, it is important to remember that predictions of the global warming associated with doubling of atmospheric CO_2 concentration are themselves only in the order 2 - 4 °K. PILPS made similar comparisons between models at monthly and daily time scales. In general, the agreement, already marginal at the annual time scale, is reduced at shorter time scales (Pitman *et. al.*, 1993).

The difference in performance between land surface parameterization schemes used in PILPS is not strongly correlated with the complexity of representation used in the models, except that some of the most significant outliers are bucket models. The origin of these discrepancies is not know at this time, but the results of PILPS must raise the question as to which matters most, the complexity of process representation used in these models, or the specification and calibration of the parameters they contain?

5. Calibration of Soil-Vegetation-Atmosphere Models for Simple Cover

In selected situations it is possible to define worthwhile studies of the potential consequences of land cover change when currently uniform biomes are at risk of large scale removal and replacement with alternative vegetation types. The potential removal of tropical forests for use as pastureland is an obvious example of dramatic (but not totally unrealistic) human-induced change, which as been much studied using General Circulation Models (*e.g.* Lean and Warrilow, 1989; Nobre *et. al.*, 1991). Since it is change in the soil-vegetation-atmosphere interactions which is the essence of these studies, it is the calibration of models of these reactions which is arguably a most relevant issue in this case. Fortunately, this is one case in which provision of data for the calibration of models has been pursued with vigor (Shuttleworth *et. al.*, 1991), at least in the case of the Amazon basin.

The Amazonian rainforest is still 90% intact, though deforestation proceeds quickly, mainly around the southern and western edges. Detailed micrometeorological measurements at a single study site in central Amazonia made more than a decade ago (Shuttleworth *et. al.*, 1984; Shuttleworth, 1989) provided an at that time unique, set of data relevant to the calibration of models of soil-vegetation-atmosphere interactions. They included routinely available above-canopy weather variables measured on a scaffolding tower (equivalent to the "driving variables") for 25 months, together with directly measured surface radiation and latent and sensible heat exchanges for four extended (typically 2 months) periods selected to sample the seasonal precipitation regime. The routine measurements also included the net loss of rainfall resulting from interception by the forest canopy, and measurements of soil moisture status to a depth of 2 m, sampled at two weekly intervals.

Early comparisons between these data and then existing, uncalibrated land surface parameterization schemes suggested that these were in substantial error (Shuttleworth and Dickinson, 1989). But, subsequently these data were used to re-calibrate the parameters in one advanced scheme, SiB (Sellers *et. al.*, 1989), and this substantially improved the descriptive ability of that model with respect to simpler, uncalibrated bucket models (Sato *et. al.*, 1989).

Although the above described experimental study provided useful calibration data of the "before" in deforestation studies, data for calibrating the pastureland "after" for such studies has only recently become available (Shuttleworth *et. al.*, 1991). Ongoing research under the Anglo-Brazilian Amazonian Climate Observational Study (ABRACOS) is providing sustained comparative measurements of near surface climate over adjacent extensive areas of undisturbed forest and cleared pastureland at three sites, which sample the ecoclimatological (precipitation) gradient across Amazonia. These measurements are being complemented by comparative measurements of surface exchanges for forest and pasture during selected periods.

Although data from ABRACOS is only now being used to calibrate GCMs, some of the early results already provide categorical evidence of distinct differences in key parameters which are bound to have effect in deforestation studies. The difference in albedo between forest and pastureland is quantified and confirmed, and this, together with the characteristic difference in surface temperatures for the two vegetation types, leads to a measurable difference in the net capture of radiant energy (Bastable *et. al.*, 1993). While because of the difference in aerodynamic interactions of forest and pasture lands, the climate over pasture shows much greater daily variation than that over forest (Bastable *et. al.*, 1993).

There is now known to be a very substantial difference in the rooting depth between forests and grassland, much larger than hitherto thought, with most of the root-accessible water in the top meter under pasture, but forest able to access water to a depth of several (perhaps 6 - 8) meters. As a result Amazonian rainforest rarely shows evidence of water stress in current climates, but the replacement grasslands respond noticeable to dry spells of a few weeks. In the course of the next few years the ABRACOS data will transform the credibility of presently available predictions of the potential consequences of large scale Amazonian deforestation, though not perhaps alter the predicted magnitude of the likely change greatly. Currently the predicted regional changes are for about a 20% reduction in evaporation, about a 30% reduction in precipitation, and about a 2 °C rise in temperature.

6. Soil-Vegetation-Atmosphere Interactions in Weather Forecast Models

Until recently weather forecast models have assigned limited importance to their description of soil-vegetation-atmosphere processes, the reasoning being that forecasts are primarily determined by the data input at the beginning of a predictive cycle, and that surface inputs have little time to modify the ensuing predictions. This is largely true, and is particularly true for short term forecasts, but recent evidence suggests that the form of land surface parameterization schemes can be important since data to define the initial status of soil moisture stores is limited.

In general terms, it is the realism of the process representation which matters most in climate (*i.e.* long term) prediction. Faults in equations or parameter values are eventually revealed at "climate" time scales, but the initiation of moisture and energy stores at the beginning of climate prediction runs ultimately has limited effect. On the other hand, it is the initiation of such stores, particularly the initiation of the soil moisture store, which has most effect in weather prediction since this has impact immediately. In practice soil moisture status is not a data variable which is currently amenable to measurement, so the initiation of weather forecast models is derived from the predictive history of the model itself. In this way, gross defects in land surface parameterization schemes can propagate to calculate an erroneous initiation of the soil moisture store, which has an immediate and detrimental impact on the predicted weather.

A recent change in the formulation of land surface parameterization used at the European Centre for Medium Range Weather Forecasts illustrates this subtle effect well. Improvements in the realism of the land surface scheme between "Version 47" and "Version 48" of this model resulted in more realistic, wetter soils in the continental interiors during summer months. The resulting improved initiation of predictive runs brought dramatic improvements in forecasts of the 1993 floods in the Mississippi basin (Beljaars et al., 1993).

Recent detailed studies of the model mechanism which allow better prediction (Viterbo, 1994) show that this is not due to greater evaporation and recycling of water locally, rather that the greater evaporation upwind (in Texas) allows a vertical temperature profile in the central Mississippi which allows the model to predict greater convective precipitation.

7. Investigation of Heterogeneous Vegetation Covers

Regardless of their complexity, most models of soil-vegetation-atmosphere processes and virtually all land-surface parameterization used in GCMs represent surface exchanges as vertical flows, often assuming uniform land cover over extensive areas. In reality for large portions of continents surfaces are heterogenous, and this is particularly the case for land areas where human intervention is greatest. Heterogeneity occurs in the overlying vegetation cover and in the underlying soil, and this is further complicated by topography, which can generate movement in surface and subsurface water, and which may in part determine the spacial position of vegetation.

In addition, atmospheric processes can themselves impose (albeit perhaps transient) surface heterogeneity by modifying surface radiation in response to cloud cover, and by distribution precipitation non-uniformly. Over the past decade most research into heterogenous covers has focussed on investigating the representation of mixed vegetation and, in particular, on determining the importance (or otherwise) of atmospheric advection in determining the values of area-average, effective parameters to be used in models of soil-vegetation-atmosphere processes.

One approach to representing heterogenous surfaces is to decrease the intergrid distance used in representational models, but this rapidly generates huge demands for computational resources. The alternative approach is to seek to provide effective values for parameters within the existing land surface scheme such that these allow the scheme to describe area-average surface exchanges with adequate accuracy. This last approach is, in effect, a direct extension of the philosophy already used in providing plot-scale calibration as described earlier. To pursue it requires that we postulate "aggregation rules", that is hypothetical and hopefully simple equations, which can be used at large scale to calculate effective area-average values of parameters relevant to the individual covers from the values present in the heterogeneous landscape.

In practice the aggregation rules relevant to a particular land surface parameterization will tend to be scheme specific, but Shuttleworth (1991) suggests that a general approach to specifying the required rules might be to define the effective parameter as the weighted average for component covers using that function of the parameter which most nearly links it to a linear average of the associated flux. This suggestion reflects the fact that it is the area-average flux which is required.

When applying this general rule it is necessary to have some basic understanding of how each parameter within a given model changes the calculated flux. When defining this it is sometimes useful to frame the models as comprising components which directly relate to the availability of energy and water to the atmosphere. Using this approach, Shuttleworth (1991) postulates hypothetical aggregation rules for certain key parameters which control the capture of energy and water from the atmosphere such as albedo, emissivity and infiltration rate; for the magnitude of energy and moisture stores in the model; and for parameters which control the rate of access by the atmosphere to these stores, namely aerodynamic and stomatal resistance.

Much recent research has sought to investigate the validity or otherwise of the aggregation rules as described above. The methodology has been to combine experiment and representative models to synthesize area-average values of near-surface weather variables and surface exchanges, and then to compare these synthetic surface "data" with values derived using a single point-model in which parameters have been calculated using the hypothetical aggregation rule under test. Large scale field experiments (see Shuttleworth, 1991a, 1991b for reviews) have been carried out in which plot scale models of component covers in the selected study area are calibrated, and which allow the validation of meso-scale atmospheric models from aircraft and boundary layer sounding systems.

Analysis and interpretation of the first of these experiments, HAPEX-MOBILHY (André *et. al.*, 1988) is most advanced. The primary land cover distinction in this case was between an extensive area of pine forest on the one hand, and an area of mixed agricultural crops on the other. Model simulations of the observed data has been very successful, and suggest that coupled surface-atmosphere models operating at meso-scale with a grid mesh of about 10 km, can indeed adequately represent both the surface exchanges and ensuing atmospheric response well.

Similar large scale field experiments have been carried out elsewhere. The First ISLSCP Field Experiment (FIFE: Sellers *et. al.*, 1988) investigated a 15 km x 15 km area of comparatively uniform prairie grass in Kansas, USA in 1987 and 1989. More recently field experiments have been carried out at the 10,000 km^2 scale in desertification-threatened areas with heterogenous vegetation in central Spain, in 1992 (EFEDA: Bolle *et. al.*; 1993), and in the Sahel in 1993 (HAPEX-Sahel: Goutorbe *et. al.*, 1994). Analysis and interpretation modeling of the data from these experiments is well in hand, and is already contributing to define the need for a change in focus within this subject area, as overviewed in the next section.

8. Current Status and Future Emphasis

Recently, the Interactional Geosphere-Biosphere Programme's Cure Project *Biospheric Aspects of the Hydrological Cycle* and the *International Satellite Land Surface Climatology Project* jointly sponsored a workshop, the *Tucson Aggregation Workshop*, whose purpose was to provide a state of the art review of progress in the area of providing aggregate description of mixed land cover. In practice this workshop provided better definition of the current status of our understanding of the following questions:

a) is atmospheric advection important when formulating area-average descriptions of mixed cover?
b) is topography important to area-average descriptions?
c) can we remotely sense important aspects of vegetation controls on exchanges?
d) what are the primary future research priorities?

We overview the results below.

Is Atmospheric Advection Important?

Of the several question just posed, this is the question which has received most attention and for which there has been most progress. As outlined above, the method used to address this question is to use coupled surface atmosphere models to synthesize "true" area-average fluxes, and to compare these values with those calculated from a single equivalent point model, whose parameters have been defined using hypothetical aggregation rules. Experiments provide reassurance that the coupled model is a reasonable representation of reality. Based on observations made during HAPEX-MOBILHY, which suggest that coherence in surface cover can effect atmospheric processes (Shuttleworth, 1988), researchers have been careful to implement this methodology using models which represent atmospheric response at two different spatial scales, *i.e.* mesoscale circulations at greater than about 10 km, and plot scale movement at much finer scale.

Noilhan (1994) made a model synthesis of the data from HAPEX-MOBILHY using a meso-scale model with a grid mesh of 10 km, this being consistent with the length scale of the main organizational structure is land cover in this experiment. The model was parameterized using soil/vegetation maps for the study area, some of the information in these maps, particularly that relating to vegetation, being derived from satellite data. Figure 2 shows that the synthetic area - average fluxes calculated with this representational model agree with those calculated from a single point model with aggregation rule derived parameters to within about 10%, both in terms of mean daily trend and in terms of the integrated evaporation over several weeks. The implication is that mesoscale atmospheric advection only has limited impact at this level, even for the dramatic land cover distinction between forest and agricultural crops, and where substantial mesoscale structure is observed (and modeled) in the atmosphere.

Similar studies have been made using coupled surface-atmosphere models to investigate the influence of atmospheric advection at patch scale, that is with vegetation patches in the order of a kilometer, or less. In fact valuable initial insight had already been gained using coupled models operating at this scale with idealized and/or extreme mixes of surface covers. Mason and co-workers (Mason, 1988; Woods and Mason, 1991) suggested

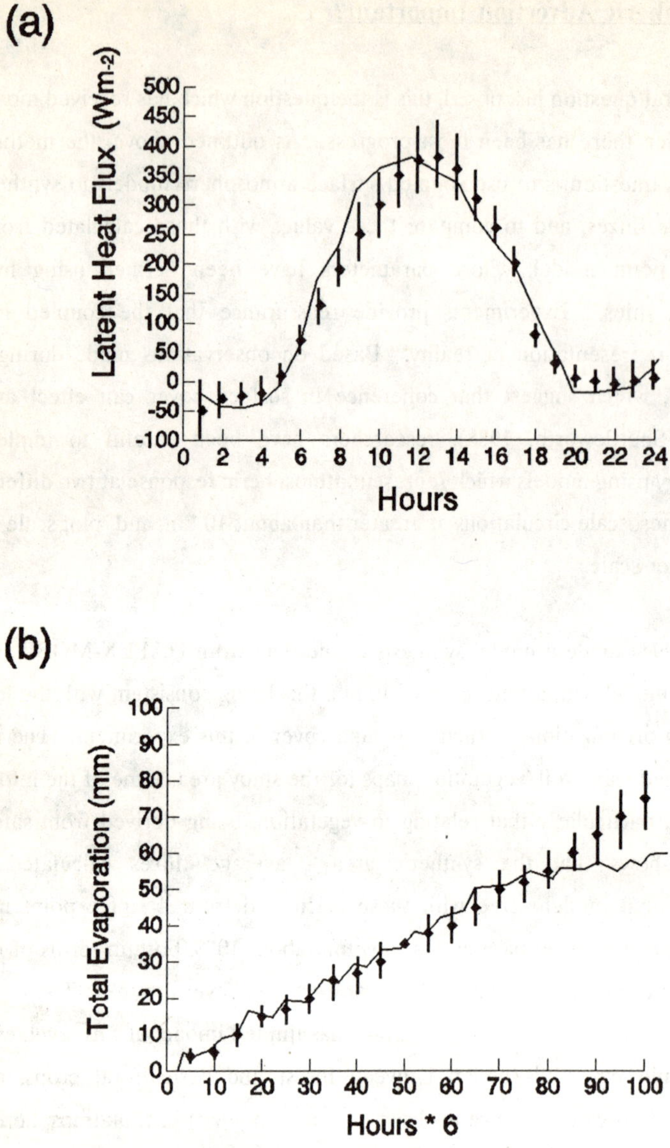

Figure 2. *(a) Latent-heat-flux for the HAPEX-MOBILHY site calculated using a land surface scheme with "aggregate" parameters (full line) compared with the mean and standard deviation for grid squares in a meso-sclae meteorological model; (b) integrated evaporation for the HAPEX MOBILHY site given with "aggregate" parameters (full line) compared with measurements for sample sites in the study area.*

that an aggregation rule based on the square of logarithms is relevant for calculating area-average values of aerodynamic roughness. Blyth and co-workers (Blyth *et. al.*, 1993), using a development of Mason's model which incorporated a simple Penman-Monteith description of surface energy balance, suggest an aggregation rule which assumes parallel summation of surface resistance was adequate to 5 - 10 % for reasonable mixtures of dry vegetation, but that more complex schemes may be needed if surface water (from rainfall) is present in small patches within the landscape.

More recently, Arain (1994) has investigated patch scale aggregation rules relevant to the BATS model for plausible mixes of vegetation relevant to the FIFE experimental site. The philosophy believed his approach is that, in practice, ecoclimatological limitations mean that the range of cover types relevant to real life, real world situations is likely to be a restricted subset of all those possible within a particular land surface parameterization scheme. In general, the results of this study confirm that simple aggregation rules work extremely well to describe area-average behavior even with an advanced land surface parameterization such as BATS. But studies with artificially wetted (irrigated) patches in an otherwise dry landscape reveal an interesting exception. In this case atmospheric advection remains a small effect, but the non-linearity of the relationship between soil moisture and plant wilting point in BATS means that the true area-average energy balance actually alters in a non-linear way in a landscape with extreme differences in soil moisture.

This last result points the way for a future emphasis in research attention, away from the current focus on atmospheric advection processes, towards a focus on subsurface variability in accessible soil moisture in the rooting zone, and towards possible topography-related movements in this.

Is Topography Important?

Research into the importance of topography in the formulation of area-average soil-vegetation-atmosphere processes is much less advanced than that into the importance of vegetation cover. However important initial progress has been made in understanding the effect of topography-induced changes in near-surface weather variables.

Raupach and co-workers (Raupach, 1994) made modeling studies in which they used a coupled surface-atmosphere model to synthesize area-average exchanges in a landscape of moderate (about 100 m) topography, but with surface cover and soil water availability was prescribed as constant across the landscape. The model therefore allowed for topography generated changes in the surface radiation balance, and in near-surface temperature, humidity and wind speed. The results of these studies are valuable in that it demonstrates that area-average exchanges are little effected by spatial variability in these meteorological variables alone.

Not withstanding this last simple result, very substantial unresolved questions remain regarding topography which are high on the agenda of the global research community for the coming decade. Specifically:

a) what are the consequences of topography - generated differences in soil moisture on area-average exchanges?;
b) what are the consequences of topography - generated differences in vegetation cover?;
c) do the topography-generated differences in soil moisture and vegetation, and those in meteorological variables interfere with each other?; and
d) how do all of the above processes alter in a region with marked (as opposed to moderate) topography.

Can We Remotely Sense Aspects of Soil-Vegetation-Atmosphere Relations Independently of Cover Class?

Certain aspects of soil-vegetation-atmosphere relations may be amenable to indirect measurement from satellite systems. Most obvious amongst these parameters are the area-average albedo, the fractional vegetation cover and the level of photosynthetic activity. Clearly there are substantial problems of a technical nature to be resolved before these can be measured - related to the calibration of sensors and signal contamination by atmospheric processes for example - but these are not the subject of attention here.

There has been a great deal of research over the last decade focussed on building possible relations between surface parameters and remotely sensed radiances (see for example, Sellers *et al.*, 1992 and associated papers). Progress has been difficult and problematic, but one more general result has emerged which has value in directing future research endeavor.

Remote sensing data can be used in two main ways. It can be used in its now well-established mapping mode, to define the location and areal extent of different land covers. Alternatively we can hope to build algorithms which relate specific canopy features to particular combinations of surface radiance as measured from satellite. Are these last relationships universal and independent of cover class?: some years ago remote sensing scientists hoped so.

Current evidence suggest that if we are to use remote sensing data to monitor parameters relevant to soil-vegetation-atmosphere relationships, we will need to exploit both satellite mapping capability and calibrated algorithms simultaneously. To take the example of albedo, satellites can measure reflected radiant energy over the wavelength relevant to solar radiation, but they are inherently directional in nature and measure the reflectivity at a particular place and a particular time only in a particular direction. The directional reflectivity of vegetation cover is determined by the structure of the particular vegetation class, normally expressed in terms of the angle of observation in the form of bi-directional directional reflectance functions. It seems likely that even to make the simple measurement of hemispherical reflectance for solar radiation at a particular location will require at least some knowledge of the vegetation class prevalent there.

In a similar way information on the fractional cover of green vegetation derived from the relative ratio of reflectances at different wavelengths perhaps is also influenced by the type and structure of the vegetation, and by the spectral properties of the underlying soil. Again, any algorithms derived to provide fractional vegetation cover from satellite data are likely to be sensitive to view angle, and so to be cover class dependent to some degree.

Finally, there is evidence that the net absorption of photosynthetic radiation by plant stands provides some information on the surface resistance of the stand (see Sellers *et. al.*, 1992). However, this relationship is strongly determined by basic physiology used by the plant to effect photosynthesis. Figure 3 illustrates an example for the FIFE site, where the relative proportion of C_3 to C_4 plants at a particular prairie grass site substantially alters the relationship between an important radiation control on surface resistance, and the "simple ratio" between reflectances in the near infrared and visible regions of the spectrum.

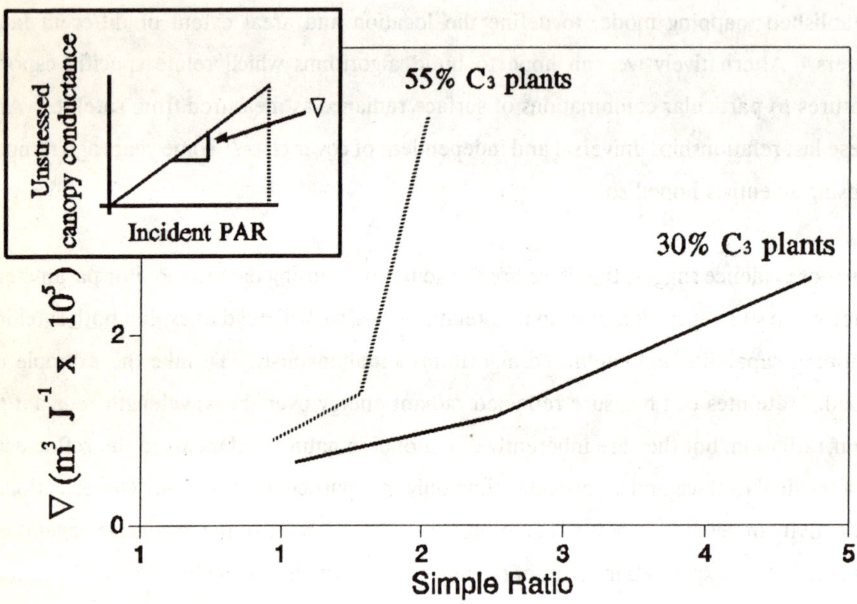

Figure 3. *The sensitivity of unstressed canopy conductance to incident Photosynthetically Active Radiation (PAR) observed at two FIFE sites with different mixes of grass cover in relation to the measured simple ratio at those two sites.*

<u>Future Research Priorities?</u>

On the basis of discussion at the *Tucson Aggregation Workshop*, what is the required shift in research emphasis?

Clearly there is a need to refocus studies of heterogeneity away from atmospheric advection processes, towards better defining the relation between vadose zone hydrology and biospheric processes, and towards the influence of subterranean variability on area-average interactions. Similarly, there is need for some shift in observational emphasis away from the current preoccupation with short-term sampling in the spatial domain, towards longer term periods of data collection, so as to better measure and understand seasonal and inter-annual variations for important vegetation cover classes.

Studies of the effect of topography are needed, and experimental and modeling studies in regions with marked topography, and studies to investigate the interaction between topography-related changes in meteorology and those in vegetation and soil moisture are a high priority.

In the area of remote sensing, we need to recognize and use the general result that it is necessary to simultaneously exploit both the mapping capability of satellites, and their capability to provide globally available fields of surrogate measures of parameters important in soil-vegetation-atmosphere relations.

Acknowledgements

This paper was prepared while the author was supported under NOAA Grant NA46GP0247. I am pleased to thank Chris Wenger for the care with which she typed the camera-ready manuscript.

References

André J-C, Goutorbe J-P, Perrier A, Becker F, Besselmoulin P, Bougeault Ph, Brusaert Y, Carlson T, Cuenca R, Gast JHC, Gelpe J, Hildebrand P, Lagouard P, Lloyd CR, Mahrt L, Mascart P, Mazaudier C, Noilhan J, Ottle C, Payen M, Phulpin T, Stull R, Shuttleworth WJ, Schmugge T, Taconet O, Tarrieu C, Thepenier RM, Valencogne Ch, Vidal-Madjar D, Weill A (1988) HAPEX-MOBILHY: First results from the special observing period. *Ann Geophys* 6(5): 477-492

Arain AM (1994) Spatial aggregation of vegetation parameters in a coupled land surface-atmosphere model. Masters Thesis, The University of Arizona

Bastable HG, Shuttleworth WJ, Dallarosa RLG, Fisch G, Nobre CA (1993) Observations of climate, albedo and surface radiation over cleared and undisturbed Amazonian forest. *Int J Clim* 13: 783-796

Beljaars ACM, Viterbo P, Miller MJ, Betts AK, Ball JH (1993) A new surface-boundary layer formation at ECMWF and experimental continental precipitation forecasts (July 1993). GEWEX News No. 3(3)

Blyth EM, Dolman AJ, Wood (1993) Effective resistance to sensible and latent heat flux in heterogeneous terrain. *Quart J Roy Met Soc* (in press)

Bolle H-J, André J-C, Arrue JL, Barth HK, Bessemoulin P, Bruin HAR de, Cruces J, Dugdale G, Engman ET, Evans DL, Fantechi R, Fiedler F, Griend A van de, Imeson AC, Jochum A, Kabat P, Krazsch T, Lagouarde J-P, Langer IJ, Oliver HR, Roth R, Saatchi SS, Sanchez Diaz J, Santa Olalla M de, Shuttleworth WJ, Sqaard H, Stricker H, Thornes J, Vauclin M, Wickland D (1993) EFEDA European field experiments in a desertification-threatened area. *Ann Geophys* 11: 173-189

Clapp RB, Hornberger GM (1978) Empirical equations for some hydraulic properties. *Water Resour Res* 14-4: 601-604

Deardorff J (1978) Efficient prediction of ground temperature and moisture with inclusion of a layer of vegetation. *J Geophys Res* 83: 1889-1903

Dickinson RE, Henderson-Sellers A, Kennedy PJ, Wilson MF (1986) Biosphere-Atmosphere Transfer scheme (BATS) for the NCAR Community Climate Model. *Tech Note NCAR/RN-275+STR*, Natl Cent for Atmos Res Boulder CO 69 pp

Goutorbe J-P and 15 others (1994) Hapex-Sahel: a large scale study of land-atmosphere interactions in semi-arid tropics. *Ann Geophys* 12: 53-64

Henderson-Sellers A, Dickinson RE (1992) Intercomparison of land surface parameterizations launched. *EOS* 73: 195-196

Lean J, Warrilow DA (1989) Climatic impacts of Amazon deforestation. *Nature* 342: 11-413

Manabe S (1969) Climate of the ocean circulation, Part I. *Mon Weather Rev* 97: 739-774

Mason PJ (1988) The formation of area-averaged roughness lengths. *Quart J Roy Met Soc* 114: 399-420

Monteith JL, Unsworth MH (1990) *Principles of Environmental Physics*. Edward Arnold London

Nobre CA, Sellers PJ, Shukla J (1991) Amazonian deforestation and regional climate change. *J Climate* 4: 957-988

Noilhan J (1994) Private communication, CNRM, Toulouse, France

Noilhan J, Planton S (1989) A simple parameterization of land surface processes for meteorological modsl. *Mon Weather Rev* 117: 536-549

Pitman A and 38 co-authors (1993) PILPS: results for intercomparison of land-surface parameterization schemes. GEWEX Publication No. 7, International GEWEX Project Office Washington

Raupach M (1994) Private communication, CSIRO, Canberra, Australia

Ross J, Monteith JL (ed) (1975) Radiation transfer in plant communities. *Vegetation and the Atmosphere*. Academic Press, 13-52

Rutter AJ, Kershaw KA, Robins PC Morton AJ (1971) A predictive model of rainfall interception in forests. I: Derivation of the model from observations in a plantation of Corsican pine. *Agric Met*, 9: 367-384

Rutter AJ, Morton AJ, Robins PC (1975) A predictive model of rainfall interception in forests. II: Generalization of the model and comparison with observations in some coniferous and hardwood stands. *J Appl Ecol* 12: 367-380

Sato N, Sellers PJ, Randall DA, Schneider EK, Shukla J, Kinter JL III, Hou Y-T, Albertazzi E (1989) Effects of implementing the simple biosphere model in a general circulation model. *J Atmos Sci* 46(18): 2757-2782

Sellers PJ, Hall FG, Asrar G, Strebel DE, Murphy RE (1988) The first ISLSCP Field Experiment (FIFE). *Full Met Soc* 69: 22-27

Sellers PJ, Hall FG (1992) Relations between surface conductance and spectral vegetation indices at intermediate ($100m^2$ to $15km^2$) length scales. *J Geophys Res* 97(D17): 19 091-19 109

Sellers PJ, Mintz Y, Sud YC, Dalcher A (1986) A simple biosphere (SiB) for use within general circulation models. *J Atmos Sci* 43: 505-531

Sellers PJ, Shuttleworth WJ, Dorman JL, Dalcher A, Roberts JM (1989) Calibrating the simple biosphere model for Amazonian tropical forest using field and remote sensing data, Part 1, Average calibration with field and remote sensing data. *J Appl Met* 28: 727-759

Shuttleworth WJ (1975) The concept of intrinsic surface resistance: energy budgets at a partially wet surface. *Bound Lay Met* 8: 81-99

Shuttleworth WJ (1988) Macrohydrology - The new challenge for process hydrology. *J Hydrol* 100: 31-56

Shuttleworth WJ (1989) Micrometeorology of temperature and tropical forest. *Phil Tran R Soc Lond* B324: 299-334

Shuttleworth WJ (1991a) Insight from large-scale observational studies of land/atmosphere interactions. In: Land surface atmosphere interactions: parameterization and analysis for climate modelling. Wood EF (ed), *D Reidel* 3-30 New York

Shuttleworth WJ (1991b) The Modollion Concept. *Reviews Geophys* 29(4): 585-606

Shuttleworth WJ, (1993) Evaporation. in *Handbook of Hydrology*, Maidment DR (ed), McGraw-Hill New York

Shuttleworth WJ, Dickinson RE (1989) Comments on "Modeling tropical deforestation: A study of GCM land-surface parameterizations" by Dickinson RE and Henderson-Sellers A. *Quart J Roy Met Soc* 115: 1177-1179

Shuttleworth WJ, Gash JHC, Lloyd CR, McNeil DD, Moore CJ, Wallach JS (1988) An integrated micrometeorological system for evaporation measurement. *Agric and For Met* 43: 295-317

Shuttleworth WJ, Gash JHC, Lloyd CR, Moore CJ, Roberts JM, Marques A de O, Fisch G, Silva Filho V de P, Ribeiro MNG, Molion LC B, Nobre C, de Sa LDA, Cobral OMR, Patel SR, Mordas JC (1984) Observations of radiation exchange above and below Amazonian forest. *Quart J Roy Met Soc* 110: 1163-1169

Shuttleworth WJ, Michaud J, Kabat P, Sellers PJ (1994) The aggregate description of heterogeneous land covers: progress and prospect. submitted to the J Hydrology

Shuttleworth WJ, Schmugge TJ and Andre J-C (eds) (1991) Evaporation models in hydrology. in *Land Surface Evaporation*. Springer-Verlag New York

Stewart JB (1988) Modelling surface conductance of pine forest. *Agric and For Met* 43: 19-35

USSR Committee for the International Hydrological Decade (1978) World Water Balance and Water Resources of the Earth. English translations, *Studies and Reports in Hydrology* 25 UNESCO Paris

Verstraete MM, Dickinson RE (1986) Modeling surface processes in atmospheric general circulation models. *Ann Geophys Ser B* 4: 357-364

Viterbo P (1994) Private communication. European Center for Medium Range Weather Forecasts Reading UK

Warrilow DA (1985) Indications of the sensitivity of European climate to land use variations using a one-dimensional model. *Proc ISLSCP Conference* Rome, Italy 2-6 December 1985. ESA SP-248, 159-166

Wood N, Mason PJ (1991) The influence of static stability on the roughness lengths for momentum and temperature. *J Quart Roy Met Soc* 117: 536-549

Xue Y, Sellers PJ, Kinter JL, Shukla J (1991) A simplified biosphere model for climate studies. *J Clim* 4: 345-364

Effects of CO_2-fertilization on Evapotranspiration

Willem Bouten
Landscape and Environmental Res. Group
University of Amsterdam
Nieuwe Prinsengracht 130
1018 VZ Amsterdam
The Netherlands

Jan Goudriaan
Dep. Theoretical Production Ecology
Wageningen Agricultural University
Bornsesteeg 65
6708 PD Wageningen
The Netherlands

SUMMARY

Concentrations of greenhouse gases are increasing rapidly. It is most likely that the present CO_2 concentrations will be doubled by about the middle of the next century. An important question to answer, when considering the role of water and the hydrological cycle in global change, is to what extent the stomata of plants, whole plants or entire plant communities adapt to changing CO_2 concentrations. Plant response could alter the temporal or spatial distribution of evaporation from terrestrial systems.

Changes in the ambient CO_2 concentrations primarily affect the net photosynthesis of C3-plants, the majority of the plant species. Higher ambient concentrations also increase the concentration gradient which is the driving force for diffusion of CO_2 into the stomatal cavities. This larger gradient and the concurrent higher diffusion rate are counteracted to some extent by a partial closure of the stomata, thus reducing the transpirational loss of water from the leaves. The instantaneous transpiration efficiency, the ratio between assimilation and transpiration rates, thus increases both by enhanced carbon assimilation and simultaneous reduced transpiration.

An enhanced carbon assimilation rate induces a stimulated growth. For annual

NATO ASI Series, Vol. I 31
The Role of Water and the Hydrological Cycle
in Global Change
Edited by Howard R. Oliver and Sylvia A. Oliver

crops increased assimilation causes a faster canopy development in spring thus reducing soil evaporation and increasing plant transpiration. However, the carbon allocation in the plant not only depends on the growth stage but also on stress factors that limit growth. Plants with limited water or nutrient availability show increased root:shoot ratios when grown at higher CO_2 levels. Plants tend to adjust their carbon allocation so that limitation of growth by different resources is equalized as much as possible.

Long term changes of natural ecosystems are difficult to predict. It is the combination of stress factors, and their spatial and temporal variability, on the one hand and differences of sensitivity among plant species to these factors on the other hand that determine the outcome of competition and the resulting biological diversity.

THE GLOBAL CARBON CYCLE

The role of water and the hydrological cycle in global change cannot be detached from that of other factors which act at the same temporal scale. Considering the global energy balance, greenhouse gases in the atmosphere (CO_2, N_2O, CH_4) reduce the thermal emittance thus increasing the earth surface temperature to equilibrate the incoming energy fluxes. Concentrations of the greenhouse gases have been increasing at a high rate. Of all atmospheric changes during the past century, the increase of the atmospheric CO_2 concentration has been most clearly established. A rise in temperature, attributable mainly to changes in atmospheric CO_2 and CH_4 concentrations, is not so easy to distinguish from the background of normal fluctuations. Other trends in rainfall, storms, etc., are even far more difficult to assess.

Atmospheric CO_2 increases at a rather steady state, at least on a time scale of decades. During the past 50 years its rate of increase has grown exponentially with about 2.4% per year to a concentration level of 350 vpm at present (1994), possibly leading to a doubling (700 vpm) around 2050. On longer time scales, there has been a rather stable level at about 280 vpm for the last thousands of years until the effects of the industrial revolution became noticeable since the mid 1800's. Further back in time, during the last glacial period (about 50000 years ago), atmospheric CO_2 was much lower and even dropped to about 200 vpm.

The change of the global atmospheric CO_2-concentration is a result of the non-balancing exchange between the carbon pools (fig. 1).

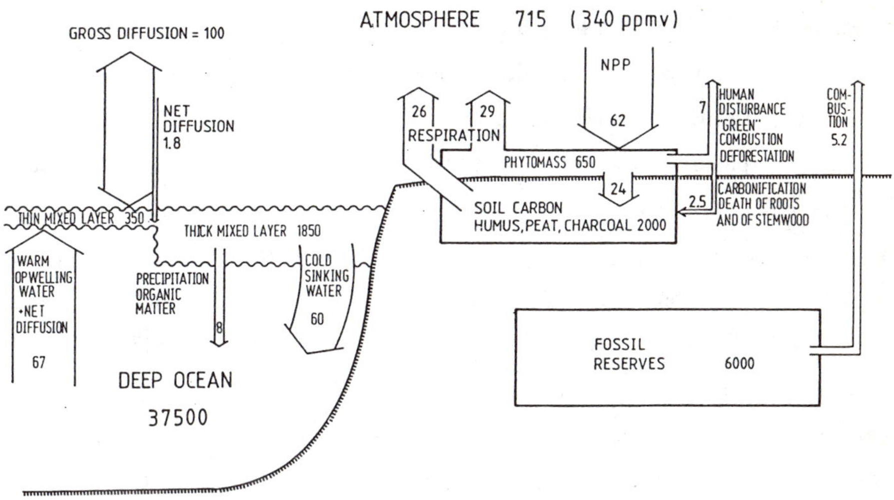

Figure 1. Simulated pools and fluxes of the global carbon cycle in 1980 A.D. Area of pools and width of fluxes are proportional to their sizes (reprinted from: Goudriaan and Ketner, 1984)

On a geological time scale the largest carbon reservoir on earth is the rock carbonate. On the time scale of present interest, exchange processes with this pool can be expected to equilibrate and therefore rock carbonate is not incorporated into calculations, although this reservoir may eventually absorb all CO_2 from fossil fuel at a time scale of thousands of years. On the whole, terrestrial and marine annual fluxes of biotic carbon uptake and release are of the same order of magnitude at about 100 Gt per year in total. With a global atmospheric amount of 700 Gton of carbon this means that the content of the atmosphere is recycled once per 7 years. Also considering the huge pools of carbon in the oceans and the biosphere, one would expect that any disturbance of injecting CO_2 from fossil fuel should be absorbed without difficulty. However, this is not the case. In oceans a chemical equilibrium exists in the chain CO_2 <-> HCO_3^- <-> $(CO_3)^{2-}$, which shifts to the right with increasing pH (Butler, 1982). When the CO_2 concentration in the air increases, more CO_2 gets dissolved and an equilibrium shift occurs towards HCO_3^-, with a release of H^+ ions. Therefore the pH decreases, offering a strong negative feed-back for further absorption. Moreover, due to poor mixing of water, especially in the deep Pacific ocean, the uptake rate of CO_2 in oceans is limited. The role of the terrestrial biosphere is even more complex. The fixation of CO_2 by net primary production is counterbalanced by a consumption by insects or various kinds of mammals or left as litter to be decomposed by fungi and microbes. Differences in fluxes of formation and consumption lead to characteristic steady state terrestrial carbon pools which differ largely for the vegetation types over the world and which have turnover times that can range from 1 year to 500 years (fig. 2). Many ecosystems in the world, however, are not in steady state. They accumulate carbon during a (large) number of years and are then disrupted by deforestation, fire or other (man-induced) drastic events.

All together, the lack of strong negative feed back mechanisms implies that relatively small amounts of CO_2 from fossil fuel burning cause an exponential increase of the atmospheric CO_2 concentration. Whatever the exact rate may be, it is to be expected that the increase of the global CO_2 concentration will have impacts on plant water relations as CO_2-assimilation through stomata goes with a water vapour flow in opposite direction. It will also affect plant growth and competition between species. As plants play a dominant role in the control of evapotranspiration at the earth's surface, the atmospheric CO_2 concentration is of vital interest for the role of water and the hydrological cycle in global change.

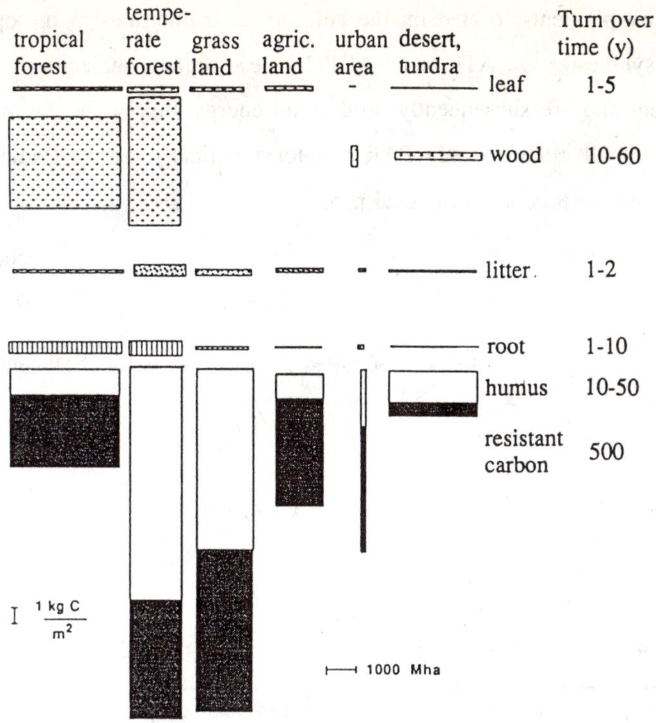

Figure 2. Simulated carbon areal density (heights of the columns) and land area for each vegetation type (widths of the columns). The areas are proportional to the amounts of carbon.

PLANT PHOTOSYNTHESIS AND TRANSPIRATION

CO_2 and photosynthesis

CO_2 is the primary substrate for photosynthesis, the basic process of formation of plant material. It is the process in which light, CO_2 and water are converted to sucrose and oxygen:

$$CO_2 + H_2O + \text{light} \rightarrow CH_2O + O_2 \quad (1)$$

Photosynthesis consists of two reactions (fig. 3). In the light reaction, light energy is absorbed by pigments located on the chloroplast membranes in mesophyll cells and used to synthesize O_2, ATP and NADPH. These organic molecules have a high energy content and are subsequently used as an energy supply for the synthesis of plant tissue. Of all radiation only PAR (Photosynthetically Active Radiation: 400-700 nm) is used for this photochemical process.

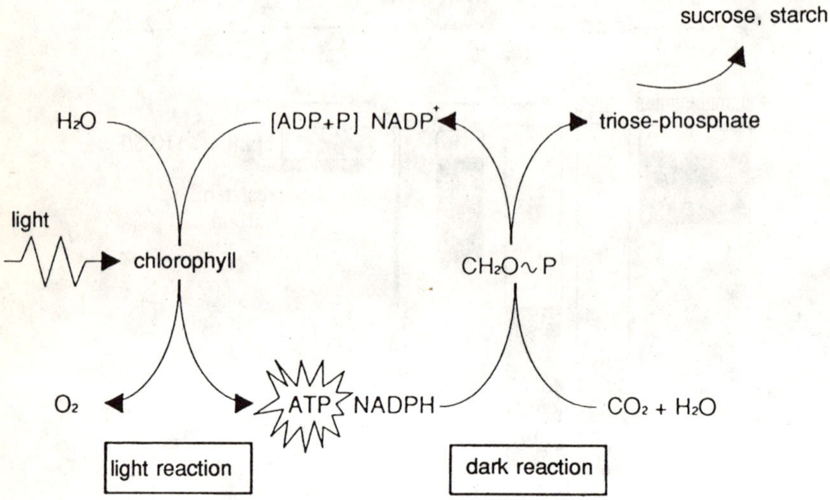

Figure 3. Schematic representation of the photosynthesis process (after: Ammerlaan and de Visser, 1993)

In the dark reaction (fig. 4a), a biochemical process which also occurs in the mesophyll cells, CO_2 is bound with Ribulose-bis-phosphate (RubP), catalized by the enzyme Rubisco to form 3-phosphoglyceric acid and then triose-phosphate, the primary substrate for the synthesis of sucrose, starch. ATP and NAPH are used to regenerate RubP. Besides carboxylation, Rubisco also catalyzes oxygenation in which O_2 is bound and which leads to the release of CO_2. In this apparently counterproductive process a part of the energy gained in the light reaction is lost as photorespiration without net production of carbohydrates. The ratio between the rates of CO_2 binding in the Calvin-cycle and the CO_2 release in the photorespiration process depends among others on the ratio of O_2 and CO_2 concentrations. At enhanced CO_2 concentrations, the competition between O_2 and CO_2 for binding at Rubisco is changed in favour of CO_2, thus increasing the carboxylation and decreasing the oxygenation.

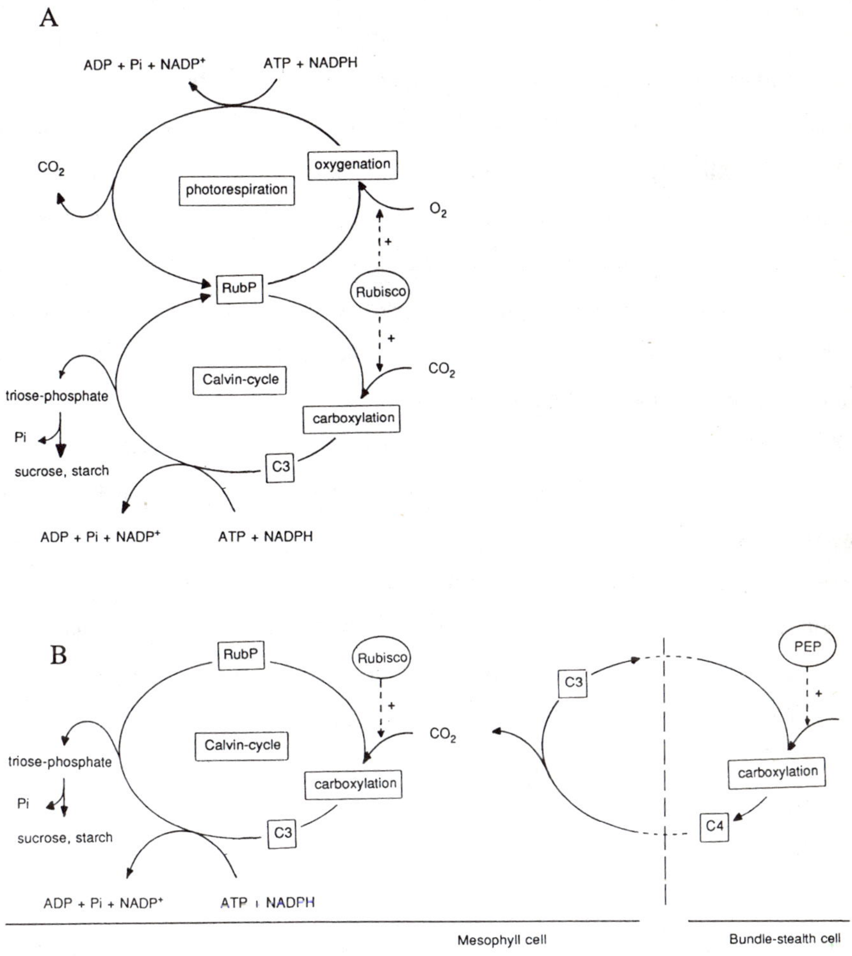

Figure 4. Schematic representation of the dark reaction of photosynthesis of A) C3 plants and B) C4 plants (after: Ammerlaan and de Visser, 1993)

Although the photosynthesis process of most plants acts as described above, we need to distinguish between plants of the so-called C3-type (most plants of the temperate climates) and the C4-type (some of the important tropical agricultural crops such as maize, sorghum, sugar cane and some tropical grasses). C3 plants are named after the carboxilation product 3-phosphoglyceric acid which has three C-atoms. C4 plants do not use RubP as a primary acceptor of CO_2 from the air. The initial carboxylation uses PEP carboxylase (PhosphoEnolPyruvate) instead to produce an organic acid with four C-atoms (oxaloacetate) as a first product of fixation (fig. 4b). After being transferred from the mesophyll cells to the bundle-sheath cells the oxaloacetate releases the CO_2 again. Then this CO_2 is bound by RubP in the Calvin-cycle, again catalyzed by Rubisco but now with hardly any competition of O_2. The initial fixation by PEP thus acts as a CO_2 concentrating mechanism improving the conditions for the rather inefficient RubP carboxylation. Photorespiration therefore does not occur in C4-plants.

These reactions are cyclic reactions in which CO_2 and H_2O are transferred into organic molecules and O_2, using the energy of harvested light. Any disturbance of the cyclic process as caused by a shortage of light, substrate, intermediate products or catalyzing enzymes reduces the assimilation rate.

Details of the biochemistry and physiology of photosynthesis can be found in appropriate text books such as those by Trebst and Avron (1977), Gibbs and Latzko (1979), Clayton (1981) or Hatch and Boardman (1981).

Exchange of CO_2 and water vapour by leaves

The supply of CO_2 for carboxylation is determined by diffusion through pathways from the ambient air to the chloroplasts in mesophyll cells. This diffusion process can be described in analogy with Ohm's law, in which the resistance stands for the ratio between diffusion length (m) and diffusivity (m^2 s^{-1}):

$$N=\frac{[CO_2]_a-\Gamma}{1.6(r_b+r_s)+r_m}=\frac{[CO_2]_a-[CO_2]_i}{1.6(r_b+r_s)}=\frac{[CO_2]_i-\Gamma}{r_m} \qquad (2)$$

where N is the net diffusion flux of CO_2 into the leaf (g m^{-2} s^{-1}), $[CO_2]_a$ is the ambient CO_2 concentration (g m^{-3}), $[CO_2]_i$ is the intercellular CO_2 concentration, Γ is the apparent CO_2 concentration in mesophyll cells, r_b, r_s are resistances to diffusion of water vapour (s m^{-1}) of the boundary layer and stomata respectively (fig. 5), the term 1.6 comes in because of the ratio of the binary diffusion coefficients of CO_2 and water vapour, r_m is a chemical reaction constant which here is written in the form of a resistance. The difference in the concentrations of the ambient air (350 μmol mol^{-1}) and in the mesophyll cells (30-80 μmol mol^{-1} for C3 plants and < 10 μmol mol^{-1} for C4 plants) causes a diffusion flux into the leaf. Characteristic values for resistances are 5-50 s m^{-1} for r_b decreasing with increasing wind speed, 150 - 4000 s m^{-1} for r_s decreasing with increasing stomatal aperture, and 300 - 600 s m^{-1} for r_m of C3 plants and 50 - 200 s m^{-1} for C4 plants, decreasing with increasing radiation.

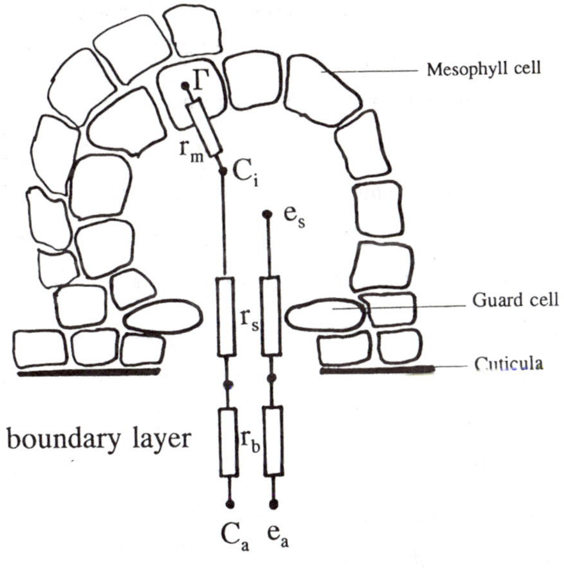

Figure 5. Schematic representation of a leaf stoma with resistances in the pathways for vapour and CO_2. The symbols are explained in the text.

The pathway through the stomata and the boundary layer around the leaf is also followed by water vapour but in the reverse direction. The transpiration flux thus can be described in the same way:

$$T=\frac{[H_2O]_i-[H_2O]_a}{(r_b+r_s)}=\frac{(e_{s,i}-e_a)\rho c_p}{\gamma\lambda(r_b+r_s)} \tag{3}$$

where $e_{s,i}$ is the interior saturated vapour pressure and e_a is the ambient vapour pressure (Pa),ρ is the air density (g m^{-3}) and c_p is the specific heat of air (J $g^{-1}K^{-1}$), γ is the psychrometer constant (Pa K^{-1}) and λ is the latent heat of vaporization of water (J g^{-1}).

Carbon allocation in plants

The allocation and partitioning of assimilated carbon provide resources for growth and for acclimation to environmental stress. Allocation comprises the processes determining the biochemical fate of carbon that has become available for distribution or partitioning among plant parts. This carbon can either be newly assimilated carbon or carbon set aside to be mobilized and used later.

Sucrose, synthesized in active leaves (sources), is actively loaded into the phloem causing a locally decreased osmotic water potential (fig. 6). Driven by the difference of osmotic potential, water from the neighbouring xylem passes the semi-permeable phloem wall, causing a locally high turgor pressure. At the sink, sucrose is actively unloaded from the phloem and water leaves the phloem resulting in a lower pressure. As a result of the difference of turgor pressure, or hydraulic potential, water and dissolved solutes thus move by mass flow from the source to the sink (Lang and Thorpe, 1986). The long distance transport of carbon in the sieve tubes of the phloem may be regarded as a passive process and thus is fully determined by the metabolic activity of sources and sinks.

Sink strengths change with the stages of development of plants. In periods of leaf development, carbon allocation is largely regulated by the sink strength of developing leaves, which compete with other sinks throughout the plant. As leaves expand and mature, they change from sinks to sources. Upper mature leaves supply assimilates to the developing leaves, lower mature leaves supply to the branch while excess of assimilate is transported to the main stem and roots. The leaves of any particular branch are responsible for growth and maintenance of its own functions. All plant parts need certain amounts of assimilates for the regeneration of disintegrated components. This carbon use is known as the maintenance respiration. Upon shading in a closed canopy, the lower branches cannot produce enough assimilates for maintenance. They are soon isolated from the water and nutrient flow from the main stem. The leaves then rapidly senesce and the whole branch dies.

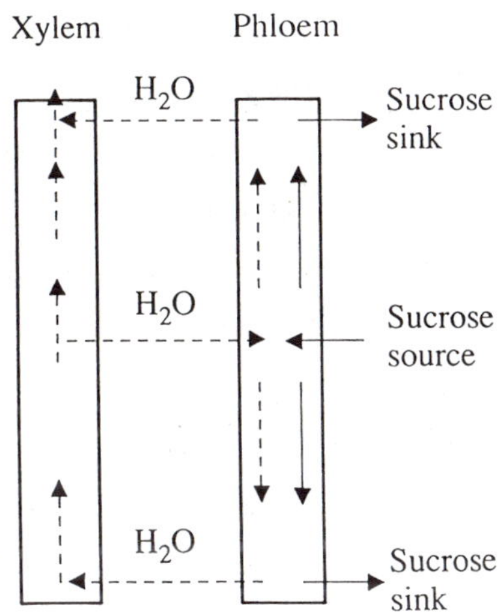

Figure 6. Transport of water (dashed arrows) and sucrose (solid arrows) from sources to sinks.

The details about metabolic activity of sources and sinks and the resulting allocation and partitioning of assimilated carbon are complex and difficult to study. Bloom et al. (1985) suggest that plants respond to imbalances in resource availability by adjusting their carbon allocation so that their limitation for growth is equalized for all resources as much as possible.

Water requirements of plants

Together with CO_2 assimilation, plants loose water through their stomata. An ideal strategy for controlling stomatal conductance is a trajectory that minimizes water loss, subject to the constraint that carbon assimilation satisfies the plant's carbon need for growth, maintenance and reproduction.

The water use efficiency (WUE) is defined as the dry matter production divided by the amount of water used during the growth of plants. It is a parameter of interest as an index of productivity versus water requirements. However, several different definitions of the water use efficiency have been proposed. Experimental data from literature are difficult to compare as water use sometimes stands for transpiration, sometimes for transpiration and soil evaporation but sometimes also includes irrigation or interception losses. Dry matter production can stand for total standing dry matter, for total produced dry matter including died leaves or for economic yield (eg grain yield).

WUE is sometimes also used as the ratio of instantaneous rates of leaf net photosynthesis to transpiration here called the instantaneous transpiration efficiency. This ratio was sometimes expressed in terms of molar ratios [mol CO_2 / mol H_2O] and sometimes as mass ratios [g CO_2 / g H_2O]. In this form it is used in studies of CO_2 and vapour exchange through stomata in response to changing environmental conditions.

Water Use Efficiency

The water requirements of crops has been subject of many studies in the past.

The design of irrigation systems, land evaluation etc. depends on the knowledge of the water use for crop production. Most of these water requirement studies related crop yields to meteorological conditions. With data on yields of plants grown in containers, de Wit (1958) showed that crop yield and transpiration are very well correlated. For dry, high-radiation climates he found the following relation:

$$\frac{Y}{T} = m \,/\, E_o \quad or \quad T = \frac{1}{m} Y E_o \tag{4}$$

where Y is the total dry mass production per area, T is the total transpiration per area during growth to harvest, E_o is the mean daily free water evaporation for the same period, and m is a parameter governed mainly by plant species and independent of soil nutrition and water availability except for cases with serious nutrition limitations or extremely wet soils. Fisher and Turner (1978) concluded that the variation in WUE in arid and semi-arid zones was best described with this relation if the root production was included in the total dry matter production. They found a characteristic value for the m factor of 110-140 kg ha^{-1} d^{-1} for wheat and many other C3 species, and an m-factor of about 210 kg ha^{-1} d^{-1} for corn and many other C4 species.

For humid regions where solar radiation was limiting, de Wit proposed an even simpler relation:

$$Y = n * T \tag{5}$$

Later the same data and similar data were reconsidered and various other empirical relations were proposed (Table 1). All these functions are useful to relate the ratio of biomass production and water requirements to meteorological conditions and some empirical constant which is dependent on species and common agricultural practice.

In natural ecosystems and some third-world agricultural systems, primary production is not only limited by solar radiation or the availability of water but also by the availability of nitrogen or phosphorus. An approximate lower limit for the requirements is 1% of the total dry matter production for nitrogen and about 0.05% for phosphorus. When the availability of nutrients is the limiting factor for growth water use efficiencies are lower.

Table 1. Relations for Water Use Efficiencies (Y: dry matter production; T: transpiration; RH: relative humidity; α, k: empirical parameters; e_s and e: saturated and actual vapour pressure; E: evaporation; ET: evapotranspiration), (Tanner and Sinclair, 1983)

$$\frac{Y}{T} = \frac{\alpha}{(1-RH)} = \frac{\alpha e_s}{(e_s-e)}$$ (Arkley, 1963)

$$\frac{Y}{T} = \frac{k}{(e_s-e)}$$ (Bierhuizen and Slayter, 1965)

$$\frac{Y}{Y_{max}} = \frac{T}{T_{max}}$$ (Hanks, 1974)

$$\frac{Y}{ET} = \frac{k_d}{(e_s-e)}\left[1-\frac{E}{ET}\right]$$ (Tanner and Sinclair, 1983)

k is calculated from C_i/C_a; biomass conversion coefficients, effective k is calculated transpiration area, LAI, ..etc.

Instantaneous Transpiration Efficiency

The fluxes of CO_2 (eq. 2) and water vapour (eq. 3) can be related producing the instantaneous transpiration efficiency ω:

$$\omega = \frac{N}{T} = a\frac{kC_a}{\Delta e} \tag{6}$$

where $k=(1-C_i/C_a)$ and a is a constant.

In general, photosynthetic rate and stomatal conductance appear to be very well correlated for changing conditions of light, temperature, nutrient and water availability. This suggests that the intercellular CO_2 concentration C_i is maintained at a constant ratio of C_a. There is much experimental evidence that treating C_i/C_a as a conservative is reasonable (Farquhar and Sharkey, 1982; Ball et al., 1987; Morison, 1993) with values of about 0.7 for C3 plants and 0.4 for C4 plants.

EFFECTS OF CO_2 FERTILIZATION ON WATER USE

For most plants the genetic potential for productivity is much higher than the actual productivity, due to all kinds of environmental stress, such as light, temperature, and nutrient or water availability. In fact the present CO_2 concentration is one of these stress factors, at least for C3 plants as their productivity can strongly increase with higher CO_2 levels. However, plant responses do not only depend on CO_2 concentration, but also on other environmental qualities. Moreover, most responses are not expressed in isolated systems but occur as an integrated response of the plant as a whole. The acclimation of plants to changing environmental factors involves both short-term physiological response and long-term physiological, structural, and morphological modifications and can be expected to act at different time scales. Plant responses upon enhanced CO_2 concentrations have been studied in single leaf gas exchange cuvettes, in open top chambers, in greenhouses or in FACE experiments. Long term effects due to a changed net primary production, changed C/N ratios in plants, changed litter and soil organic matter quality finally leading to changed competition between species in natural ecosystems are far more difficult to study. The complexity of the plant's response to a combination of environmental factors and the variety of time scales of response complicates the interpretation of response measurements. It is therefore not surprising that there are many apparent contradictions in published plant response to elevated CO_2 concentrations.

In trying to understand the plant's response to enhanced CO_2 concentration and for the interpretation of measurements we recall two rules which we use as starting points for our discussion: A) that an ideal strategy for controlling stomatal conductance is a trajectory that minimizes water loss, subject to the constraint that carbon assimilation satisfies the plant's carbon need for growth, maintenance and reproduction and B) Bloom's (1985) statement that plants respond to imbalances in resource availability by adjusting their carbon allocation so that their limitation for growth is as equal as possible for all resources. In our discussion of effects of CO_2 fertilization we will start with instantaneous changes at the stomatal level and continue with longer time scales through the effects of the plant's changed carbon allocation to change of competition of plant communities in natural ecosystems. All effects that we will discuss below are summarized in figure 7.

Stomatal response to enhanced CO_2 concentrations

There are many environmental factors that regulate the assimilation rate and with it the stomatal aperture and transpiration rate. At low light, the gross photosynthesis rate at the cell level is linearly dependent on the irradiance of PAR, the tangent being the light efficiency, until the availability of CO_2 is the limiting factor. This availability causes a maximum rate which depends on the CO_2 concentration and temperature. This broken-line relation results in a smooth curve (fig. 8) at the leaf level when averaged for cells with different irradiance and CO_2-concentrations. Figure 9 shows the effect of the CO_2-O_2 competition for Rubisco on the CO_2 uptake rate as a function of the internal CO_2 concentration. The intersection with the x-axis is the so-called compensation point (Γ) with no uptake at all. Γ equals to about 50-80 vpm for C3-plants, depending on temperature, and to about 10 vpm for C4-plants. At the present ambient level of CO_2 (350 vpm) the internal concentration is about 240 vpm (C_i/C_a=0.7) for C3-plants and about 140 vpm (C_i/C_a=0.4) for C4-plants. For these constant C_i/C_a ratios we can see from figure 9 that a doubling of C_a and consequently a doubling of C_i results in an almost doubled CO_2 uptake for C3-plants whereas C4-plants hardly profit with their assimilates from a rise of CO_2 levels. The change in stomatal resistance follows from the change in the CO_2 uptake flow through the stomatal aperture and the change of the difference of the concentrations (eq.2). The differences in response of C3 and C4 plants also lead to differences in response of transpiration.

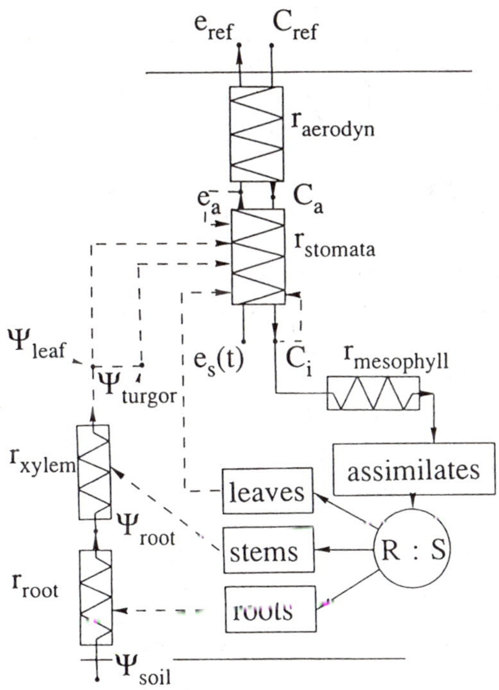

Figure 7. Effects of enhanced CO_2 concentration on water flows in the Soil-Plant-Atmosphere Continuum.

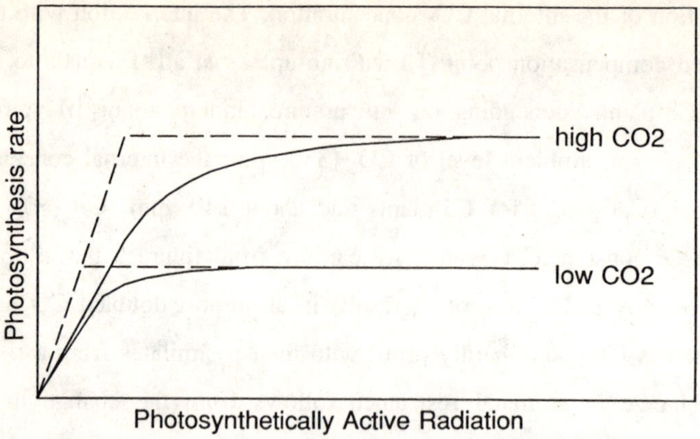

Figure 8. Light response curves at different CO_2 concentrations at cell level (dashed lines) and at leaf level (solid lines)

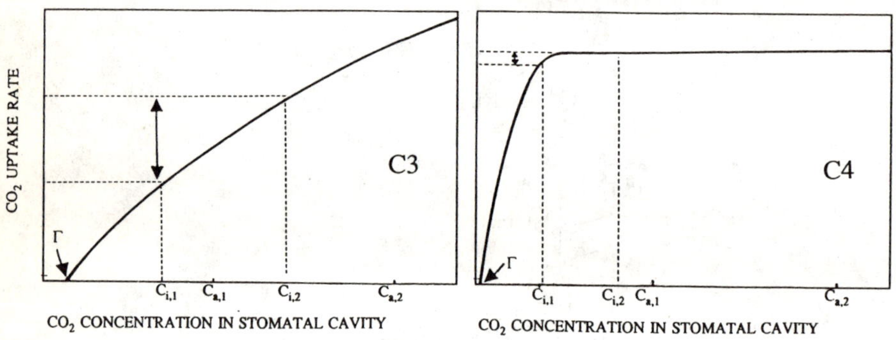

Figure 9. CO_2 response curves of C3 and C4 plants

It is commonly observed that the plant water use efficiency increases with increasing CO_2 concentrations both for C3 and for C4 plants. Many researchers have ascribed these changes to greater assimilation rates associated with higher ambient CO_2 concentrations; others mention increased stomatal resistances or a combination of the two. Morison (1993) collected measured ω values of 20 C3 species and 9 C4 species, measured in single leaf gas exchange cuvettes (fig. 10), clearly showing the validity of equation 3, also under changing ambient CO_2 concentrations in the range from 350 vpm to 700 vpm. At a doubling of the pressure gradient with C_a increasing from 350 to 700 vpm the stomatal resistance increases with 15-20% for C3 plants and with 100% for C4 plants. Instantaneous repercussions for transpiration are then in the same order of magnitude which explains the doubling of the instantaneous transpiration efficiency for both C3 and C4 plants, evaluated at leaf level. This mechanism answers our first starting point. C4 plants do not profit from a higher C_i due to their PEP carboxylation. They reduce their water consumption as an unchanged stomatal aperture would be a pure loss of water without any gain of carbon assimilation. C3 plants prioritize a full advantage of the increased CO_2/O_2 ratio in the competition for Rubisco and additionally slightly reduce the stomatal conductance, saving some water. These results of instantaneous change of stomatal resistance effectively cause the same change in transpiration as the reduction of stomatal density which was reported for plants grown at elevated CO_2 concentrations (Evens et al., 1993). Also historical herbarium plant material of the pre-industrial period (1800-1900) with atmospheric CO_2 concentrations of about 280 μmol mol^{-1} show a decrease of this stomatal density in time (Woodward, 1987).

The reduction of water requirements has strong implications in case of drought. Several studies (Tyree & Alexander, 1993) found that at elevated C_a plants wilted less, that the plant turgor pressure was higher and that plants incurred less drought related damage. Otherwise, ω may increase in case of drought. An increase of the stomatal resistance causes the transpiration to reduce in proportion. In case of assimilation, the influence of stomatal resistance is less dominating as the mesophyll resistance can be in the same order of magnitude. This means that the difference of (C_a - C_i) increases and C_i/C_a decreases in case of drought induced stomatal closure. Indeed Jacobs (1994) found a linear reduction of $(C_i-\Gamma)/(C_a-\Gamma)$ with an increasing specific humidity deficit at the leaf level. At a larger scale micrometeorological feed-back mechanisms partly counteract these differences. The difference between effects at plant level or at field scale or even regional scale is that the aerodynamic resistance is then added to the total pathways for vapour and CO_2 (Goudriaan and Unsworth, 1990). The importance of one particular resistance such as the mesophyll resistance or the stomatal resistance is thus reduced.

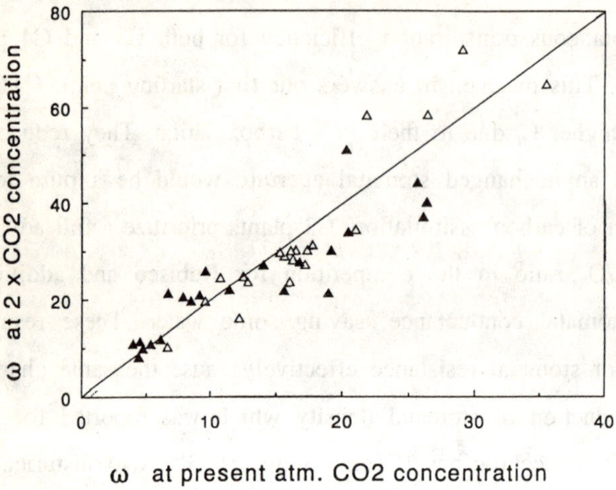

Figure 10. Relationship between the instantaneous transpiration efficiency, ω, at present and at doubled CO_2 concentrations (reprinted from: Morison, 1993)

Response of carbon allocation

On longer time scales, the increased carbon assimilation at elevated CO_2 levels causes an increase of the total biomass. Apart from possible changes in carbon allocations resulting in changed root:shoot ratios we first evaluate the effect of stimulated leaf production. For annual agricultural crops, the exponential growth in the period of development is accelerated. Standing biomass increases and soil evaporation is reduced. The increase of LAI enhances both carbon assimilation and transpiration. Due to the increase of the LAI there is more shading within the canopy. Then, the photosynthetic effects of enhanced CO_2 concentrations is lower at low light intensities, as light is the limiting factor in the light reaction; the stomata close and the transpiration is reduced. In the germination period, no large differences are to be expected. On the whole, the plant production increases although not proportional to the relative increase of CO_2 concentrations. Transpiration also increases but to a lesser extent and opposite to the results of instantaneous effects at leaf level. For the total growing season the WUE increases by 20-50% (table 2).

Table 2. Water use efficiencies (g total dry matter / kg water use) of plants grown in stands at ambient CO_2 (350 vpm) and doubled CO_2 and their ratio (summarised by Morison, 1993).

	$CO_{2,a}$	$2xCO_{2,a}$	WUE_a/WUE_{2a}
Sorghum	3.08	4.13	1.49
Wheat ww	5.1	6.3	1.40
Wheat ws	6.2	8.9	1.21
Wheat	2.62	3.45	1.52
Wheat* ww	1.58	2.14	1.43
Wheat* ws	1.27	1.86	1.33
Faba bean	4.91	7.82	1.26
Water hyacinth	1.4	2.6	1.00

ww: well watered; ws: water stressed; *: grain weight only

Not as many data exist for perennial crops and forests due to experimental problems of imposing higher CO_2 concentrations and simultaneous measurement of photosynthesis and transpiration. However, the same trends in increasing LAI and shading can be expected. In terms of water use it is important to mention that rainfall interception losses of forests can be very important for the water availability. As interception losses are strongly dependant on the canopy water storage capacity -and thus on the LAI-, these losses can be expected to be positively correlated with CO_2 concentrations.

Although an increased carbon assimilation almost always leads to an enhanced leaf production, the extent depends on the allocation of carbon. In periods of leaf development the sink strength of developing leaves is very high and most assimilates are used for shoot extension and leaf development. For several temperate tree species, wood dimensions, including stem height, stem diameter and wood density, have been reported to increase upon CO_2 fertilization (Conroy et al., 1990). With respect to root growth, both increased and suppressed root growth have been reported (Goudriaan and de Ruiter, 1983). We already discussed the extended biomass production and the enhanced transpiration. In cases of severe water or nutrient deficiency or unfavourable distribution in the soil, carbon allocation favouring root growth was reported to an extent that also the root:shoot ratio increases. We assume that the reported low root:shoot ratio at high soil water and nutrient availability was associated with increased leaf production and low assimilate requirements of the roots. Probably root:shoot ratios increase more for nutrient/water limited natural vegetation than for intensively irrigated and fertilized agricultural crops. The increase of root dimensions both in terms of the number of root tips as well as the rooting density or rooting depth reduces the resistance to water uptake for the plant as a whole. Until now it is not clear whether changed allocation of carbon also changes the xylem resistance to water flow from roots to leaves (Tyree & Sperry, 1989) and if so whether this change is relevant in comparison to the resistance to root water uptake and the stomatal resistance.

All the results discussed in this section consolidate our second starting point that the plant balances its limitations for growth as much as possible; LAI increases as long as light deficiencies due to shading do not prevail and reported changes in root/shoot ratios vary as a result of the allocation of assimilates which is focused on equalizing the limitations for growth.

As shown in figure 7, any increase of the biomass or increase of specific plant parts (root-shoot) affects the resistances of the transpiration pathway of water.

Long term response of ecosystems

The changes of growth and water use upon increasing CO_2-levels can be rather well predicted for agricultural crops without water or nutrient limitations. Due to their complexity there are hardly any natural ecosystems of which we understand the response to environmental change. Responses to elevated CO_2 concentrations are difficult to study as natural ecosystems can not easily be transported to greenhouses. FACE experiments in the coming years will hardly solve this problem.

Most ecosystems have a primary production that is far below the potential level. It is the combination of stress factors, and their spatial and temporal variability, on the one hand and the variety of sensitivity of species to these factors that rules the competition and determines the biological diversity. When we try to extend the discussion of response of stomata and carbon allocation we can expect that C3 plants are more favoured by increased CO_2 levels than C4 plants. This is specially the case in situations with a high soil water availability since C3 plants can profit from their increased carbon assimilation while C4 plants probably do not profit from their reduced transpiration. For severe drought situations the competitive effects might be reversed. Another point is the nutrient availability in poor soils. In many natural ecosystems a subtle equilibrium exists in stationary nutrient flows through the system. Plants can be extremely effective in nutrient uptake in such a way that nutrients are immediately taken up after their release during the mineralization of litter material. In this kind of system the standing biomass is dependent on its turnover time and the total amount of nutrients present in the soil-biomass system. A change of CO_2 concentration and enhanced plant growth increases the C/N or other nutrient ratios and reduces the quality of litter and soil organic matter. This change of quality slows down the mineralisation rate and thus increases the pool of soil organic matter with immobilized nutrients. It thus reduces the availability of nutrients which can have strong implications for the competition between species. In view of the complexity of ecosystem behaviour and the uniqueness of ecosystems over the world, rather than try to generalize we will return to our two starting points. The only generality which may be relevant here is one which can be deduced from both our points: the more flexible the plant in its allocation of assimilates and its adaption to resource limitations the better its competitive position in increasing stress. When we don't consider other factors of the climatic change, enhanced CO_2-levels will reduce the stress for plants may be except for nutrient limited systems.

CONCLUDING REMARKS

Until now experiments have taught us more about the behaviour of plants upon a change of environmental factors than about the numerous mechanisms that may control this behaviour. The interpretation of the measurements is often speculative and much more research is necessary to really understand the mechanisms and to build explanatory models which should be able to be used to predict the diversity changes as a result of enhanced CO_2 levels at a variety of situations. In the meantime we are able to interpret the response of plants in more general terms with only two basic rules which we used as starting points throughout the discussion. Models which are founded on these rules and on general empirical relations reported should be very useful in estimating the effects of enhanced CO_2 concentrations.

The lack of general laws for the competition between species in natural ecosystems and uncertainties about other mean climatic changes, such as increased temperature and rainfall and their changed spatial patterns over the world, prevent definite statements being made about the possibility of disappearance of specific biotopes.

REFERENCES

Ammerlaan F.H.M., de Visser A.J.C. (1993) Effects of CO_2 enrichment on photosynthesis and carbohydrate utilisation: consequences for regrowth of Lolium perenne. In: S.C. vander Geijn, J. Goudriaan, F. Berendse (eds), Climate change, crops and terrestrial ecosystems: 1-22, AB-DLO, Wageningen, the Netherlands

Arkley R.J. (1963) relationships between plant growth and transpiration. Hilgardia 34: 559-584.

Ball J.T., Woodrow I.E., Berry J.A. (1987) A model predicting stomatal conductance and its contribution to the control of photosynthesis under different environmental conditions. Prog. Photosynth. Res. 5: 221-224.

Bierhuizen J.F., Slatyer, R.O. (1965) Effect of atmospheric concentration of water vapor and CO_2 in determining transpiration-photosynthesis relationships of cotton leaves. Agric. Meteorol. 2: 259-270.

Bloom A.J., Chaplin III F.S., Mooney H.A. (1985) Resource limitation in plants - an economic analogy. Ann. Rev. Ecol. Syst. 16: 363-392.

Butler J.N. (1982) Carbon dioxide equilibria and their applications. Addison-Wesley Publ. Corp.

Clayton R.K. (1981) Photosynthesis. Cambridge: Cambridge University Press.

Conroy J.P., Milham P.J., Mazur M., Barlow E.W.R. (1990) Growth, dry weight partitioning and wood properties of Pinus raiata D. Don after 2 years of CO_2 enrichment. Plant, Cell and Environ. 13: 329-337.

de Wit C.T. (1958) Transpiration and crop yield, Versl. Landbouwk. Onderz. 64.6.

Evens L., Peterson R., Lee H.S.J., Jarvis P.G. (1993) Effects of elevated CO_2 on birch, Vegetatio, 104/105: 452-453

Farquhar G.D., Sharkey T.D. (1982) Stomatal conductance and photosynthesis. Ann. Rev. Plant Physiol. 33: 317-345.

Fisher R.A., Turner N.C. (1978) Plant productivity in the arid and semiarid zones. Annu. Rev. Plant Physiol. 29: 277-317.

Gibbs M., Latzko E. (eds.) (1979) Photosysnthesis, vol.2, Photosynthetic carbon metabolism and related processes. Berlin: Springer-Verlag.

Goudriaan J., Ketner P. (1984) A simulation study fro the global carbon cycle, including man's impact on the biosphere. Climatic change 6:167-192.

Goudriaan J., de Ruiter H.E. (1983) Plant growth in response to CO_2 enrichment, at two levels of nitrogen and phosphorus supply. 1. Dry matter, leaf area and development. Neth. J. Agric. Sci. 31:157-169.

Goudriaan J., Unsworth M.H. (1990) Implications of increasing carbon dioxide and climatic change for agricultural productivity and water resources. ASA Spec. Publ. 53: 111- 130.

Hanks R.J. (1974) Model for predicting plant yield as influenced by water use. Agron. J. 66: 660-665.

Hatch M.D., Boardman N.K. (eds.), (1981) The biochemistry of plants, vol.3, Photosynthesis. New York: Academic Press.

Jacobs C.M.J. (1994) Direct impacts of atmospheric CO_2 enrichment on regional transpiration. PhD-thesis, Wageningen Agric. Univ., the Netherlands, 179 pp.

Lang A., Thorpe M.R. (1986) Water potential, translocation and assimilate partitioning. J. Experim. Bot.37:495-503.

Morison J.I.L. (1993) response of plants to CO_2 under water limited conditions, Vegetatio 104/105: 193-209.

Tanner C.B., Sinclair T.R. (1983) Efficient water use in crop production: Research or re-search? in: H.M. Taylor, W.R. Jordan and T.R. Sinclair (eds.), Limitations to efficient water use in crop production. Am. Soc. Agron.

Trebst A., Avron M. (eds.), (1977) Photosynthesis, vol.1, Photosynthetic electron transport and photophosphorylation, Berlin: Springer-Verlag.

Tyree M.T., Alexander J.D. (1993) Plant water relations and the effects of elevated CO_2: a review and suggestions for future research. Vegetatio 104/105: 47-62.

Tyree M.T., Sperry J.S. (1989) Vulnerability of xylem to cavitation and embolism. Annu. Rev. Plant Phys. Mol. Biol., 40: 19-38.

Woodward F. (1987) Stomatal numbers are sensitive to increases in CO_2 from pre-industrial levels. Nature 327: 617-618.

Snow and Ice Cover and Climate Sensitivity

J. Oerlemans and R. Bintanja
Insititute for Marine and Atmospheric Research
Utrecht University
Princetonplein 5
UTRECHT, The Netherlands

ABSTRACT

The area of the earth covered by snow, sea ice and land ice is about 16%. According to calculations with a zonal season-resolving climate model, the earth climate would be 2.6 °C warmer without snow and ice cover. Local changes in the seasonal cycle would be much larger, however.

Climate sensitivity, i.e. temperature changes for small changes in external forcing, is significantly increased by the albedo feedback associated with snow and ice. For a 1% decrease of the solar constant, we calculate a -0.88 °C temperature change without snow and ice feedbacks. With sea ice and snow included this number is -1.25 °C. Including land ice gives a further amplification of the temperature signal, but this would occur only on very long time scales. In this paper, we discuss in some detail the seasonal and latitudinal structure of the sensitivity.

ICE AND SNOW COVER ON THE EARTH

Snow, sea ice and land ice affect the climate of the earth in several ways. Most obvious is the control on surface albedo, directly affecting the radiation budget. Other mechanisms include the insulating effect of sea ice on air-sea exchange and the production of cold bottom water underneath ice shelves. On the longer time scale, large ice sheets affect atmospheric flow also through orographic forcing, and initiate vertical land movement through glacio-isostatic adjustment.

From Table 1 it is clear that the area covered by snow, sea ice and land ice is small: about 16% of the earth surface. So one may anticipate that the direct effect of snow and ice on the present-day climate is limited, or, more carefully formulated, restricted to polar and subpolar regions. Nevertheless, when local feedbacks initiate an increase in areal coverage of snow and ice, global climate sensitivity may be significantly affected. The growth of the large Northern Hemisphere ice sheets at the start of glacial epochs is an example: here local feedbacks make ice sheets grow, which then gradually increase their radius of influence until, at a full glacial maximum, the global mean temperature is suppressed by several degrees through the albedo effect associated with those ice sheets (and by the same order of magnitude through lower atmospheric concentrations of carbon dioxide and methane).

NATO ASI Series, Vol. I 31
The Role of Water and the Hydrological Cycle
in Global Change
Edited by Howard R.Oliver and Sylvia A.Oliver

	AREA (10^6 km^2)	FRACTION
all earth	510.1	1
sea ice	24.2	0.0474
snow cover on land	28.1	0.0551
land ice	15.37	0.03013
- Antarctica	13.59	0.02664
- Greenland	1.73	0.00339
- remaining glaciers	0.0528	0.0001035

Table 1. Areal coverage of snow (annual mean), sea ice (annual mean) and land ice on earth. Data from Robinson (1993), Harvey (1988) and Haeberli et al. (1989), respectively.

Effects of snow and ice cover on the annual temperature range are of the same interest as those on annual mean temperature. In fact, in many cases, especially concerning the possibility of biological activity, summer temperatures are more important than annual means. So when studying the albedo feedback annual cycles should also be considered. Figure 1 shows the annual cycles in sea ice and snow cover (Northern Hemisphere only because snow cover outside Antarctica is very limited in the Southern Hemisphere).

Snow cover in the Northern Hemisphere apparently is a climate variable with a pronounced seasonal cycle. In late summer only a fraction of the maximum cover (in February) remains. With regard to sea ice cover, the cycles are less pronounced. The annual range is larger in the Southern Hemisphere than in the Northern Hemisphere, which is due to the different geometrical setting of land and ocean.

There is also significant interannual variation of sea ice and snow cover. The correlation between NH snow cover and annual mean air temperature north of 30°N is highly significant, namely -0.76 based on 18 years (Folland et al., 1992). In contrast, no such simple relation appears to exist between temperature and sea ice cover or any other basic meteorological variable (Folland et al., 1992).

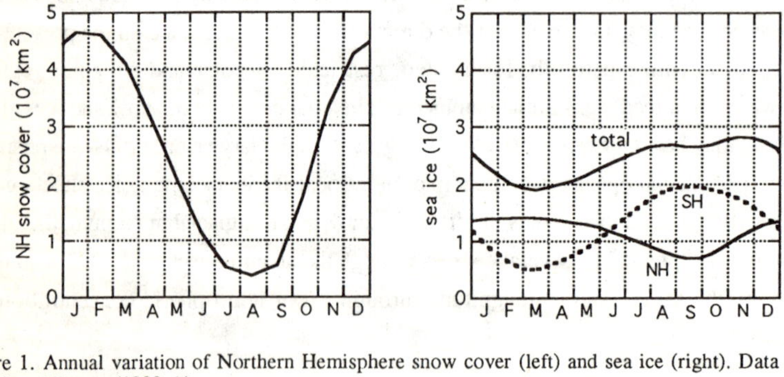

Figure 1. Annual variation of Northern Hemisphere snow cover (left) and sea ice (right). Data from Robinson (1993) and Harvey (1988a,b).

ALBEDO FEEDBACK

The concept of albedo feedback is best introduced by looking at the global energy balance, which reads:

$$(1-\alpha)\frac{S^*}{4} = \sigma\, T_{eff}^4 = \tau\, \sigma\, T^4 \qquad (1)$$

Here S* is the solar constant (about 1365 W m-2), α is planetary albedo, σ the Stefan-Boltzmann constant [$=5.67 \times 10^{-8}$ J m^{-2} K^{-4} s^{-1}], and T_{eff} an effective radiation temperature of the earth. With a planetary albedo of 0.3, a value of -15°C is found for T_{eff}, showing that the radiative temperature of the planet is not the surface temperature but the air temperature at a height of about 5 km. This reflects the greenhouse effect, of course. To express the outgoing longwave radiation in the mean surface temperature, and effective transmissivity τ has to be introduced, that appears to have a value of 0.61. As $\tau=1$ for a planet without greenhouse atmosphere, one can state that the greenhouse effect on the earth is quite significant (e.g. Raval and Ramanathan, 1989). It is well known that the most important contributors to the greenhouse effect are clouds, water vapour and carbon dioxide.

With regard to climate change, τ is not a constant. Probably τ increases with increasing temperature because total water vapour in the atmosphere would increase (the water vapour feedback). For our purpose it is thus more practical to include this in a linearized form of the right-hand side of eq. (1), i.e.

$$(1-\alpha)\, S = a + b\, T \qquad (2)$$

where T is now global mean surface temperature, $S=S^*/4=341$ W m^{-2}, and a and b are constants. The global planetary albedo feedback can be evaluated by looking at the response of the surface temperature to a change in the solar constant ($T=T_o+\delta T$; $\alpha=\alpha_o+\delta\alpha$; $S=S_o+\delta S$). A similar exercise could be done for a general perturbation of the radiation balance (i.e. associated with greenhouse gas concentration). Assuming that $\delta\alpha = -\beta\, \delta T$, $\beta \geq 0$, we find:

$$(1-\alpha_o - \delta\alpha)\,(S_o + \delta S) = a + b\,\delta T \quad \text{--->} \quad \delta T = \frac{(1-\alpha_o)\,\delta S}{b - \beta S_o} \qquad (3)$$

Higher order terms have been ignored With eq. (3) we evaluate values of δT for a 1% change in S. Without albedo feedback ($\beta=0$) and for a black body b=5.42 W m^{-2} K^{-1}), so $\delta T = 0.44$ K. We can compare this with the case of constant greenhouse effect ($\tau=0.61$ --> b=3.3 W m^{-2} K^{-1}), variable greenhouse effect (b=2.2 W m^{-2} K^{-1}) and including albedo feedback ($\beta=0.0016$; this is a mean value of six different climate model studies, only involving changes in snow and sea ice cover - *not* land ice):

no greenhouse effect:	$\delta T = 0.44$ K
constant greenhouse effect::	$\delta T = 0.71$ K
+ variable water vapour:	$\delta T = 1.09$ K
+ albedo feedback:	$\delta T = 1.46$ K

Table 2. Change in global mean temperature for a 1% increase of the solar constant for different assumptions concerning greenhouse and albedo feedback. The albedo feedback is associated with changes in snow and sea ice cover - not land ice.

EXPERIMENTS WITH A ZONAL CLIMATE MODEL

To investigate the importance of the albedo feedback on latitudinal and seasonal variation of climate we employ a zonal climate model. This model is of the seasonal energy balance type (e.g. Warren and Schneider, 1979). The structure of the model is shown schematically in Figure 2. Calculations are based on the energy budget of an atmospheric column. At the upper boundary shortwave and longwave radiative fluxes balance while at the lower boundary the atmosphere exchanges radiative and turbulent energy with land and ocean. The processes in the atmosphere (i.e. radiation, horizontal heat transport and turbulent exchange of energy) are expressed in terms of surface air temperature (T_a). The zonal mean surface air temperature is governed by:

$$\frac{\partial T_a}{\partial t} = \frac{1}{r^2 \cos\varphi} \frac{\partial}{\partial \varphi} D_a \cos\varphi \frac{\partial T_a}{\partial \varphi} + \frac{(H + LE + \Delta L_a + \Delta S_a)}{(c_p \Delta p/g)} \tag{4}$$

where r is the earth's radius, φ is latitude, D_a the diffusion coefficient for total atmospheric energy, H and LE the surface-air fluxes of sensible and latent heat, respectively, ΔL_a the divergence of longwave radiation, ΔS_a the absorbed shortwave radiation and $c_p\Delta p/g$ the thermal inertia of the atmosphere.

As in all energy balance climate models, the zonally averaged transport due to complicated, three-dimensional atmospheric motions is represented as a diffusive process in terms of the surface air temperature (first term on the right-hand side of equation (4)). According to Stone and Miller (1980), the diffusive treatment of the meridional transport is applicable when one considers long-term averaged radiative forcing while using a seasonal model. The diffusion coefficient D_a represents the diffusion coefficient for the total atmospheric flux. The meridional distribution of D_a is taken from Harvey (1988a).

The structure of the zonal mean ocean circulation is prescribed and can only vary in magnitude. Upwelling takes place in the entire basin except for two areas at 50° - 70° S and 60° - 80° N where downwelling occurs. Water flows northward in the uppermost model layer and equatorward in the layer closest to the bottom. Thus, by transporting warm water poleward and cold water equatorward, the thermohaline circulation effectively transports heat towards the poles.

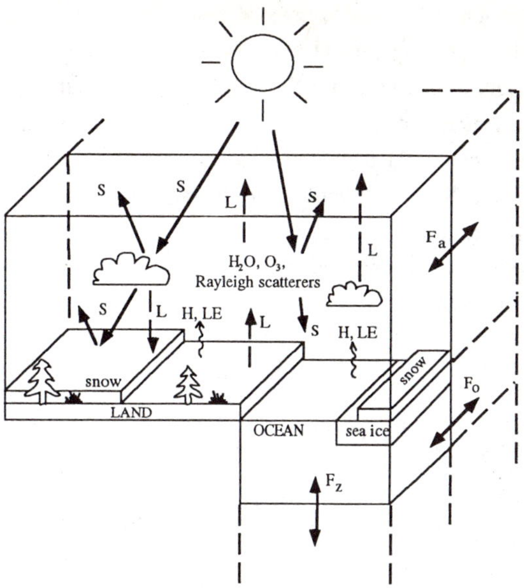

Figure 2. A picture of energy transfers dealt with in a zonal climate model. S is shortwave radiation, L longwave radiation, H sensible heat flux, L latent heat flux, F_a heat transport by atmospheric motions, F_o heat transport by ocean currents, F_z heat exchange with the deep ocean.

The land fraction at a specific latitude is subdivided into three surface types: grass, forest and permanent land ice. This makes it possible to account for the differences in albedo between the types and to incorporate the effect of masking albedo changes due to snowfall in snow-covered forests (Oerlemans and Van den Dool, 1978; Robock, 1980). Fractions of land surface types are fixed at the present-day distribution, implying that the latitudinal shift of vegetation zones under varying climatic conditions is neglected here. The fractional amount of snow present on land (a_s), equally distributed over the three surface types, is expressed as a function of surface air temperature (Harvey, 1988a):

$$a_s = \begin{cases} 1 & T'_a < 260\ \text{K} \\ 0.05\,(280 - T'_a) & 260\ \text{K} \leq T'_a \leq 280\ \text{K} \\ 0 & T'_a > 280\ \text{K} \end{cases} \qquad (5)$$

Here $T'_a = T_a + \Gamma z$ is surface air temperature corrected for zonal mean surface elevation (z), using a global mean lapse rate Γ of -0.0065 K m^{-1}

Sea-ice is incorporated in a purely thermodynamic way. Only changes in sea-ice extent are considered, not thickness. The albedo of sea-ice is taken the same as that of land-ice with respect to its dependence on snow cover, solar zenith angle, surface temperature and cloudiness. In the meridional direction, a five degree grid distance is used in the ocean and atmosphere. The ocean is subdivided into six vertical layers with increasing layer thickness towards the bottom.

The meridional distribution of the annual surface air temperature and its seasonal range are shown in Figure 3. There is overall agreement with observations, although the model somewhat underestimates temperatures in the region 10° N - 45° N (the maximum difference is about 4 °C). Temperatures over the Antarctic continent are significantly lower than over the North Polar Basin (even when corrected to sea-level temperatures), as correctly simulated by the model. Due to the large land masses in the Northern Hemisphere, the amplitude of the seasonal cycle of air temperature is largest there. The poleward enhancement of the seasonal temperature range, due to the poleward increase in seasonal amplitude of incident solar radiation, is captured by the model, although the small scale structures are not.

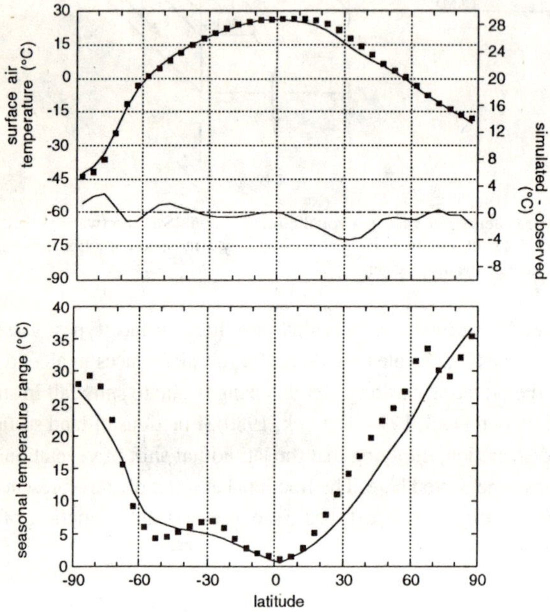

Figure 3. A comparison of calculated annual mean temperature (upper panel) and temperature range (lower panel) with observations (dots) from Warren and Schneider (1979).

The model successfully reproduces the seasonally varying snow cover on land (not shown). On the other hand, some problems exist with respect to the seasonal behaviour of sea-ice amount: in the Northern Hemisphere, the seasonal range is overestimated while in the Southern Hemisphere it is underestimated. This is a common feature of climate models with a simple representation of sea ice. However, the simulation of the latitudinal distribution of annual mean surface albedo is fairly good, with maximum discrepancies of about 0.03.

To illustrate the role of snow cover, Figure 4 shows annual mean temperature and temperature range when no snow is allowed in the model. The largest differences with the standard case occur at the high northern latitudes, of course. Averaged over the 60°-90°N latitude belt, the climate without snow would be 4 °C warmer, and the annual range would be 10 °C smaller. Here the larger differences in albedo when snow is included add to the seasonal cycle.

In Figure 5 the change in annual temperature and temperature range are shown separately for the albedo effects of snow, sea ice and land ice. Sea ice has not been removed, i.e. its effect on air-sea exchange remains the same. Without any snow or ice, the antarctic continent would be more than 15 °C warmer, the arctic basin about 10 °C. Interestingly, the effect on annual temperature range is opposite for the hemispheres. This is because at present the highly reflective surface of the Antarctic ice sheet persists through the year. If it were to be removed, the forcing of the energy balance by absorbed solar radiation would have a much larger annual amplitude. In the arctic region, the pronounced annual cycle in albedo works in the opposite way. In the tropical and subtropical regions, the temperature increase for the no ice/snow case is within a few °C. The annual range is not influenced at all.

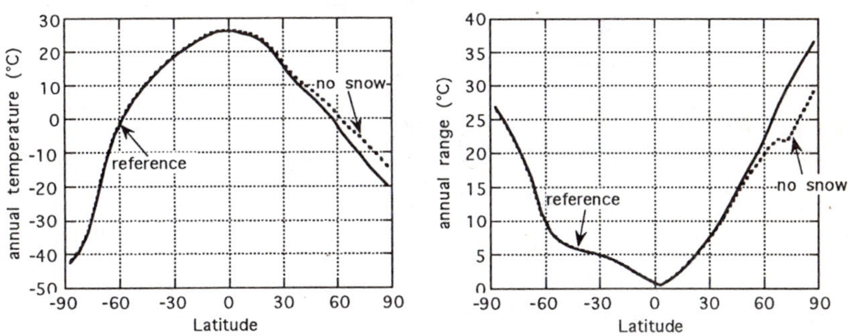

Figure 4. Annual temperature (left) and temperature range (right) in the reference case and without snow cover feedback.

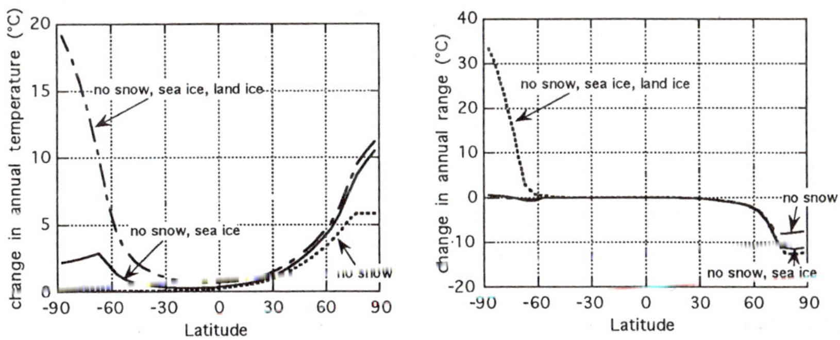

Figure 5. Change in annual temperature (left) and temperature range (right) without albedo feedback asociated with snow, sea ice and land ice, respectively.

Of course one should realize that this type of experiment is extremely schematic. The ocean circulation, in particular, is prescribed and cannot react to a changing temperature distribution and sea ice cover, which is unrealistic. Nevertheless, insight is gained in the overall effect of snow and ice cover on the reflectivity of the earth and, consequently, on the global mean surface temperature.

Of more practical interest is the question to what extent ice and snow effects climate sensitivity for small excursions of the radiative forcing, either by changing greenhouse concentration, aerosol loading of the atmosphere, or solar constant. Figure 6 shows changes annual mean temperature and temperature range in case of a 1% drop in the solar constant. With albedo feedback, the largest temperature drops are found in the polar regions. Without albedo feedback, the pattern is reversed (because the change radiative forcing decreases towards the poles). With regard to the annual temperature range, changes without albedo feedback would be very small, with albedo feedback the annual range increases by 1 to 2 °C in the polar and subpolar regions.

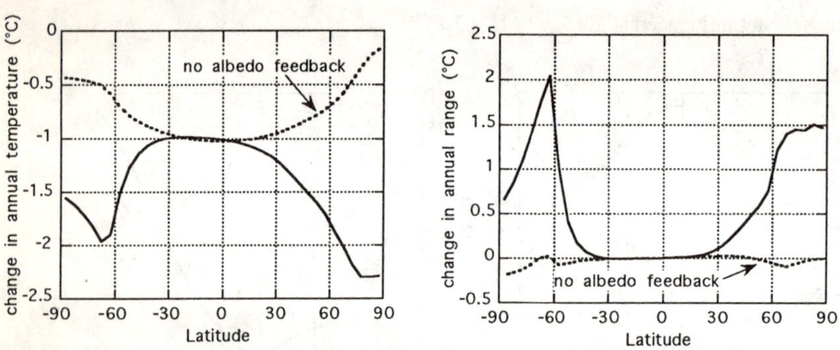

Figure 6. Change in annual temperature (left) and temperature range (right) without albedo feedback associated with snow, sea ice and land ice, respectively.

Table 4 shows some global mean values of calculated surface temperature for the experiments dicsussed above. Taking out snow would cause a 0.85 °C drop in mean surface temperature. Removing the sea ice creates a further drop of about 0.6 °C. Removing the land ice also roughly doubles the temperature decrease.

Concerning climate sensitivity, a distinction has to be made between short and long time scales. On the short time scale, decades to centuries say, the area of land ice will not change significantly and only snow and sea ice feedback is relevant. According to the results of the zonal climate model, the albedo feedback parameter β for snow and sea ice cover only is 0.0016 K^{-1}. On the long time scale, land ice area will change. When using estimated mean albedo and surface temperature for the last glacial maximum (Climap, 1976) and comparing this with the present-day values, provides an estimate β of 0.003 K^{-1}.

ROLE OF SNOW AND ICE		
reference case:	13.32	
no snow:	14.17	0.85
no snow & sea ice:	14.75	1.43
no snow, sea ice & land ice:	15.93	2.61
CLIMATE SENSITIVITY FOR S = - 1%:		
all feedbacks*:		-2.29
only snow and sea ice:		-1.25
no snow, sea ice & land ice changes:		-0.88

Table 4. A summary of the climate model experiments on snow / ice albedo feedback. For explanation of * see text. All numbers are global and annual mean surface temperatures (first column) or difference with the reference case (second column). In °C.

LAND ICE AND THE GLOBAL ENERGY BALANCE

It has been argued sometimes that the melting of the large Pleistocene ice sheets had a significant direct effect on the energy budget of the climate system. This view cannot be supported when looking at characteristic energy flows. Table 5 provides some numbers.

It appears that energy flows involved in melting of snow and ice are generally small compared to the amount of solar energy absorbed in the climate system. To melt all land ice currently present, for instance, requires 2.5 yr of absorbed solar radiation. If the melt down took 10000 years (a characteristic time scale for the glacial-interglacial transition, which reflects a very fast worldwide climatic change in the earth's history), a fraction of only 0.00025 of the absorbed solar energy would be needed. So one can safely state that energy flows associated with water phase transitions in building or melting ice and snow cover are negligible on the global scale. So whether ice sheets grow or shrink depends on the local conditions at the glacier surface, not on the global energy budget.

absorbed solar radiation in climate system:	3.8×10^{24} J/yr	1
absorbed at the earth's surface:	2.8×10^{24} J/yr	0.737
heating the oceans by 1 K requires:	5.8×10^{24} J	1.53
heating the atmosphere by 1 K requires:	5.1×10^{21} J	0.00134
present mass of land ice is 2.7×10^{19} kg		
to melt this requires:	9.3×10^{24} J	2.45
mass of ice in full glacial is 7.3×10^{19} kg		
return to interglacial requires:	1.6×10^{25} J	4.21

Table 5. Global energy flows involved in heating / melting components of the climate system. Last column gives values scaled with absorbed solar radiation.

REFERENCES

Climap Project Members (1976) The surface of the ice-age earth. Science 191: 1131-1137.

Haeberli W, Bösch H, Scherler K, Østrem G and Wallén CC (1989) World Glacier Inventory. Status 1988. IAHS(ICSI)-UEP-UNESCO

Folland CK, Karl TR, Nicholls N, Parker DE and Vinnikov KYa (1992) Observed climate variability and Change. In: Climate Change 1992 (eds.: Houghton JT, Callander BA and Varney SK), University Press Cambridge, p. 159 and 160

Harvey LDD (1988a) A semianalytic energy balance climate model with explicit sea ice and snow physics. Journal of Climate 1: 1065-1085

Harvey LDD (1988b) On the role of high latitude ice, snow, and vegetation feedbacks in the climate response to external forcing changes. Climatic Change 13: 191-224

Oerlemans J and Van den Dool HM (1978) Energy balance climate models: Stability experiments with a refined albedo and updated coefficients for infrared emission. Journal of Atmospheric Sciences 35: 371-381

Oerlemans J and Van der Veen CJ (1984) Ice Sheets and Climate (Kluwer), 217 pp.

Robinson DA, Dewey KF and Heim RR Jr (1993) Global snow cover monitoring: An update. Bulletin of the American Meteorological Society 74: 1689-1696

Robock A (1980) The seasonal cycle of snow cover, sea ice and surface albedo. Monthly Weather Review 108: 267-285

Raval A and Ramanathan V (1989) Observational determination of the greenhouse effect. Nature 342: 758-761.

Stone PH and Miller DA (1980) Empirical relations between seasonal changes in meridional temperature gradients and meridional fluxes of heat. Journal of Atmospheric Sciences 37: 1708-1721

Warren SG and Schneider SH (1979): Seasonal simulations as a test for uncertainties in the parameterizations of a Budyko-Sellers climate model. Climatic Change 8: 231-241.

The role of the atmosphere in the water cycle

José Pinto Peixoto
Instituto Geofísico do Infante D. Luís
Universidade de Lisboa
Rua da Escola Politécnica, 58
1200 Lisboa
Portugal

Abstract

The hydrological cycle is a consequence of the conservation of water substance in the climatic system in its three phases and must be regarded *in toto* with the aerial and terrestrial branches.

The atmosphere due to its mobility and its capacity to promote phase transitions which lead to precipitation, forces, in a way, the terrestrial branch through the two common boundary hydrological elements: the evaporation and precipitation.

The atmosphere is responsible for the transport of water from the oceans into the continents. The atmosphere plays an important role in modulating the precipitation and also in the radiative forcing at the land surface, through the radiative effects of clouds, trace gases including carbon dioxide, water vapour and aerosols. They profoundly alter the radiative proprieties of the atmosphere. Clouds are the prime regulator of the energy exchanges in the atmosphere. Their feedbacks are the greatest uncertainty in modelling the general circulation of the atmosphere.

Actually Hydrology is based upon the so called classical equation of hydrology, which is a balance equation for the terrestrial branch. It is only a condition equation that does not involve any dynamic principle. However, the balance equation for the moisture in the atmosphere allowing the use of meteorological quantities to define the hydrological elements is based on the dynamics of the general circulation of the atmosphere. Thus, due to the forcing of the terrestrial branch by the aerological branch the dynamics of the terrestrial hydrology can be linked to the dynamics of the aerial branch.

Using aerological data, the basic fields of precipitable water and water vapour transport were analyzed on a global scale. In order to study its intrinsic structure the transport water field was partitioned into a nonrotational part (the potential field) and a nondivergent part (the streamfunction field).

The connection between the atmospheric branch and the hydrology of the Earth's surface can be studied through the divergence field of water vapour transport, through the two dimensional mean paths of water vapour in the atmosphere and the vertical mean streamlines.

NATO ASI Series, Vol. I 31
The Role of Water and the Hydrological Cycle
in Global Change
Edited by Howard R. Oliver and Sylvia A. Oliver

Due to the release of latent heat associated with cloud formation there is a strong generation of entropy within the atmosphere, which is equivalent to the entropy generated by the absorption of solar radiation.

This work points to the importance of the atmosphere for the study of the hydrological cycle showing that the dynamics of Hydrology is, in part, ruled by the dynamics of the general circulation of the atmosphere.

I. WATER IN THE CLIMATE SYSTEM

I.1 Introduction

We will accept the conservation principle of water in its three phases: solid, liquid and gaseous, within the climatic system. The same cannot be accepted for each phase separately. Thus, an excess of liquid water is compensated by the depletion of the cryosphere (solid phase). In the atmosphere, the vapour phase has a variable concentration, which depends, not only, on the higher or lower availability of water in other phases, but also on the temperature.

The hydrological cycle is a closed circulation of the water in its three phases inside the climatic system. It is a closed sequence of natural phenomena, in which the globe provides the atmosphere with water vapour, that afterwards returns in liquid or solid phases to the globe.

The natural phenomena that make up the hydrological cycle are: a) the transfer of water, in its gaseous phase, from the globe to the atmosphere, by the evaporation of liquid or solid water of the seas, lakes, rivers, etc and the transpiration of plants at the surface; b) water transport in its gaseous and solid or liquid phases (clouds) by the general circulation of atmosphere; c) partial condensation of water vapour in small droplets and ice crystals that remain in suspension in the atmosphere, making the clouds; d) water transfer in its solid or liquid phases from the atmosphere to the globe by hydrometeor precipitation and deposition at the surface; e) runoff and storage at the surface, water infiltration at the subsurface and absorption by plant roots, the intensification of rivers on their way to the sea or lakes, and underground water storage (phreatic deposits). And then this cycle starts again.

As we can see, the hydrological cycle consists of an aerial branch, which transports water vapour and liquid and solid water of clouds by the general circulation of the atmosphere, and a terrestrial branch formed by the surface and subterranean runoff.

These two branches are linked by evaporation and precipitation. Precipitation is the main source of fresh water for the globe, feeding and maintaining the terrestrial branch of the hydrological cycle. On the other hand, evaporation, mainly terrestrial, feeds and maintains the aerial branch of this cycle.

We can say that the terrestrial branch is forced by the aerial branch, characterized by its large mobility and the ability to induce phase transitions that lead to precipitation. The dynamics of the hydrological cycle is conditioned by sources and sinks of water vapour in the atmosphere, by the thermodynamics of phase transitions and by the dynamics of the general circulation of the atmosphere.

Vast quantities of water are continuously on the move in the climatic system. Under the direct or indirect influence of solar energy, water evaporates from the oceans and the continents and is transpired by plants and animals into the atmosphere. Within the atmosphere the ascending moist air develops, by vertical adiabatic cooling, into visible moisture in the form of clouds (condensation). Thus, in the atmosphere, water is stored and transported in the condensed phase (liquid water and ice crystals) as clouds, or in the vapour phase (water vapour). Under the action of gravity it falls on the continents and oceans in the form of rain, snow, hail, or other forms of precipitation. The water that is precipitated returns partly to the atmosphere, through evaporation and evapotranspiration, infiltrates into the ground, or runs off over or under the ground to the rivers and streams, which carry it back to the oceans and seas, and becomes ready to start a new journey.

This gigantic and complex system of distillation, pumping and transporting the water substance in its many forms, originates an unending circulation that constitutes the hydrological cycle. Basically it is a consequence of the conservation of water substance. Not all water of the climatic system takes part in this continuous cycle. Some remains for shorter or longer periods of time in the atmosphere, in the biosphere, as snow accumulations in the cryosphere, in the oceans, river valleys, reservoirs and lakes, or as connate water chemically or physically bound to the lithosphere (soil and rocks), etc.

The atmosphere is the physical environment where man lives. It is essential to all phenomena responsible for the existence of the hydrological cycle, and it is more than a spectator of the physical and chemical processes of water phase transitions that provide the rainfall mechanisms.

The global hydrological cycle describes the transport and occurrence of water in all three phases from one place to another on a global scale.

The importance of water for the energetics of the Earth is not due only to the direct transport of energy in the latent heat form. In fact, the gains and losses of energy from the atmosphere and the Earth's surface by radiative processes are strongly coupled with the hydrological cycle. Clouds and water vapour modulate the incident solar radiation through reflection and absorption. On the other hand the absorption and the emission of longwave radiation by water vapour and clouds generate a greenhouse effect which influences substantially the Earth's climate.

I.2 The place of the atmosphere in the climate system

The role of the atmosphere in the hydrological cycle cannot be understood if the other components of the climate system are disregarded, since they interact strongly with the atmosphere. Thus the atmosphere must be taken within this more general framework.

The climatic system ($\mathcal{S}$) is a composite system formed by five major subsystems: the atmosphere ($\mathcal{A}$), the hydrosphere ($\mathcal{H}$), with the oceans ($\mathcal{O}$), the cryosphere ($\mathcal{C}$), the lithosphere ($\mathcal{L}$) and the biosphere ($\mathcal{B}$), this is:

$$\mathcal{S} = \mathcal{A} \cup \mathcal{H} \cup \mathcal{C} \cup \mathcal{L} \cup \mathcal{B}$$

as shown in figure 1.

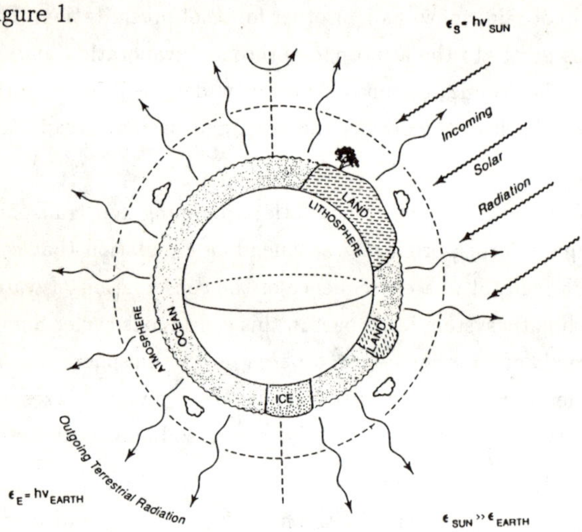

Fig. 1:Schematic diagram of the Climate System

The climatic system is closed for the mass and for the angular momentum but is open for the energy in the radiant form and for the entropy. On the other hand, all the subsystems are open for the mentioned quantities, with strong non linear feedback interactions.

The water is always present in any of the subsystems in one or various of its three phases: liquid, solid and vapour. Furthermore the water is the only substance which can exist simultaneously in earth in three phases under natural conditions.

Since the climatic system is closed for the mass and there are no significant destructive mechanisms for the water, the amount of water substance within the climatic system remains invariant. However, the same conservation principle cannot be applied to the individual phases separately.

Water must then be regarded as one of the most important links among the subsystems of the climate system, since it connects all of them. The knowledge of distribution of the water in the climatic system and its transport in any of its three phases is fundamental for full understanding of the hydrological processes which play also a crucial role as a regulator of the climate.

⋆

⋆ ⋆

The water on the continents is distributed in several reservoirs, namely in glaciers ($25\times10^{15}\,m^3$), ground water ($8.4\times10^{15}\,m^3$), lakes and rivers ($0.2\times10^{15}\,m^3$), and in the living matter of the biosphere ($0.0006\times10^{15}\,m^3$). The amount of water locked in the polar ice is impressively large, totalizing some 1.8% of all water in the hydrosphere. Of the total amount of underground water, vadose water (water present in soil) accounts for only $0.066\times10^{15}\,m^3$. The remainder is about evenly divided between reservoirs deeper than 800 m and reservoirs shallower than that level (Peixoto and Kettani, 1973). The residence time of water in the various reservoirs can be deduced from the ratio of the amount of water in the particular reservoir by the corresponding water fluxes. It is found to vary from nine days for atmospheric water to thousands of years for the polar ice and the oceans (Peixoto and Oort, 1992).

The atmosphere is the central subsystem of the global climate system. It is linked to the other subsystems through highly non-linear interactions. It has the highest mobility of them all. The water in the atmosphere can exist in the three phases. In spite of its small amount when compared with the amounts of the other subsystems, the atmospheric water is decisive for the hydrological cycle and exerts a central influence on the energy balance of the climate system. Furthermore, through the latent heat and the radiative processes the moist atmosphere is essential for the energy exchanges in the climate system.

I.3 The forcing of the hydrological cycle

The main forcing of the hydrological cycle is the incoming solar radiation. However the spherical shape of the Globe, the earth's rotation and the gravity must also be considered.

As shown in figure 1 Earth is imbedded in a sea of photons of solar and terrestrial origin. The solar photons are of higher energy and lower entropy than those emitted by the Earth.

The Earth exchanges radiant energy with outerspace. The incoming solar radiation is absorbed, reflected and scattered by the atmosphere and the clouds and only a fraction of 50% reaches the globe. On the other hand all the components of the climate system emit longwave radiation which is partly absorbed by the atmosphere and clouds and the remaining is transferred to outerspace. Due to its importance for the dynamics of the hydrological cycle and of the climate we will give a detailed discussion of the energy balance of Earth.

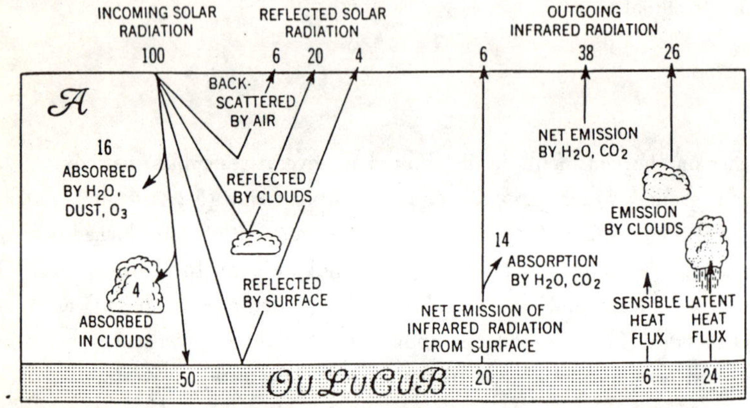

Fig. 2:Schematic diagram of the global radiation budget in the climate system. A value of 100 units is assigned to the incoming flux of solar energy.

Figure 2 summarizes the annual-mean radiation balance for the climate system as a whole, separating it into the solar and terrestrial radiation budgets on the left and right hand sides, respectively. The numbers are scaled, so that incident solar radiation flux is set at 100 units, equivalent to $342\,\mathrm{Wm}^{-2}$.

Of the 100 units of incoming solar radiation shown on the left hand side of figure 2, 16 units are absorbed by stratospheric ozone, tropospheric water vapour and aerosols; 4 units by clouds, and 50 units by the earth's surface. The remaining 30 units of solar radiation are backscattered by the air (6 units), reflected by clouds (20 units), and reflected by the earth's surface (4 units). These 30 units do not participate in the physical and chemical processes that occur in the climate system.

Of the 50 units of solar radiation absorbed at the earth's surface, as the right hand side of figure 2 shows, 20 units are emitted as longwave radiation into the atmosphere and 30 units are transferred upward into the atmosphere by turbulent and convective processes in the form of sensible heat (6 units) and latent heat (24 units). Of the 20 units of emitted longwave radiation, 14 units are absorbed in the atmosphere mainly, by water vapour and carbon dioxide, and 6 units are emitted directly into space.

Considering now the atmosphere by itself, it absorbs 20 units of solar radiation plus 44 units of energy coming from the earth's surface. This total absorbed amount of 64 units is then balanced by the emission into space of infrared radiation by water vapour and carbon dioxide (38 units) and by clouds (26 units). Adding the 6 units of radiation passing directly through the atmosphere from the earth's surface, we find a total loss of 70 units in the form of longwave radiation at the top of the atmosphere, as required to balance the 70 units of absorbed shortwave solar radiation.

I.4 The moisture of the atmosphere

I.4.1 The role of clouds

The atmosphere can be regarded as a binary system, with two main components: "dry air" and water substance. The latter can exist in the vapour phase alone or in the condensed phases (solid or liquid) forming the clouds.

The clouds may be regarded as a colloidal solution where the dispersed phase is formed by water droplets and ice crystals immersed in a dispersant phase – the saturated moist air.

The water droplets are formed when the threshold of supersaturation is attained and there are enough suspended particles that can act as condensation nuclei (Aikten). Their further growth depends upon the availability of water vapour in the surrounding air, on its stability and upon radiation. Furthermore, collisions among droplets and exchanges of water molecules between ice crystals and water droplets are also of particular importance.

The size spectrum of the droplets (dispersed phase) ranges from 0.1 μm to 50 μm. Above this limit drizzle and precipitation starts.

Besides its importance for precipitation, the clouds, its shapes and the spatial distribution, have an important role in modulating the planetary radiation budget. In fact, they are responsible for about one half of the total reflection back to space; without clouds the earth's "clear sky" albedo would range between 15 and 16%.

In the longwave part of the spectrum, terrestrial heat radiation is mainly emitted from cloud tops which are colder than the surface. Cloud bottom and the lowest layers of the troposphere emit an appreciably high amount of radiation to the surface. Thus, in summary, without clouds the earth's surface would loose more radiative energy to space than it does with clouds.

Clouds participate in the atmosphere's greenhouse effect by reducing it considerably. While without clouds the earth's surface temperature could assume a mean value of more than 300 K (assuming the present state of the climate system unchanged) its present value is around 288 K. In the shortwave bands of the spectrum the clouds contribute to cooling the earth's surface, whereas in the longwave radiation bands they contribute to a warming.

The average cloudiness of the Earth is around 45% to 55%. Deep convective clouds contribute with 10 to 15%, low stratus with 30–60%, medium level clouds with 40–50% and high level clouds (cirrus) with 25–40%.

Thus, clouds can considerably shield the ground surfaces from heating by solar radiation, but simultaneously they prevent their cooling by remitting heat radiation back to the ground. They absorb appreciable amounts of solar radiation and reflect very effectively large fractions of the incident radiation back to space, thus affecting the solar heating of atmospheric layers above their tops. High and cold clouds emit much less thermal radiation than low clouds and the ground, thus they may even "heat" the climate system. Furthermore, many feedback processes due to clouds occur in the climate system.

In the present state of our climate, clouds cause a general "net cooling" over most parts of the earth's surface. However, over some other areas they even may enlarge slightly the radiative energy gain of the planet. These latter regions are known to be covered mostly by high level cirrus only.

I.4.2 The role of water vapour

The most important link connecting the various parts of the hydrological cycle is the water vapour of the atmosphere. Furthermore the water vapour is the predominant greenhouse gas of the atmosphere. Thus the knowledge of the distribution of the water vapour concentration and of its transport is fundamental for a complete understanding of the hydrological processes and the energy processes involved which are important in regulating the climate.

The transport of water in the vapour phase is by far much larger than the transport of water by clouds in the condensed phases. The last component is almost irrelevant for the total transport on the global scale. As we stated before, the same cannot be said, however, when we deal with energy. Thus the water vapour is the main source of water to keep the clouds going on and therefore to feed the precipitation mechanisms.

The transport of water vapour in the atmosphere, being the leading component of the aerial runoff, brings the moisture deeply inside the continents. Its incoming flux is equivalent to the runoff of all the rivers into the sea.

The quasi-horizontal water vapour transport within the atmosphere is dominantly a large scale process, moving water vapour from its primary source regions over the subtropical and tropical oceans to feed the precipitation occurring in tropical and middle latitude weather systems, as we will see. However, the time and spatial scales of the atmospheric circulations actually involved in cloud formation, latent heat release and precipitation are, by comparison, relatively small.

All the fields of water vapour content and transport in the atmosphere can be evaluated using the aerological quantities that characterize the instantaneous state of the atmosphere, as we will shown in the next chapter.

II. THE GENERAL CIRCULATION OF THE ATMOSPHERE AND THE WATER CYCLE.

II.1 The balance equation for the water

II.1.1 The hydrological cycle

As we mentioned the hydrological cycle is basically a consequence of the general principle of the conservation of water substance in its three phases on the Earth. We will see that this conservation leads to a generalized balance equation for the water substance.

For a closed circuit $[\gamma]$ this equation can be expressed as:

$$\Delta \oint \vec{Q}_t \cdot d\vec{\gamma} = 0 \tag{1}$$

where $\vec{Q}_t$ is the total transport field of water substance in its three phases:

$$\vec{Q}_t = \vec{Q}_{\text{liq}} + \vec{Q}_{\text{sol}} + \vec{Q}_{\text{vap}}. \tag{2}$$

The integral taken along the closed circuit $[\gamma]$ can be decomposed in two line integrals along the two complementary paths, one $\hat{\gamma}_A$ for the atmosphere and the other $\hat{\gamma}_0$ for the Globe, so that:

$$\Delta \int_{\hat{\gamma}_A} \vec{Q}_t \cdot d\vec{\gamma} + \Delta \int_{\hat{\gamma}_0} \vec{Q}_t \cdot d\vec{\gamma} = 0 \tag{3}$$

In the upper path, which constitutes the aerial branch of the hydrological cycle, the dominant phase is the vapour phase ($\vec{Q}_t \approx \vec{Q}_{\text{vap}}$). The lower path defines the terrestrial branch, confined to the hydrosphere and lithosphere, and so the transport of water is in the liquid and, to a smaller extent, solid phase. For the terrestrial branch the transport field reduces to the classical surface and subsurface runoff ($\vec{Q} = \vec{R}$).

Then, expression (3) can be written under the form:

$$\int_{\hat{\gamma}_A} \vec{Q}_{\text{vap}} \cdot d\vec{\gamma} = - \int_{\hat{\gamma}_0} \vec{R} \cdot d\vec{\gamma} \qquad (4)$$

which shows that the aerial runoff is exactly compensated by the terrestrial runoff.

II.1.2 The classical equation of Hydrology

For a given interval of time the fundamental equation of Hydrology is as follows:

$$(P + N + D) - (R_0 + R_u + E) = \Delta(S_s + S_g + S_u) \qquad (5)$$

where:

- P, N and D represent, the amounts of water received on the Globe, by precipitation (P), interception of fog (N) and deposition of dew and frost (D);
- ΔS_s, ΔS_g and ΔS_u represent the water stored on the surface of the Globe, in the soil and at the subsurface;
- R_0 and R_u represent the surface and subterranean runoff;
- E is the evaporation and evapotranspiration.

For a large region, the net interception, deposition and subterranean transport are usually small so that they can be disregarded, and the classical equation (5) can be written:

$$P - E = \Delta S_s + R_0 \qquad (6)$$

For long periods of time and large areas ΔS tends to be small, compared to the other terms, and equation (6) can be reduced to:

$$P = R_0 + E \qquad (7)$$

which shows that the excess of precipitation over evaporation is compensated by the runoff ($R = P - E$).

II.1.3 The balance equation for the water vapour

The formulation of the atmospheric branch of the hydrological cycle is based on the balance requirements for the water vapour in the atmosphere.

The amount of water vapour, W, contained in a unit area column of air which extends from the earth's surface to the top of the atmosphere, is given by the expression:

$$W = \int_0^{p_0} q \frac{dp}{g}, \tag{8}$$

where q is the specific humidity, g the acceleration of gravity and p the pressure. It is called the precipitable water in the atmosphere. It represents the amount of liquid water that would result if all the water vapour in the unit column of the atmosphere were condensed. It is usually expressed in units of $\mathrm{kg\,m^{-2}}$ or mm. This field, represents the atmosphere's capacity to store water vapour.

The horizontal transport of water vapour is obtained by multiplying the wind vector velocity by the specific humidity (amount of water vapour per kg of moist air). The total horizontal transport vector, integrated along the vertical, is the "aerial runoff", $\vec{Q}$, defined as:

$$\vec{Q} = \int_0^{p_0} q\vec{v} \frac{dp}{g} = Q_\lambda \vec{i} + Q_\varphi \vec{j}, \tag{9}$$

where $\vec{v}$ is the wind vector, Q_λ the zonal component (parallel to the latitude circles), Q_φ the meridional component and $\vec{i}$ and $\vec{j}$ the unit vectors tangent to a parallel and to a meridian, respectively. Terms Q_λ and Q_φ represent the fluxes of water vapour across unit strips of latitudinal and meridional walls, respectively.

From the balance requirements it is easy to show that the excess of evaporation over precipitation $(E - P)$, at the earth's surface, is balanced by the local rate of change of water vapour storage, $\partial W/\partial t$ and by the net inflow, or the net outflow of water vapour given by the divergence of $\vec{Q}$ and represented mathematically by $\operatorname{div}\vec{Q}$. Thus, we obtain a simplified general balance equation for the water vapour in the atmosphere:

$$\frac{\partial W}{\partial t} + \frac{\partial W_c}{\partial t} + \operatorname{div}\vec{Q} + \operatorname{div}\vec{Q}_c = E - P, \tag{10}$$

where W_c and $\vec{Q}_c$ are the amount of condensed water in a unit column of air and the horizontal vector transport, respectively.

In general $\partial W_c/\partial t \ll \partial W/\partial t$ and $\vec{Q}_c \ll \vec{Q}$ so that both the rate of change of condensed water in clouds and its horizontal transport can be disregarded. Thus, we can write equation (10):

$$\frac{\partial W}{\partial t} + \operatorname{div}\vec{Q} = E - P, \tag{11}$$

The former equations can be applied on a regional scale, by defining an average space operator for a surface A: $\langle(\cdot)\rangle = (1/A)\int\int(\cdot)\,dA$ and the time average operator, for an interval τ: $\overline{(\cdot)} = (1/\tau)\int(\cdot)\,d\tau$. The equation (11) assumes the form:

$$\left\langle \frac{\partial \overline{W}}{\partial t} \right\rangle + \left\langle \mathrm{div}\overline{\vec{Q}} \right\rangle = \langle \overline{E-P} \rangle \tag{12}$$

Using the Gauss theorem, this equation may be written under a different form which is more useful for regional studies:

$$\left\langle \frac{\partial \overline{W}}{\partial t} \right\rangle + \frac{1}{A}\oint \overline{\vec{Q}} \cdot \vec{n}\, d\ell = \langle \overline{E-P} \rangle, \tag{13}$$

where $\vec{n}$ denotes the outward normal unit vector to the boundary $[\ell]$ of the region with an area A.

These equations describe the atmospheric branch of the hydrological cycle.

Except for the case of severe storms and for short intervals of time, the rate of change of precipitable water, $\partial W/\partial t$, is very small compared to the other terms. For sufficiently long periods of time, this term can be disregarded, and the equation (12) yields:

$$\left\langle \mathrm{div}\overline{\vec{Q}} \right\rangle = \langle \overline{E-P} \rangle \tag{14}$$

Thus, the divergence of water vapour, is found over those regions of the globe where evaporation exceeds precipitation ($\langle \overline{E-P} \rangle > 0$), whereas convergence is found where precipitation is greater than evaporation ($\langle \overline{E-P} \rangle < 0$).

The term $(\overline{E} - \overline{P})$ establishes the connection between the terrestrial (equation 6) and the atmospheric (equation 11) branches of the hydrological cycle. Elimination of $\overline{E} - \overline{P}$ between these two equations yields:

$$\langle \Delta S_s \rangle + \langle R_0 \rangle = -\left\langle \frac{\partial \overline{W}}{\partial t} \right\rangle - \left\langle \mathrm{div}\overline{\vec{Q}} \right\rangle \tag{15}$$

which shows how the two branches of the hydrological cycle are linked together. This equation, which combines both branches of the hydrological cycle, can be regarded as a general equation of Hydrology. If, besides the aerological terms, $\langle R_0 \rangle$ and $\langle \overline{P} \rangle$ are known over a certain region, the rate of change in ground water and the rate of evaporation can be estimated.

Over long periods of time, such as a year, changes in storage in the ground and in the atmosphere become small so that, for example, for a continent the surface and subsurface runoff have to be exactly balanced by the "aerial runoff" into that continent from the surrounding ocean areas. Furthermore, when the global atmosphere is considered over a long period of time, all transport and storage terms vanish, and we can conclude that the global-mean evaporation equals the global-mean precipitation.

II.1.4 Data

The basic data necessary to evaluate the various fields, W, are the daily values of specific humidity and wind components at various levels in the atmosphere. This was done using the rawindsonde data for the ten year period, May 1963 through April 1973, by the Geophysical Fluid Dynamics Laboratory / NOAA, Princeton University. The rawindsonde network covers the entire globe, from pole to pole.

II.2 The importance of the basic fields for the dynamics of the water cycle

II.2.1 Precipitable water and specific humidity fields

The mean fields of precipitable water for the year, summer and winter seasons, are presented in figure 3. These figures show the general decrease of $\overline{W}$ from Equator to the Poles, with maxima located slightly to the north of the Equator. This location is associated with the mean position of the thermal equator. The distribution is almost zonal with an accentuated decrease in medium latitudes in both hemispheres. The northern hemisphere presents a larger zonal asymmetry than the southern hemisphere, due to different areal coverage by continents in both hemispheres.

In order to show the annual variability of precipitable water, fields for the two seasons were analyzed and are presented also in figure 3.

During DJF, $\overline{W}$ increases in the southern hemisphere and decreases in the northern hemisphere and there is a displacement of the maxima, located on the equatorial region, to the south, which follows the thermal equator movement into the south of the geographical equator.

During JJA, $\overline{W}$ increases in the northern hemisphere and decreases in the southern hemisphere. As expected, the equatorial maximum follows the thermal equator, located to the north of the geographical equator.

Larger variations occur in middle and high latitudes where there is an increase of $\overline{W}$ in the summer hemisphere and a decrease in the winter hemisphere. These seasonal shifts are much more pronounced over the northern hemisphere.

During summer, over the northern hemisphere, zonal asymmetries are larger, reflecting not only the orography effects, but also the thermal contrast associated with ocean-continent differences. The significant increase of W over Southeast Asia during JJA is associated with the monsoon and is an important case of the direct and indirect continental

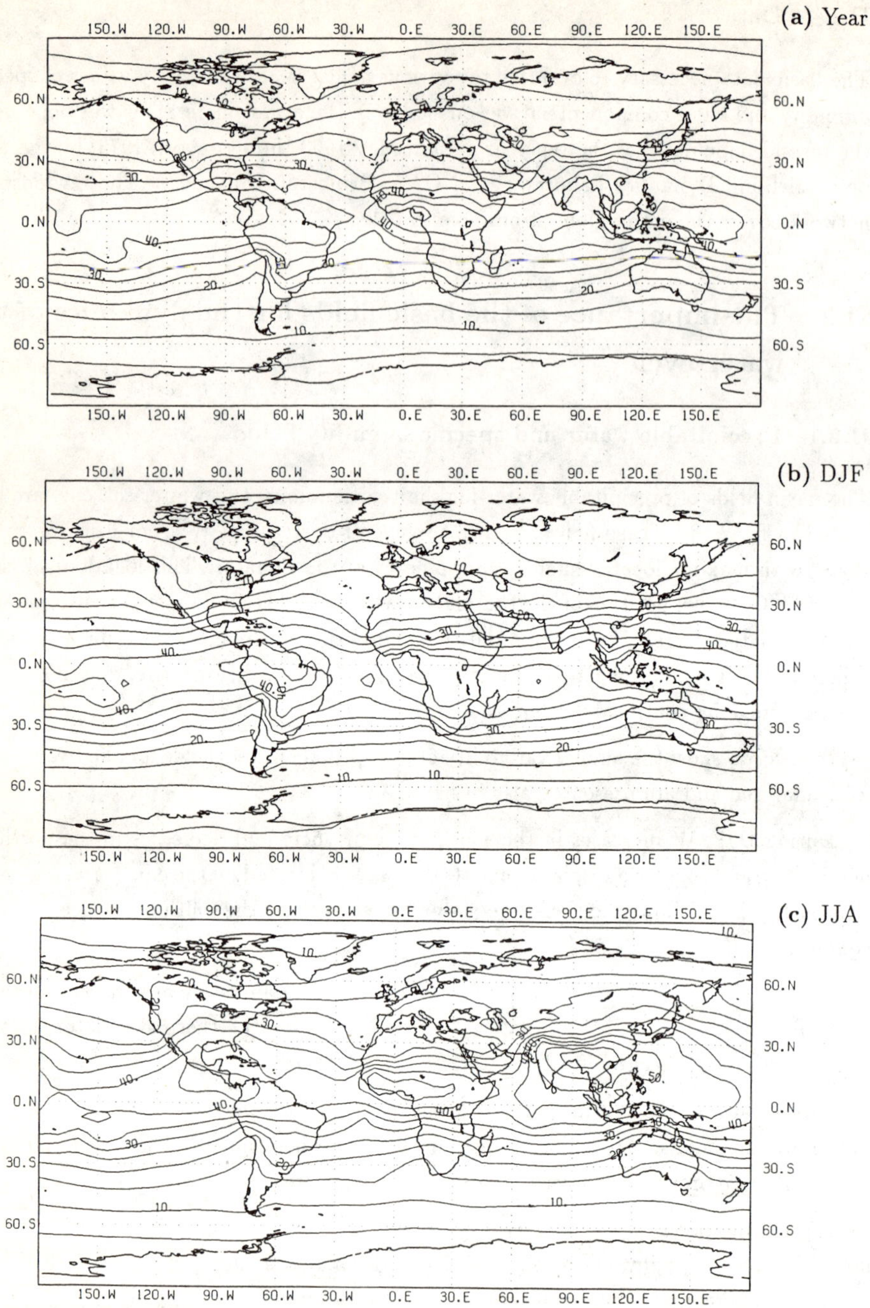

Fig. 3:Global distribution of the vertically integrated of mean water vapour abundance (precipitable water) for (a) annual, and (b) DJF, and (c) JJA seasons. Contour interval 5kg/m^2.

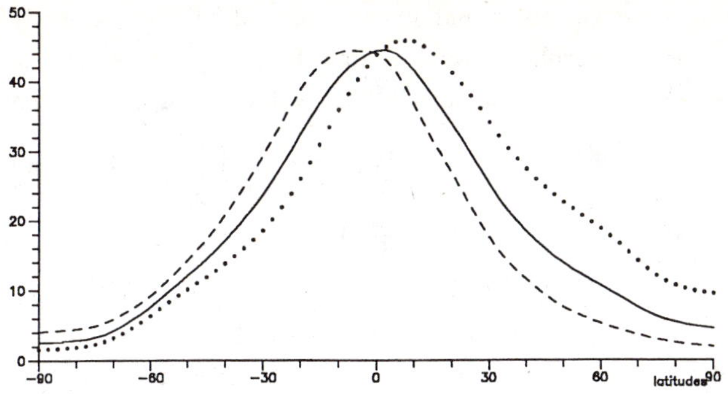

Fig. 4:Mean meridional profiles of the mean zonally averaged precipitable water $[\overline{W}]$ in kg/m^2. The full line profile is for annual conditions, the dashed for DJF and the point line for JJA seasons.

influence upon the precipitable water field.

The lower increase of $\overline{W}$ in higher latitudes over the southern hemisphere, during summer, is associated with orography of the Antarctic where the surface pressure is very low.

Due to the Earth's rotation and its spherical shape, meteorological fields present a mainly zonal distribution. It is then possible to define a zonal mean operator:

$$[(\cdot)] = \frac{1}{\ell}\oint(\cdot)\,d\ell = \frac{1}{2\pi}\int_0^{2\pi}(\cdot)\,d\lambda, \tag{16}$$

such that $\left[\frac{\partial(\cdot)}{\partial\lambda}\right] = 0$.

From these fields, mean zonal values $[\overline{W}]$ were computed at various latitudes. The profiles are presented in figure 4. This figure shows, in a compact way, all the features just discussed about the $\overline{W}$ field. A maximum is located in the equatorial region, slightly north of the equator, moving during JJA to the north and during DJF to the south. $[\overline{W}]$ increases in the summer hemisphere and decreases in the winter hemisphere; the seasonal variation are more pronounced over the northern hemisphere.

Numerically, the integration of these profiles shows that the annual mean value is 25.9 kg m^{-2} in the northern hemisphere and 24.3 kg m^{-2} in the southern hemisphere (due to the area of an hemisphere, 1 kg m^{-2} is approximately equivalent to 25×10^6 tons of water).

$\overline{W}$ charts depict also the average water vapour contribution to the total pressure of the humid atmosphere.

Global precipitable water has an annual average value of 2.6 cm or 1.3×10^{13} tons of water for the whole of the atmosphere, which is almost equal to the water content of all lakes. This value is small when compared with the water stored in the other climatic subsystems.

If we take the global average value of precipitation as 100 cm/year, we see that all the water vapour, would be consumed in 10 days (W/P). This number gives an idea of the average life (residence time) of water vapour in the atmosphere.

As the precipitable water does not suffer sizeable local fluctuations, from day to day, we must conclude that the continuous replacement of the condensed water vapour can only result from local evaporation or advection of water vapour. This advection must have a dominant role since, locally, a balance between precipitation P (water vapour consumption) and evaporation E (water vapour supply) is not observed showing the importance of the water vapour transport fields $\vec{Q}$ for the dynamics of the water cycle.

The precipitable water decreases, as expected, from Equator to the Poles, with the strongest gradient over the equatorial subregions.

Bands of intense precipitation in middle and higher latitudes, associated with the polar front, occur in regions where the values of total water vapour are not high. This shows that the necessary water must be supplied by water vapour that comes from other latitudes. The maximum seasonal values occur near the equator, and are always located in the summer hemisphere.

We can ask what is the contribution of the clouds to the water content of the atmosphere. Clouds are formed by an aerosol, where the dispersed phase is made up of water droplets and ice crystals and the dispersing phase is saturated humid air. The water droplets and ice crystals have very small dimensions ($20\,\mu$) and are separated by much larger distances, leading to concentrations on the order of 100 million/(cubic meter) of humid air. Thus, the amount of water in condensed phases is approximately $1\,\mathrm{g/m^3}$. This value is much smaller than the absolute humidity of the atmosphere (ρW). On other hand, the cloud cover is on the average around 50%, and the cloud thickness barely exceeds 3 km. So, at the global scale, the contribution of the condensed water of clouds to precipitable water is not significant. It can be important, however, at the local scale, and under certain synoptic situations.

During JJA, the value of $[\overline{W}]$ increases by $6.8\,\mathrm{kg\,m^{-2}}$, over the northern hemisphere, and decreases by $3.6\,\mathrm{kg\,m^{-2}}$, over the southern hemisphere. During DJF, this situation reverses: $[\overline{W}]$ increases over the northern hemisphere by $3.6\,\mathrm{kg\,m^{-2}}$, and decreases by $5.7\,\mathrm{kg\,m^{-2}}$ over the southern hemisphere.

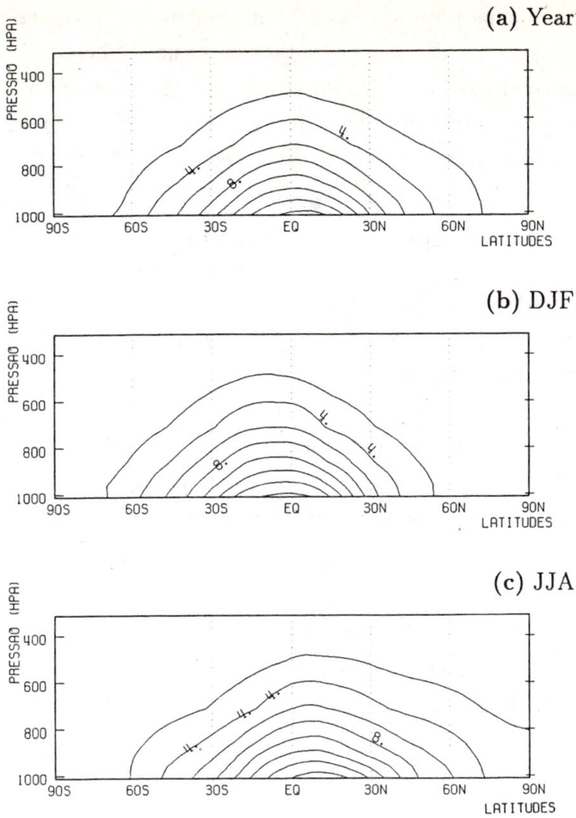

Fig. 5:Cross sections of mean specific humidity $[\overline{q}]$ for year, and DJF, and JJA seasons. The contour interval is 2 g/kg.

The analysis of the precipitable water field allows us to infer the total water vapour content but it does not give any information about the vertical distribution in the atmosphere. This can be obtained from the mean specific humidity profiles $[\overline{q}(p, \varphi)]$ as presented in figures 5. They show that the specific humidity decreases with altitude, with greater concentrations on the lower levels of the atmosphere (34% water vapour is below 900 hPa, 48% below 850 hPa and almost 95% below 500 hPa). There is also a decrease from the Equator towards Poles, moving southward during DJF and northward during JJA.

These changes are, in part, associated with the temperature increases over the summer hemisphere and temperature decreases over the winter hemisphere. They are much less pronounced in intertropical regions. On the other hand seasonal variability is much more prominent over the northern hemisphere, mainly in middle and high latitudes where significant increases of specific humidity are observed.

II.2.2 Water vapour transport in the atmosphere

We will present now the mean fields $\overline{\vec{Q}}(\varphi, \lambda)$ of the aerial runoff of water vapour in the atmosphere, which are important for the development of precipitation and evaporation. They are presented in fig 6 for the year and for DJF and JJA seasons. Water vapour transport depicts the characteristics of the general circulation of the atmosphere in the troposphere. As we can see from the inspection of maps the fields are almost zonal showing a minimum transport over the subtropical belts. In the intertropical regions, a strong easterly transport is observed, whereas over the mid–high latitudes a more westerly transport occurs.

Latitudinal averages of $\overline{\vec{Q}}_\lambda$ field are presented in fig 7, in a more compact way. The profiles show that in the latitudinal band of 21^0S to 22^0N, there is an easterly transport associated with the trade winds and outside this band a westerly transport is observed in both hemispheres. Maxima westerly transports occur around 40^0N in the northern hemisphere, and 45^0S in the southern hemisphere. The larger values in the southern hemisphere are connected with the strong westerly winds that blow in this hemisphere ("roaring forties").

The direction of the transport reverses, in both hemispheres, at the average latitudes of the centers of the major subtropical anticyclones.

During the DJF period $[\overline{\vec{Q}}_\lambda]$ the maxima move to the south, while during JJA they move to the north, following the migration of the subtropical anticyclones. The easterly transport in intertropical regions presents a maximum in the winter hemisphere. The $[\overline{\vec{Q}}_\lambda]$ fields present a strong seasonal variability.

Figure 8 presents the zonally averaged values of the mean meridional transport of water vapour. From this figure we see that, on the average, for latitudes higher than 20^0N and 17^0S, the transport is directed towards the Poles with a maximum transport around 40^0 latitude. This transport is mainly accomplished by the eddies associated with the polar fronts. The southern hemisphere maximum is stronger than the one in the north, with a weak seasonal variation.

In intertropical regions the meridional transport is mainly associated with the lower branch of the Hadley cell. It is directed towards the equator, feeding the ITCZ rain-

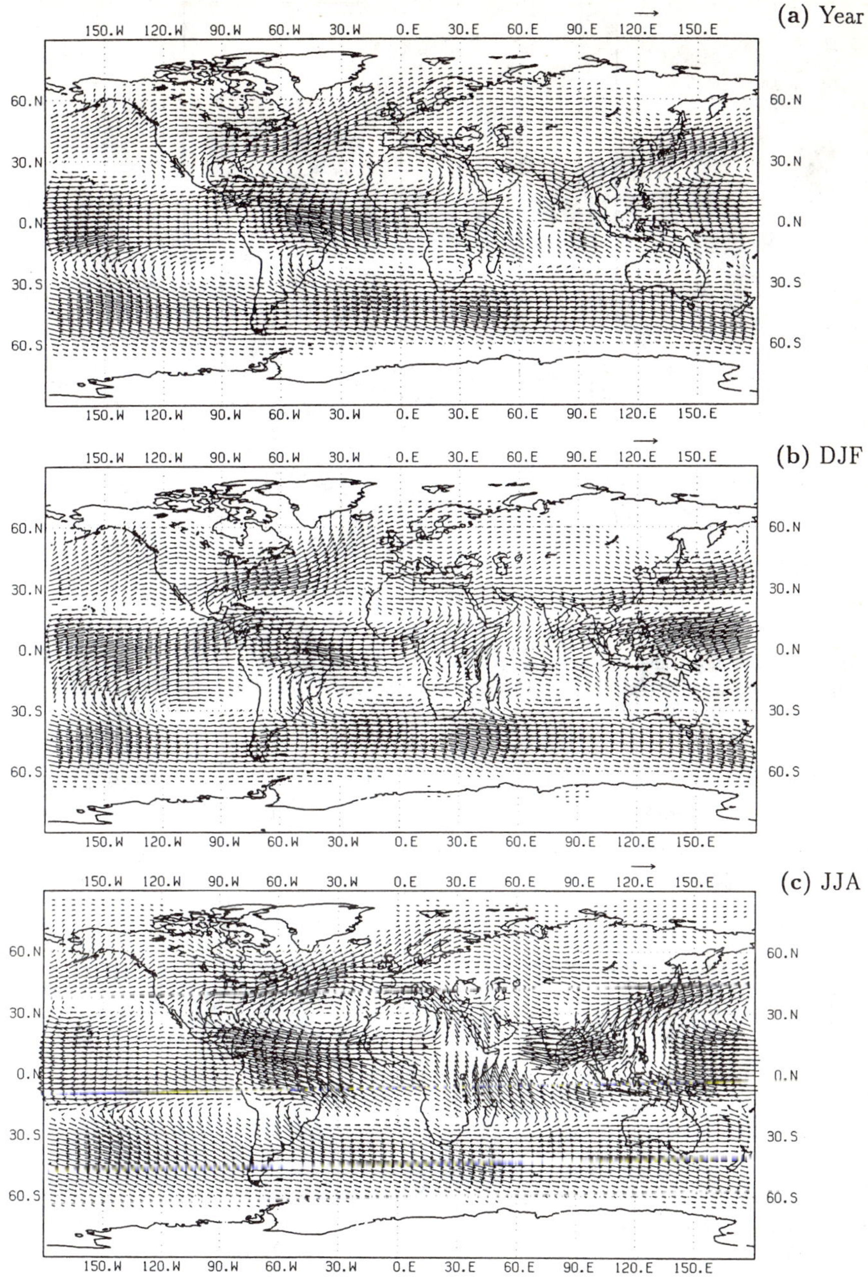

Fig. 6:Global distribution of the total mean transport of water vapor $\vec{Q}$, for (a) annual, and (b) DJF, and (c) JJA seasonal conditions. The sample arrow on the top of the right hand side corner represents a transport of 200 kg/(m s).

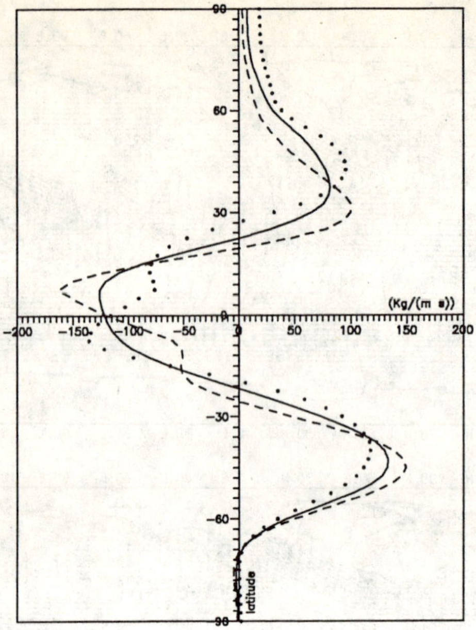

Fig. 7:Meridional profiles of the total mean eastward transport of water vapour $[\overline{Q}_\lambda]$ in units of $kg\,m^{-1}s^{-1}$, for annual (full line), DJF (dashed) and JJA (point line) conditions.

fall. The corresponding southern hemisphere values are larger than those of the northern hemisphere. It is important to point out that there is a net annual transport from the southern into the northern hemisphere, which explains the excess of rainfall observed in this hemisphere.

The larger seasonal differences occur between latitudes 23^0S and 30^0N. During the DJF there is a maximum flux towards the southern hemisphere around 10^0N, exporting water vapour from the northern into the southern hemisphere ($30.8\ kg\,m^{-1}s^{-1}$). During JJA, this flux reverses the direction. A maximum is observed around 5^0S, with a much stronger transport across the equator into the northern hemisphere, which explains the mean annual excess of rainfall, as already mentioned.

II.2.3 The importance of the water vapour basic fields for the dynamics of the hydrological cycle

We have previously described and analyzed the basic fields of water vapour in the atmosphere, v.g. the $\overline{W}$ and $\vec{\overline{Q}}$ fields. Now we will analyze some of their implications for the water cycle, on a global scale.

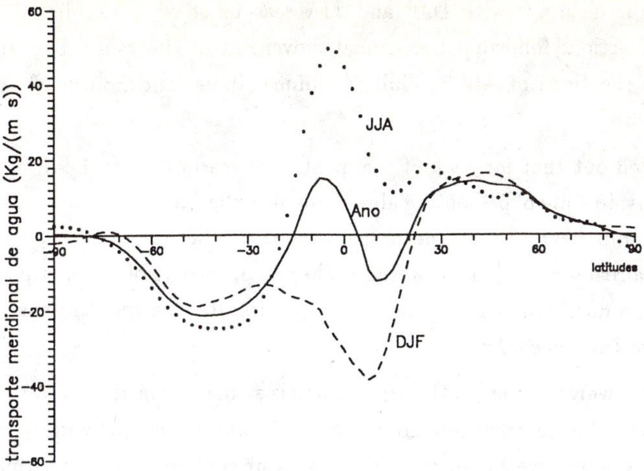

Fig. 8:Meridional profiles of the mean meridional transport of water vapour $[\overline{Q}_\varphi]$, in kg/(m s), for annual, DJF and JJA mean conditions.

Let us discuss now the importance of water vapour transport fields, $\vec{Q}$, in the atmosphere for the water balance in the climate system. We must keep in mind that the vectors, $\vec{Q}$, which are expressed in $\mathrm{kg\,m^{-1}\,s^{-1}}$ represent the aerial runoff of the water, forming the atmospheric branch of the hydrological cycle.

The structure of these fields is mainly conditioned by the characteristics of the general circulation of the atmosphere. We can notice that the great anticyclone gyres in subtropical latitudes, separate an easterly transport on the equatorial side from an westerly transport on the opposite side. The fields are more intense over the oceans (Pacific, Atlantic, Indian, etc) than over continents. In general, they show that there is a water transfer from the oceans into the continents. Even when the vectors are almost zonal their intensities are larger over the oceans than over the neighbor continents, leading to an aerial runoff of water from the oceans to the continents. In some cases, the ocean-continent transport becomes more apparent as with the flux from the Gulf of Mexico to North America or over the monsoon regions. This aerial runoff towards the continents and some reverse outflow from them must be compensated by the river's runoff into the oceans, since, for large time intervals, on a global scale we have a steady regime.

Winds increase, in general, with altitude, while specific humidity decreases exponentially. So, the transport fields of water vapour reflect, in altitude, these two effects, reaching a maximum between levels 850 and 800 hPa.

Due to the large decrease of water vapour with altitude, the transport is not significant above 200 hPa in the tropics and above 400 hPa in middle and high latitudes.

The inspection of the maps with DJF and JJA seasons show a meridional shift of the large centers of action, following the annual movement of the Sun. The monsoon effects, mainly over the Horn of Africa, Gulf of Guinea, India, and Southeast Asia are also prominently shown.

It must be pointed out that for each of the mentioned seasons, over the intertropical regions, the $\vec{Q}$ vectors, in spite of presenting almost zonal configurations, have a sufficiently strong meridional component to feed the precipitation that occurs along the ITCZ. In particular, we have already seen that, on average, there is an export of water vapour from the southern into the northern hemisphere, where precipitation is more intense due to orography and convection mechanisms.

Although the total water vapour in the atmosphere is a small quantity, its transport by the general circulation of the atmosphere does have significant values, showing a very high mobility. In spite of the relative smallness of water vapour content by the atmosphere its influence on Earth's climate and water resources, exceeds by far what could be expected when only its relative proportions are considered.

II.3 Derived fields of horizontal water vapour transport

II.3.1 Helmohltz components

The analyzes of the basic W and $\vec{Q}$ fields lead already to some conclusions on the contribution of the water vapour for the global water cycle. There are, however, some particular aspects that can be better understood, when we consider some other derived fields. Through these fields it is possible to identify, on the Globe, the sources and sinks of water vapour for the atmosphere, as well as their distributions and intensities. This can be accomplished by studying the field of divergence of $\vec{Q}$, obtained by applying the differential operator, div, to the $\vec{Q}$ vector field.

Other derived fields will allow subsequent derivation, in some detail, of the main trajectories of water vapour flow from its origins to its sinks (precipitation). Using the Helmohltz decomposition theorem, it is possible to identify and isolate a non divergent background flow of water vapour, derived from the stream function field ψ that does not participate in the precipitation phenomena, and another flow component derived from the potential function field χ. This is the only component responsible for the divergence–convergence phenomena, which are responsible by evaporation and precipitation.

Using these two components, specific details can be given about the dynamics of the aerial branch of the water cycle, having important implications for the dynamics of the terrestrial branch, specially in the assessment of water resources. In spite of having different intrinsic properties for the transport, they should not be studied separately, since each one of them represents only a fraction of the total transport.

In fact, the $\vec{Q}$ field can be written as:

$$\vec{Q} = \vec{Q}_\psi + \vec{Q}_\chi, \tag{17}$$

where $\vec{Q}_\psi$ and $\vec{Q}_\chi$ represent the solenoidal and irrotational components of $\vec{Q}$, so that:

$$\vec{Q}_\psi = \vec{k} \wedge \operatorname{grad} \psi \tag{18}$$
$$\vec{Q}_\chi = \operatorname{grad} \chi. \tag{19}$$

Applying the divergence operator, and the operator $\vec{k} \cdot \operatorname{rot}$ to these 2 equations, the following Poisson equations are obtained:

$$\operatorname{lap} \chi = \operatorname{div} \vec{Q}, \tag{20}$$
$$\operatorname{lap} \psi = \vec{k} \cdot \operatorname{rot} \vec{Q}, \tag{21}$$

These two equations allow the evaluation of the streamfunction,ψ, and potential function,χ.

★

★ ★

The transport components associated with the potential function (equation 19) follow the orthogonal trajectories to the isolines $\chi = const.$ Zones of minimum χ are regions of divergence (sources of water vapour) and zones where χ is maximum are convergence regions (sinks of water vapour).

For a closed region, we see (equation 19) that

$$\oint \vec{Q}_\psi \cdot \vec{n}\, d\ell = 0, \tag{22}$$

and so the flux of the total transport across the boundary of an region with an area A is

$$\oint \vec{Q} \cdot \vec{n}\, d\ell = \oint \vec{Q}_\chi \cdot \vec{n}\, d\ell \tag{23}$$

or using the Gauss theorem:

$$\frac{1}{A} \oint \vec{Q}_\chi \cdot \vec{n}\, d\ell = \langle \overline{E} - \overline{P} \rangle . \tag{24}$$

It is easy to show that the mean zonal and the mean meridional transports are related to the ψ and χ functions by the expressions:

$$[Q_\lambda] = -\frac{1}{a}\frac{\partial [\psi]}{\partial \varphi}, \tag{25}$$

$$[Q_\varphi] = \frac{1}{a}\frac{\partial [\chi]}{\partial \varphi}. \tag{26}$$

where a is average radius of the earth.

II.3.2 Streamfunction and potential fields

As we have said previously, the ψ field represents the solenoidal (rotational) component of the transport. The water vapour aerial runoff between two isopleths is invariant and given by the difference of the values of the streamlines. Furthermore they follow rules similar to the Buys-Ballot laws.

The ψ field has no sources or sinks. It represents the permanent average flux of water vapour in the atmosphere. Being divergenceless its isopleths must be closed lines without a start or an end.

The streamfunction field was obtained solving numerically the Poisson equation (21) on a sphere. The resulting analyses are presented in figures 9. They depict the wide anticyclonic circulations over the Pacific and the Atlantic, in the subtropical regions of both hemispheres. There is a dominant easterly transport in the intertropical regions, and a westerly transport in middle latitudes. Over the equatorial zone and over middle and high latitudes the flow is almost zonal and very strong. On a global scale, this field presents a marked zonal symmetry which is associated with the dominant circulation at the lower levels of the atmosphere. As expected, due to large extent to the continents and to orography, the westerly flow is less regular over the northern than over the southern hemisphere.

The seasonal variability of the ψ field does not differ much from that observed in other water vapour fields. During JJA the effects of the Asian monsoon extend to both hemispheres with the occurrence of a cyclonic circulation over the Equator (according to the northern hemisphere convention) breaking the dominant easterly transport normally observed at these latitudes.

Figure 10 shows the average profiles, $[\overline{\psi}]$, of the $\overline{\psi}$ field. The derivative of these profiles with respect to the latitude, give the average profiles of the zonal component of the water vapour transport, as seen in figure 7.

⋆

⋆ ⋆

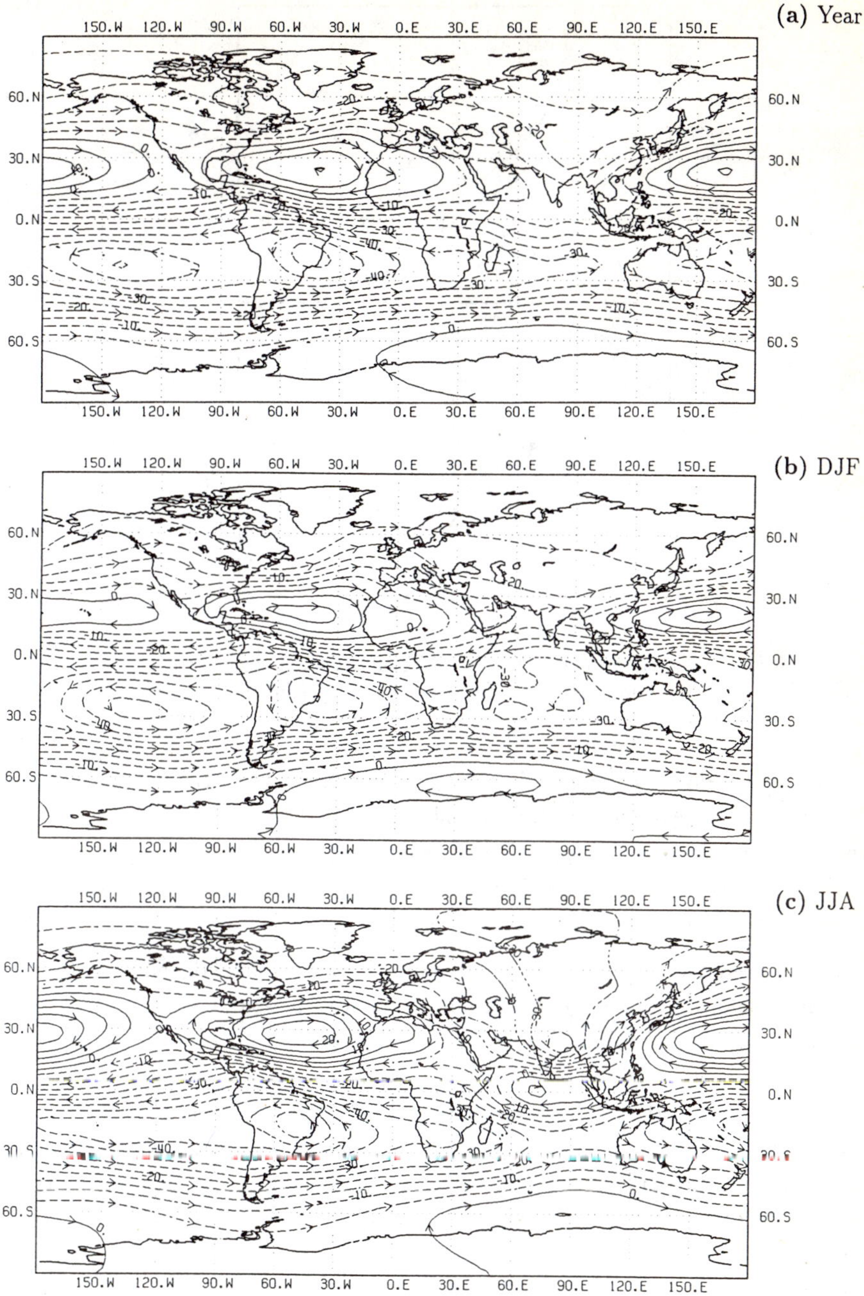

Fig. 9:Global distribution of the mean stream function fields in units of 10^7kg/s; for annual, DJF and JJA conditions. The contour interval is 5×10^7kg/s.

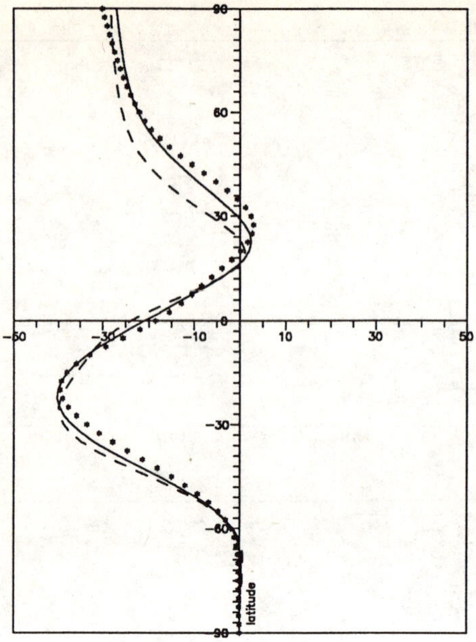

Fig. 10:Mean meridional profile of the mean zonal stream function values, $[\overline{\psi}]$, in units of 10^7kg/s; for year (full line), and DJF (dashed), and JJA (point line) seasons.

Let us consider now the χ field. The figure 11 presents the mean field of the potential function for the year and for DJF and JJA seasons. The χ fields were computed from the transport field $\vec{Q}$, solving the Poisson equation (20) on a sphere. Contrary to the ψ fields, which show an almost zonal symmetry, the χ fields present a cellular structure with alternating maximum and minimum centers.

The annual map shows centers of minimum (divergence) over the northwestern Pacific and the equatorial Atlantic, slightly displaced towards the southern hemisphere. It shows also centers of maximum (convergence) on the equatorial region of the Indian Ocean, over the western Pacific, slightly located to the north of the equator, and a less intense maximum, centered over the south Pacific. The seasonal maps show considerable differences in the location of the centers, their intensities and number. The large center over East Asia intensifies and moves into the north, during the JJA season. Small centers over the Horn of Africa and the Malabar Sea, due to the monsoon, during the JJA season are worth mentioning.

The radial transport components which are normal to the χ lines (orthogonal trajectories) are presented in figures 12. The corresponding intensities are given by the gradient of the χ field. These fields show, in a very nice way, zones of convergence alternating with zones of divergence.

II.3.3 Divergence field

Among the derived fields of $\vec{Q}$, its divergence field, $\operatorname{div}\overline{\vec{Q}}$, is, by far, the most important to show the connection between the aerial and terrestrial branches of the hydrological cycle.

The basic fields, W and $\vec{Q}$, do not have any direct connection with the terrestrial branch of the hydrological cycle. In fact, regions where the aerial runoff $\vec{Q}$ reaches larger values, are not those where higher precipitation occurs. On the other hand, precipitation and evaporation are not sensitive to the precipitable water W.

It is the spatial variation of the $\vec{Q}$ field, that contributes to precipitation P and evaporation E, while the variations of precipitable water have almost no effect on the precipitation.

A first picture of the space distribution of the water vapour sources and sinks was, already, presented in discussing the potential field. This field shows regular spatial features with two or three, well defined centers. However, we should not take these centers as the only water vapour sources and sinks, since they only give a gross representation of the divergence field. We should point out that it is the variability of the χ field, along its trajectories expressed by its laplacian (equation 20) that defines the convergence and divergence centers (the variability of the spatial derivatives of χ, i.e. the laplacian, is not always easy to see immediately). Indeed, the analysis of the planetary divergence field is much more detailed than the χ field. The average divergence for a limited area, defined by a closed χ isoline, is the circulation of $\operatorname{grad}\chi$ for this closed circuit, divided by the enclosed area (equations 20 and 24).

Figures 13 present the divergence field computed from the $\overline{Q}_\lambda$ and $\overline{Q}_\varphi$ basic fields, for yearly and seasonal conditions.

The maps show centers of divergence ($E - P > 0$) alternating with centers of convergence ($E - P < 0$). On a global scale the distribution of these centers shows a certain regularity. In fact, a belt of convergence is observed in the equatorial region, which is associated with the ITCZ, while in subtropical latitudes, with the large anticyclones, divergence centers dominate. Convergence centers appear again at higher latitudes, associated with the stormtracks in both hemispheres. Over the polar latitudes divergence occurs again.

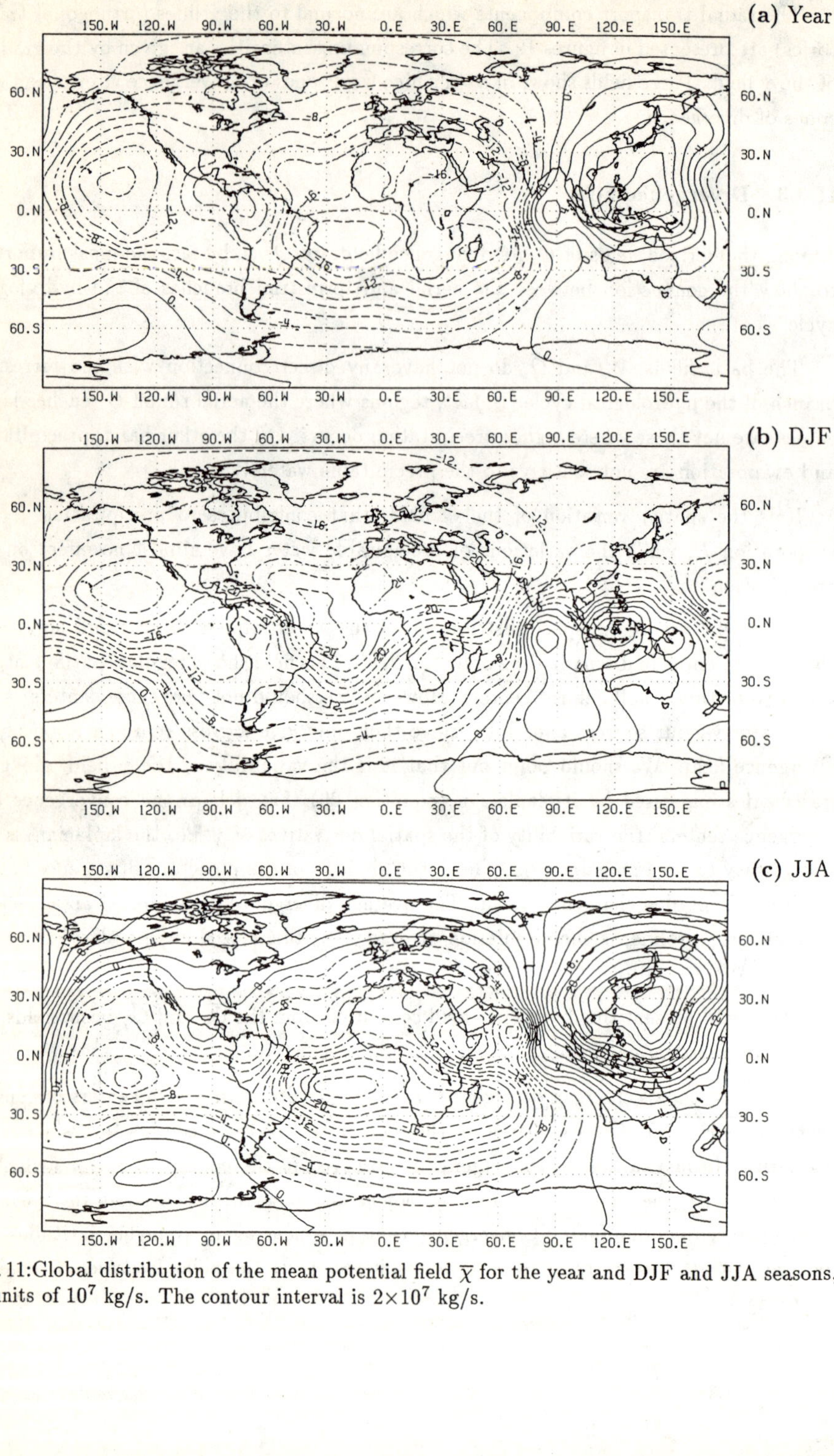

Fig. 11: Global distribution of the mean potential field $\overline{\chi}$ for the year and DJF and JJA seasons, in units of 10^7 kg/s. The contour interval is 2×10^7 kg/s.

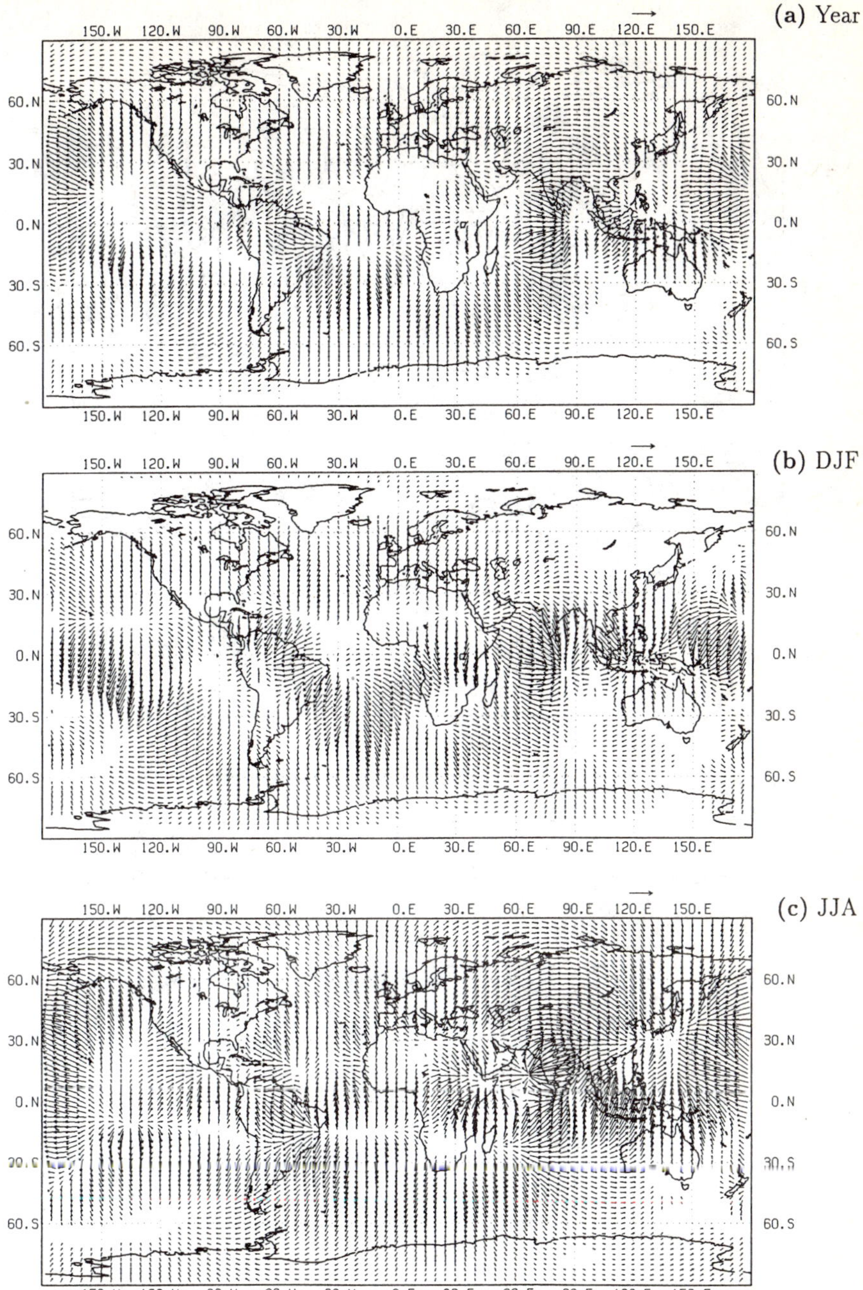

Fig. 12:Global distribution of the divergent component $\vec{\overline{Q}}_\chi$ of the total water vapour transport. The sample arrow on the top of the right hand corner corresponds to a transport of 10^2 kg/(m s).

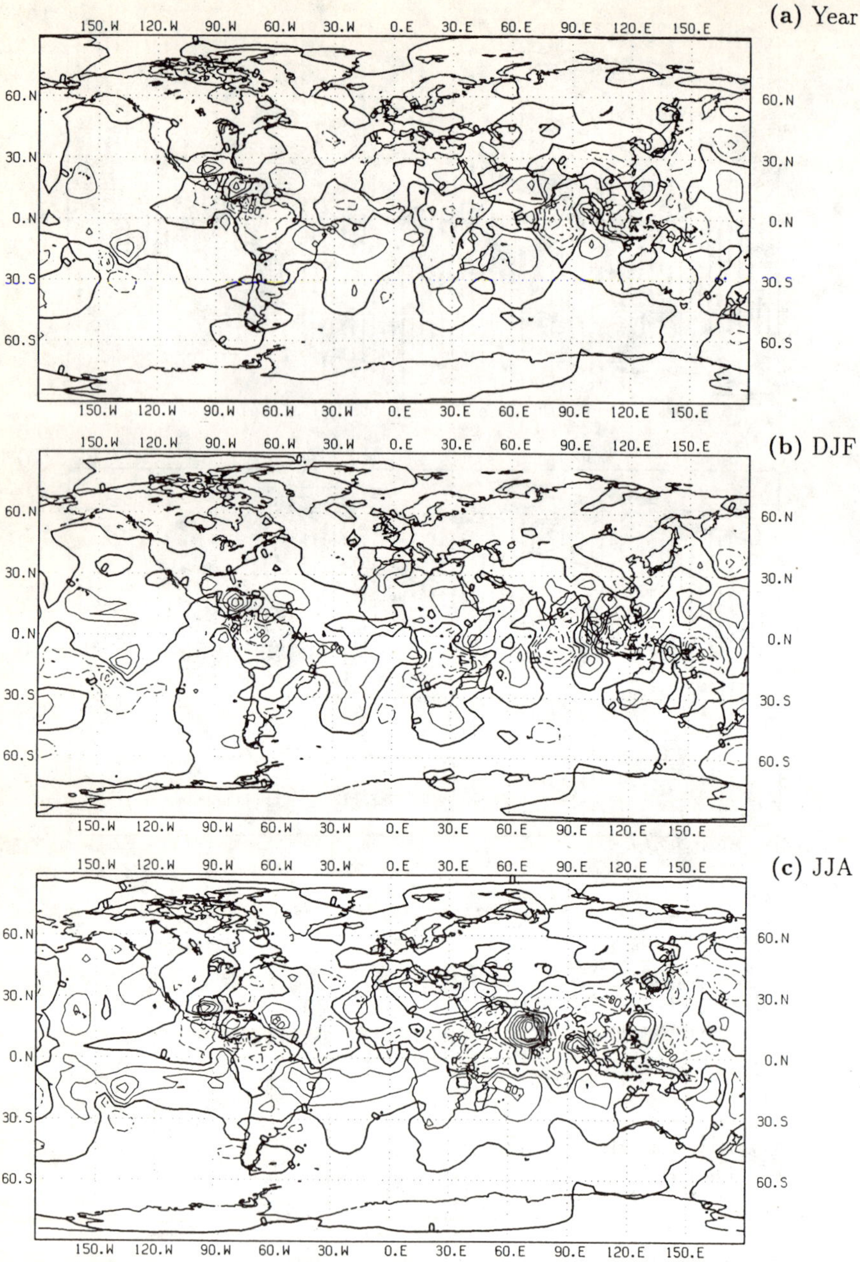

Fig. 13:Global distribution of the horizontal divergence for total water vapour transport, div $\vec{Q}$, in units of 10^{-6}kg m^{-2} s^{-1}, for annual (a), DJF (b) and JJA (c) mean conditions. The contour interval is 40×10^{-6}kg m^{-2} s^{-1}. Negative values (convergence) are dashed.

The fields of divergence reproduce, with fidelity, the climatological distribution of evaporation and precipitation. In fact, a rainy equatorial zone (convergence) is separated by two subtropical arid or semi–arid zones (divergence); these are followed by temperate and rainy zones over middle and higher latitudes, and by arid regions (divergence) near the poles.

A detailed inspection of the maps shows convergence centers over the great river basins and divergence centers over arid, or desert regions.

Divergence centers are particularly strong over the subtropical oceanic anticyclone areas which constitute the main sources of water vapour for the atmosphere. Without oceans, these regions would be real deserts.

Over the continents divergence centers are more dispersed and less intense. In this case, it is the terrestrial runoff (surface and subsurface) and the soil moisture that must supply the water to compensate for the excess of evaporation over precipitation. In general, continental areas, that present a positive annual mean divergence, correspond to arid or desert areas, as those that occur over north of Africa (Sahara desert), south of Africa (Kalahari and Namibia deserts), Mexico and the south of the United States, in Central Asia (Tabkla Makan, Gobi deserts), South America (Patagonia), Australia and so on.

Figure 14 shows the mean zonal profiles of div $\vec{Q}$ for the year and the two extreme seasons. The profiles summarize our previous findings, showing that there is always

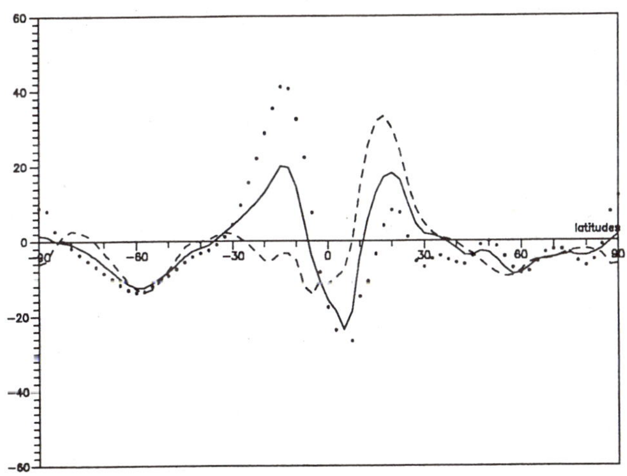

Fig. 14:Meridional profile of the zonal divergence field, $[\operatorname{div}\overline{\vec{Q}}]$ in units of $10^{-6}\mathrm{kg\,m^{-2}\,s^{-1}}$ for annual, DJF and JJA mean conditions.

a strong convergence in equatorial regions, a strong divergence in subtropical latitudes followed by convergence in the middle and high latitudes.

II.4 Paths of the water vapour in the atmosphere

By and large the horizontal paths of water vapour in the atmosphere are fairly well depicted by the stream function ψ. This field, however, does not allow the identification of the sources and the sinks of water vapour. Thus, this representation does not give a complete description of the dynamics of the hydrological cycle. Even when the potential is considered it is difficult to come out with that identification.

It is possible, however, to develop techniques to obtain, as a first approximation, bidimensional representations (horizontal maps and vertical cross sections) of the average paths that connect the major sources and sinks of the water vapour in the atmosphere. A first technique is based on the computation of fictitious horizontal trajectories of the water vapour. A second technique is based on the vertical integration of the water balance equation, assuming, as boundary condition, zero vertical transport at the top of the atmosphere.

II.4.1 Fields of horizontal path lines

By definition and construction, streamlines are tangent at each point to the water vapour transport vector $\vec{Q}$, which give an immediate indication of the paths of the water vapour at a given point of the atmosphere. As mentioned before, these fields are not the same as the solenoidal fields of the stream function. Path lines do have a beginning and an end, in opposition to the ψ isolines. On the other hand, the concentration of path lines in a given region does not give any indication of the strength of the transport, as can be inferred from its construction.

Path lines begin in areas of divergence of water vapour transport (source areas) and end in areas of convergence (sink areas), so that, under steady conditions, path lines indicate the average trajectories of the water vapour in the atmosphere from source areas (evaporation) to sink areas (rainfall).

Figure 15 show the path line maps.

From the inspection of these maps, we see again that the major source regions of water vapour are over the subtropical oceans, which coincide with the large subtropical anticyclones. Moreover, the anticyclonic curvature of these path lines shows that they agree with the potential and divergence fields. Path lines do not give any indication of the strength of the water vapour transport, but show very nicely the dominant trajectories of the water vapour flux in the atmosphere.

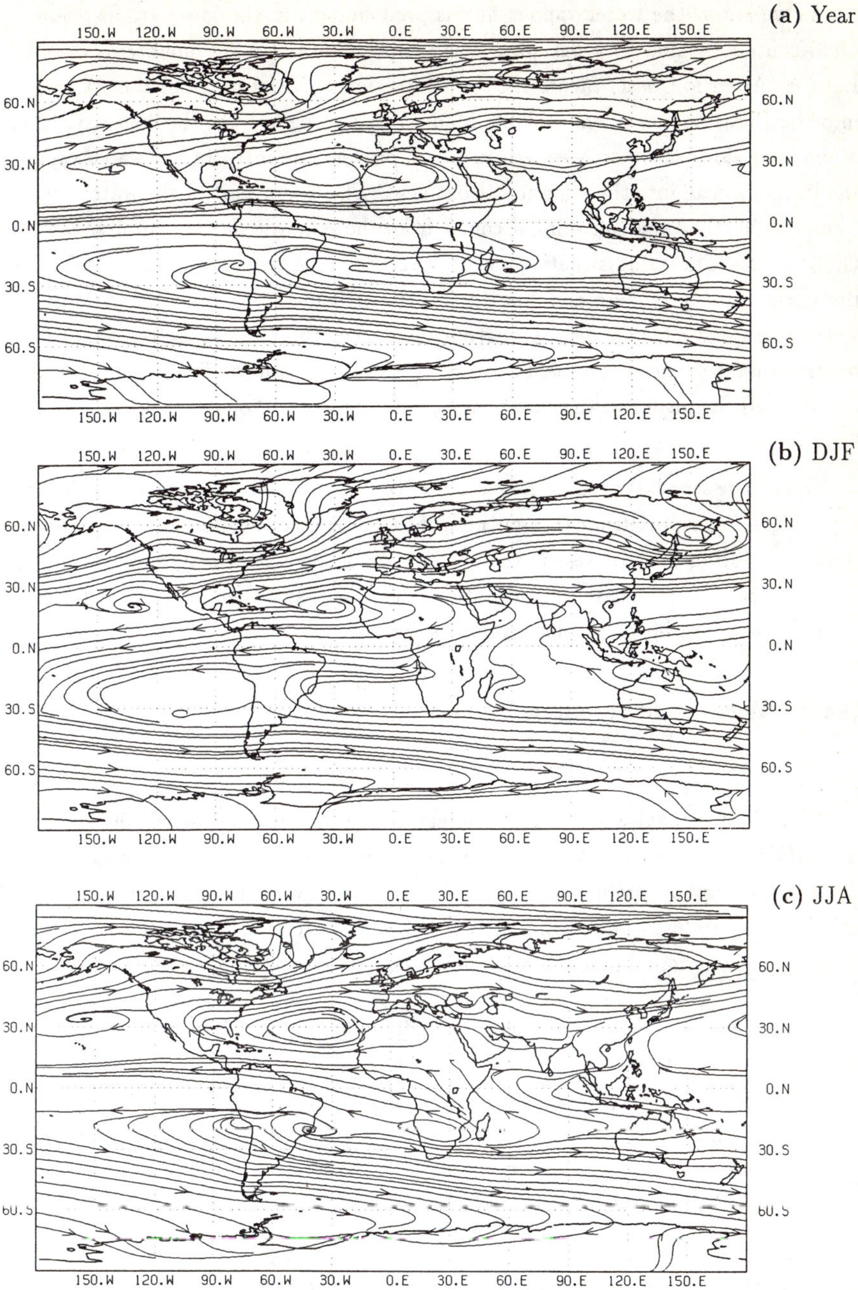

Fig. 15:Global distribution of the total aerial runoff $\vec{Q}$.

As we know, the water vapour flux is predominant in the lower troposphere and, as such, can be affected by orography. The non existence of large mountains over Western Europe (Atlantic Coast) favors the deep intrusion of moisture from the Atlantic Ocean into the Eurasian Continent and the Mediterranean region. On the contrary, the Rocky Mountains along the North American Coast prevent the intrusion of water vapour from the Pacific Ocean into the American Continent. The major part of the water vapour that feeds the North American rainfall comes from the Atlantic Ocean through the Mexican Gulf Coast. The same situation occurs over South America with the Andes, so that the major supply of humidity for this continent comes also from the Atlantic Ocean. This orographic effect is also strongly evident over the Himalayan region. Due to the relative smaller area of continents and to the less pronounced effects of topography over the southern hemisphere, the path lines are more regular (zonal) than over the northern hemisphere.

As we have seen before, and the path lines confirm, there is on the average, for the year, a net water transport from the southern into the northern hemisphere. The path lines show that during the JJA season the penetration deepens, and more northern regions are reached. For the DJF season the situation is reversed, and there is a net water transport from the northern into the southern hemisphere.

II.4.2 Meridional stream functions

The bidimensional and vertically integrated fields, just discussed, do not show the trajectories of the water vapour in altitude. This can, in part, be solved considering the average vertical profiles of $[\overline{qv}]$ and $[\overline{qu}]$. However, those trajectories are better highlighted, building a meridional stream function Ψ from the $[\overline{qv}]$ values.

The local balance equation for the water vapour can be written as

$$\frac{\partial q}{\partial t} + \mathrm{div}_p\, q\vec{v} + \frac{\partial}{\partial p}(\omega q - g\mathrm{D}) = 0 \tag{27}$$

where $\omega = dp/dt$ is the vertical velocity and D is the vertical water vapour diffusion rate. If we apply the zonal average operator, using spherical coordinates, we get the following expression

$$\frac{\partial\,[q]}{\partial t} + \frac{1}{a\cos\varphi}\frac{\partial\,[qv]\cos\varphi}{\partial\varphi} + \frac{\partial}{\partial p}[q\omega - g\mathrm{D}] = 0. \tag{28}$$

Considering the time mean for a long period of time the local rate of chage of $[\overline{q}]$ is zero, so that equation 28 becomes:

$$\frac{1}{a\cos\varphi}\frac{\partial\,[\overline{qv}]\cos\varphi}{\partial\varphi}+\frac{\partial}{\partial p}\left[\overline{q\omega}-g\overline{\mathrm{D}}\right]=0. \tag{29}$$

From this equation a meridional stream function Ψ in the (φ, p) plane can be defined, and we have then:

$$\frac{\partial\Psi}{\partial p} = \frac{2\pi a}{g}\cos\varphi\,[\overline{qv}] \tag{30}$$

$$\frac{\partial\Psi}{a\partial\varphi} = -\frac{2\pi a}{g}\cos\varphi\left[\overline{q\omega}-g\overline{\mathrm{D}}\right], \tag{31}$$

Assuming that the water vapour flux on the top of atmosphere is zero and integrating downward, the function Ψ is given by:

$$\Psi = \int_0^p 2\pi a\cos\varphi\,[\overline{qv}]\,\frac{dp}{g}. \tag{32}$$

At each isobaric level p the stream function Ψ is then the integrated meridional flux of water vapour between the top of the atmosphere and the pressure p.

On the (φ, p) plane the water circulation follows the Ψ isolines leaving the larger values to the right. Also the Ψ lines start at the surface ($\operatorname{div}\vec{Q} > 0$), over the regions of excess of evaporation over rainfall, and end, also at the surface ($\operatorname{div}\vec{Q} < 0$), over areas where the rainfall exceeds the evaporation.

The difference between the Ψ values of two isolines, represents, on the (φ, p) plane, the total flux of water vapour that crosses a normal unit surface, per unit of time.

With this stream function the average meridional transport of water vapour from the mean sources to mean sinks can be computed, as well as the vertical distribution of this transport (see figure 16).

From the first cross section, we get the confirmation of the subtropical sources for the atmosphere. It is interesting to compare the latitudes where Ψ lines start and end with the $[\operatorname{div}\vec{Q}]$ profiles. The Ψ lines are concentrated in the planetary boundary layer (below 500 hPa). The global pattern reveals a meridional tricelullar structure of the water vapour circulation.

The Annual cross sections of the meridional flux patterns of the water vapour show some similarity and symmetry in both hemispheres with respect to the thermal equator (slightly to the north of the geographical equator).

Source regions of water vapour ($[\operatorname{div}\vec{Q}] > 0$) are located between 12^0 and 30^0N and 10^0 and 35^0S, and sinks between 10^0S and 12^0N and again at latitudes higher than 40^0 in both hemispheres.

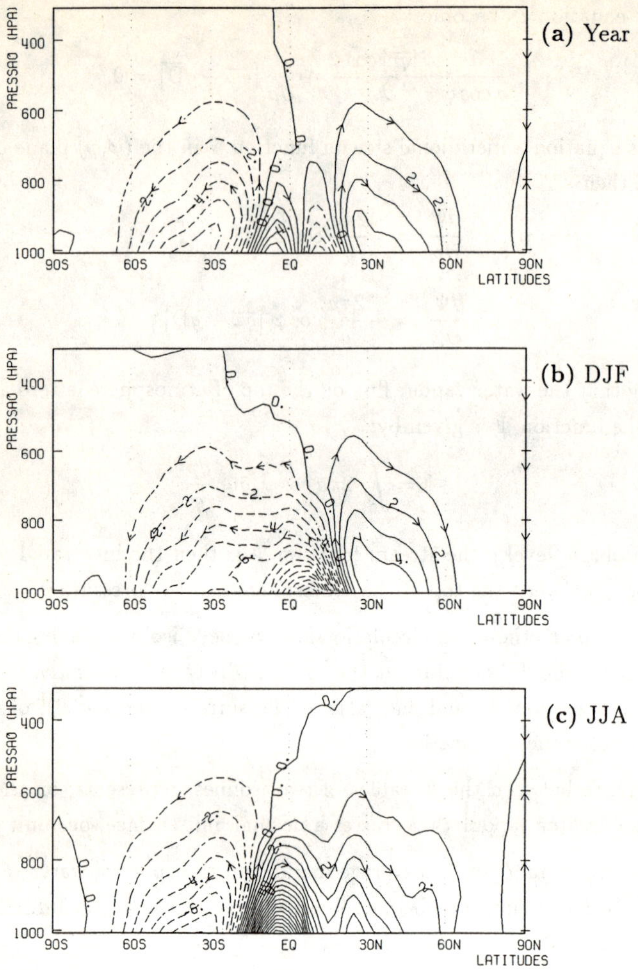

Fig. 16:Stream lines cross sections of the zonal mean transport of water vapour for annual, DJF and JJA mean conditions in units of 10^8kg/s. The contour interval is 1×10^8kg/s.

The water vapour which is released mainly in the subtropical oceans of each hemisphere, divides itself in two parts, one moving towards the equator and the other towards the poles. The equatorialward flow feeds the ITCZ and the others flows feed the rainfall of the lows associated with the polar fronts, in each hemisphere. The annual regimes are almost symmetric with the axis slightly located to the north of the equator, as a result of the net annual export of humidity by the southern hemisphere into the northern.

The mean circulations in the intertropical region are associated with the lower branches of the Hadley Cells, whereas the circulations over the middle and high latitudes are mainly associated with transient eddies.

A comparison of the seasonal cross sections shows strong asymmetry for both seasons. However, the winter and summer displays are like mirror images of each other. The ψ fields are more intense in summer in both hemispheres.

A more detailed analysis shows that during DJF the downward branch of the circulation moves to the south of the equator with a net transport of water vapour, from the northern into the southern hemisphere. It is interesting to notice that the water vapour transported into middle latitudes of the south hemisphere comes from the tropical region of the north hemisphere.

During the JJA season, over the northern hemisphere the circulation from the tropics disappears but the northward circulation increases, and the corresponding downward branch goes further north. Similarly to the DJF season, part of the water vapour northward flux into middle and high latitudes originates in the tropics of the southern hemisphere. During this same period, a small circulation between 15^0 and 25^0N, due to Asian monsoon can, also, be observed.

II.5 The water vapour aerological fields and the dynamics of the hydrological cycle

II.5.1 Some introductory remarks

In previous sections, after having identified the large sources and sinks of water vapour for the atmosphere, we have already made some qualitative and interpretative comments. However, we must present a more objective analysis of the connections of water vapour transport and its derived fields with the hydrology of the Earth.

Let us start with some quantitative implications of those fields and compare them with results of the Classical Climatology, Synoptic Meteorology and of Hydrology.

In the present study we will deal mainly with the connections of the water vapour fields with the global hydrology. Of course, the same connection exists for any spatial scale, namely, for regional hydrology (polar regions, Mediterranean Sea, etc) for large river basins (Mississippi, Danube), etc. The discussion of these results are, however, outside the scope of the present study.

For comparison we will show in figure 17 a map with the global rainfall, as presented by Jaeger (1976), and in figure 18 meridional profiles of zonal averages of the mean rainfall (cm/year) over oceans, continents and the whole Globe (Peixoto *et al* 1992). Figure 19 shows the distribution of the global evaporation, as given by Baumgartner and Reichel (1975).

In table 1 the zonally averaged mean annual values of $[\overline{W}]$, $[\overline{\vec{Q}}_\varphi]$ and $[\operatorname{div}\overline{\vec{Q}}]$ are also presented.

	$[\overline{W}]$	$[\overline{\vec{Q}}_\varphi]$	$[\operatorname{div}\overline{\vec{Q}}]$
80 S	2.9	−0.0	−0.8
70 S	4.2	−3.0	−6.5
60 S	7.6	−11.5	−12.5
50 S	12.0	−19.8	−8.3
40 S	17.1	−21.1	−2.0
30 S	23.6	−18.0	4.1
20 S	32.4	−7.6	12.9
10 S	40.5	12.3	14.4
Equ	44.4	8.7	−15.9
10 N	41.5	−12.2	−3.7
20 N	33.8	0.2	18.1
30 N	25.1	12.3	1.8
40 N	18.3	14.5	−2.0
50 N	13.8	13.1	−3.1
60 N	10.6	8.3	−8.6
70 N	7.5	4.0	−4.5
80 N	5.4	1.5	−4.1

Table 1: Zonal averages of precipitable water $[\overline{W}]$ kg/m^2, meridional transport $[\overline{Q}_\varphi]$ kg/(m s) and divergence for the year, in $10^{-6} kg\, m^{-2}\, s^{-1}$.

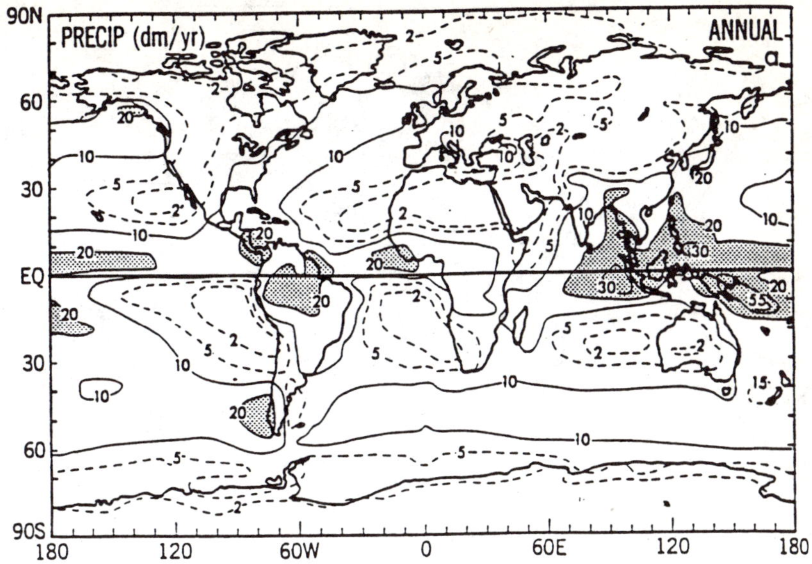

Fig. 17:Global distribution of the precipitation for annual mean conditions in units of dm/yr (Jaeger 1976).

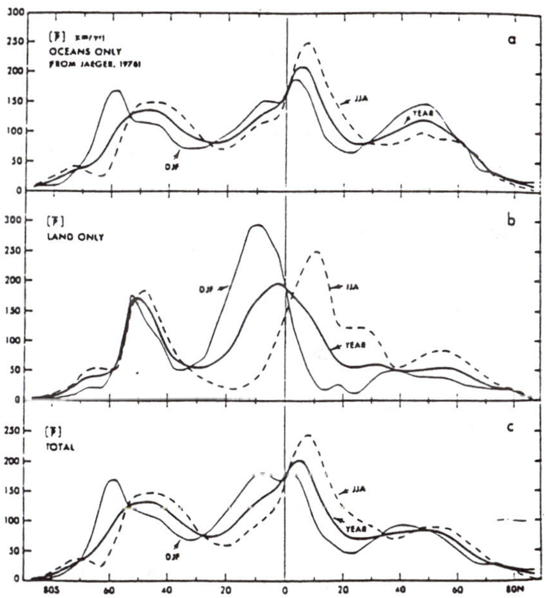

Fig. 18:Meridional profiles of the zonal mean precipitation rate in cm/yr for the ocean area (a), the continents (b) and the total land plus oceans area (c) for yearly, DJF and JJA mean conditions (Jaeger 1976).

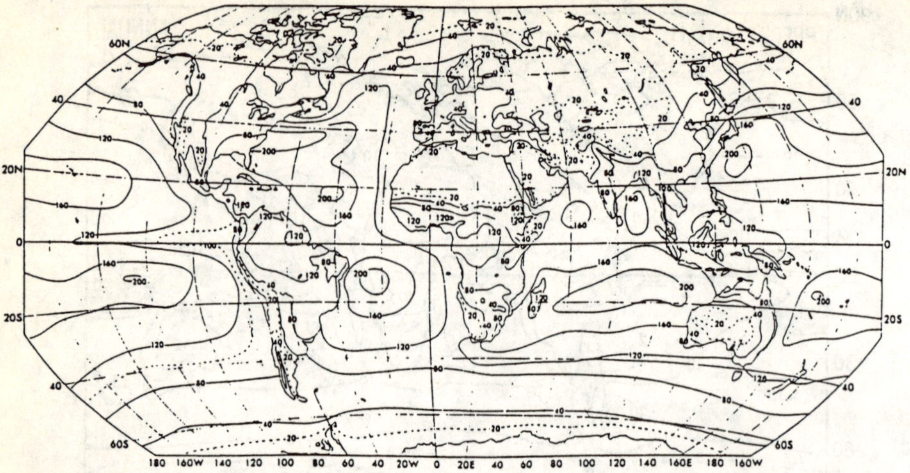

Fig. 19: Global distribution of the anual mean evaporation rate in cm yr^{-1} (Baumgartner e Reichel 1975).

II.5.2 Global hydrological balance

From table 1 we see that there is a a net northward transport of water vapour flux across the equator ($[\overline{\vec{Q}}_\varphi] = 8.7\,\mathrm{kg\,m^{-1}s^{-1}}$), which is equivalent to an excess of rainfall over evaporation over the northern hemisphere of the order of 42 mm/year. This value is a bit lower than that obtained by Jaeger's climatological data (a net balance of $P - E =$ 73 mm/year).

Similarly, during the DJF season, the northern hemisphere exports water vapour across the equator ($[\overline{\vec{Q}}_\varphi] = 30.8\,\mathrm{kg\,m^{-1}s^{-1}}$) which contributes to a surplus of rainfall ($P - E =$ 152 mm/year) over the southern hemisphere. This value can be compared with the influx that the northern hemisphere receives during the JJA season ($P - E = 221$ mm/year).

Similar exercises, with values of $[\overline{\vec{Q}}_\varphi]$, could be extended to other latitudes. They should lead to comparable results with those inferred from the distribution of the meridional stream function Ψ. The intra annual variability of the transport components can be transformed into equivalent climatological values of $E - P$. These results give support to the feasibility of the aerological method.

From the $\vec{Q}$ fields we obtained the values (not presented here) of the net flux of water vapour across continental boundaries. The results show a net inflow which leads to a surplus of rainfall over evaporation on the continents. Larger values were found for South America and for Europe, and lower values for Australia, Africa and Antarctic, etc.

The net inflow of water vapour is compensated for by river and glacial discharges into the oceans and by evaporation. The same conclusion is reached when the divergence field is integrated over the continents. The integrated spatial divergence over the continents is, of course, negative. However, over the oceans the areas of divergence are so wide and intense, that they are not compensated for by the convergence areas, resulting then in a net divergence of water vapour from the oceans.

When the various oceans are taken separately, the corresponding integrated values of divergence, not reproduced here, lead to a variety of values, some positive, some negative. These integrals converted in $E - P$ values show that they are highly positive for the Atlantic and Indian Oceans, and negative for the Arctic and Pacific Oceans.

If we take into consideration the river runoff into the oceans, there is no compensation for the loss of evaporation for the Atlantic ($R_0 = 197\,\mathrm{mm/year}$) and for the Indian Ocean ($R_0 = 37\,\mathrm{mm/year}$). Being so, there must be a continuous input of fresh water from the Arctic and Pacific Oceans into the Atlantic and Indian Oceans. These results agree well with oceanographic studies.

II.5.3 Water vapour paths in the atmosphere

Let us present some results concerning the water vapour paths in the atmosphere. When we discussed the potential χ and stream function ψ fields, we mentioned the large water vapour gyres in the atmosphere. These fields give only a static representation of those paths. Once the sources and sinks are identified, through the fields of divergence, of potential and of meridional stream function, it is only natural to look for the preferred destinations of water vapour in the atmosphere.

The path lines (fig 15) give a very good representation of the bidimensional mean total water vapour trajectories in the atmosphere. They identify very well the large subtropical sources and the preferred pathways of water vapour, associated with the dislocation of the humid air masses.

These pathlines are influenced by the orography, and they show how the water vapour can travel long distances from its sources. They allow us also to get an idea of the long journeys of the water vapour in the atmosphere and its capacity for penetration into the continents. Furthermore, they show, in a simpler way, the transfer of water vapour between the two hemispheres and the patterns of this transfer.

In order to get an order of the magnitude of its strength, we could superimpose the pathline maps on the streamline ψ charts (fig. 9).

*

* *

In analyzing the meridional transfers we can pursue a little further.

Let us consider again to the Ψ meridional stream lines cross sections, fig 16. These Ψ lines start (sources) and finish (sinks) at the Earth's surface. Source and sink regions correspond exactly to latitudes where mean divergence and convergence are respectively predominant (figure 14).

If we start with annual average conditions, the sources present some symmetry with respect to the equator. In the northern hemisphere, from the nine drawn lines, five of them are confined to the lower troposphere and extend towards the equator with short paths that end between 5^0 and 10^0N. The other four lines, have longer paths ending at latitudes between 38^0 and 70^0N. It is important to point out that this last transport, directed into higher latitudes, have significant values in the middle troposphere. The patterns show that almost half of the released water vapour in subtropical latitude is going to feed the ITCZ rainfall and that its transport is mainly confined to the lower branch of the Hadley Cell. On the other hand the water vapour export to middle and high latitudes is accomplished by eddies, and its flow is more diffuse and attains higher altitudes.

It is interesting to see that values computed by the aerological method agree with climatological values, obtained from Baumgartner and Reichel (1975).

The cross sections (figure 16) prove that most of the water vapour that feeds middle and higher latitudes comes from this subtropical source.

Over the southern hemisphere the Ψ lines have a similar pattern, but the sources of water vapour are wider. Over this hemisphere the moisture transport is separated into two parts, one going towards the equator and the other towards polar regions. The part of the transport directed towards the north crosses the equator and penetrates into the northern hemisphere.

Profiles for DJF and JJA seasons show that in the winter hemisphere the sources intensify, being strongest during the winter over the southern hemisphere. During the summer the evaporation area is not as wide.

III. GENERATION OF ENTROPY BY PHASE TRANSITIONS IN THE ATMOSPHERE

III.1 Preliminary remarks

As we have already mentioned the atmosphere is an open system that exchanges mass, energy, momentum and entropy with the globe and energy plus entropy with outerspace.

The dominant forcing of the climate system is, by far, provided by the incoming solar radiation. In absence of radiating forcing the earth's atmosphere would tend towards an equilibrium state that would be *isothermal*, *hydrostatic* and *motionless*. Under the constraint of mass and total energy conservation, the entropy would monotonically increase to its maximum value at the state of equilibrium.

The internal production of entropy is positive definite for an adiabatic and closed system (second law of thermodynamics). For systems that are neither closed nor adiabatic the entropy variation will also be affected by the flux of entropy across the system boundary, so that the total entropy change (dS) is given by the sum of the external exchange term (d_eS) and the internal production term (d_iS):

$$dS = d_eS + d_iS \tag{33}$$

Unlike the internal term that is always positive ($d_iS \geq 0$), the external exchange (d_eS) can be of either sign.

For steady state systems not in equilibrium the internal entropy production must balance exactly the net entropy flux leaving the system through the boundary.

As viewed from the Earth the Sun emits energy as a black body at a temperature near 6 000 K from a relatively small solid angle in the sky ($\approx 0.5^0$). The solar radiation intercepted by the Earth is *reflected*, *scattered* and *absorbed*. Finally the absorbed energy will be remitted isotropically at a lower temperature of about 255 K to the space. Thus, the Earth receives high quality short wavelength radiation from a small solid angle source and emits it to longer wavelength radiation of lower quality over a larger solid angle, to the space (see figure 1).

According to Planck's equation, the energy, ϵ, of a photon is given by $h\nu$:

$$\epsilon = h\nu \tag{34}$$

where h is the Planck constant and ν is the frequency of radiation which is inversely proportional to the wavelength ($\nu \propto 1/\lambda$). The solar photons are rich in energy and poor in entropy, contrary to the terrestrial photons. Thus, since the energy of both radiations are equivalent, the entropy associated with the incoming solar radiation is much lower than the amount of entropy associated with the emitted terrestrial radiation. Therefore, the climate system receives high quality energy and returns the same total amount, but of lower quality to the space. These radiative exchanges at the top of the atmosphere lead to the export of the generated entropy within the climate system. This cleaning operation allows the solar radiation to revitalize the meteorological phenomena, to feed the hydrological cycle and to renovate the biosphere.

There are numerous irreversible processes in the climate system that lead to an increase of entropy, namely the absorption of solar radiation and terrestrial radiation, melting of snow and ice, condensation, evaporation, erosion, turbulent and molecular diffusion of matter, energy and momentum transformations, etc.

The most important generating processes of entropy in the atmosphere are the absorption of solar radiation and release of latent heat associated with the phase transitions of the water substance, mainly the condensation of the water vapour, which leads to cloud formation.

The clouds play a fundamental role in the hydrological cycle. Furthermore, they are important because they are strong absorbers of infrared radiation and excellent reflectors of solar radiation. These two opposite effects make the clouds an essential modulating factor of the climate. They are also strong factories of entropy within the atmosphere, due to the release of latent heat of condensation and to the precipitation mechanisms.

III.2 Balance equation of entropy

The rate of change of entropy per unit mass is given by

$$\frac{ds}{dt} = \frac{\dot{Q}}{T} \tag{35}$$

where $\dot{Q}$ is the rate of change of diabatic heating per unit mass.

The diabatic heating of atmosphere includes radiational heating (shortwave and long-wave), the release of latent heat, and the heating by conduction and frictional dissipation, so that:

$$\frac{ds}{dt} = -\frac{1}{\rho}(\mathrm{div}\vec{F}_{rad} + \mathrm{div}\vec{F}_{LH} + \mathrm{div}\vec{J}_H) - \frac{1}{\rho}\boldsymbol{\tau} : \mathrm{grad}\,\vec{c}. \tag{36}$$

where ρ is the density of air. $\vec{F}_{LH}$ is the latent heat flux, $\vec{F}_{rad}$ is the net radiative flux, $\vec{J}_H$ is the sensible heat flux due to conduction, $\boldsymbol{\tau}$ is the wind stress tensor, and $\vec{c}$ is the wind velocity.

After some mathematical manipulations (see Peixoto et al, 1991) we can write the following equation for the entropy balance of the atmosphere:

$$\frac{\partial S}{\partial t} = \frac{G_{rad}}{\overset{\star}{T}_{rad}} + \frac{G_H}{\overset{\star}{T}_H} + \sigma_{rad} + \sigma_{LH} + \sigma_H + \sigma^{dis} \tag{37}$$

where S is the total entropy, the G's are the integrated energy fluxes, the $\overset{\star}{T}$ are characteristic temperatures of the processes involved and the σ's are the entropy generation terms.

Using the observed data in the atmosphere all the terms of equation (37) can be evaluated and analyzed (Peixoto *et al*, 1991).

III.3 Observed entropy budget of the atmosphere

Figure 20 shows the entropy budget of the atmosphere for the annual mean conditions.

The three terms of the upward flux of entropy at the top of the atmosphere are associated with the fluxes of infrared radiation emitted by the earth's surface (through the atmosphere spectral window, 8.5-11μm), the atmosphere and by the clouds. The total outgoing flux of entropy is 930 $mWm^{-2}K^{-1}$, whereas the incoming of entropy associated with the solar radiation is only 39.2 $mWm^{-2}K^{-1}$. This negative difference of the fluxes of entropy (891 $mWm^{-2}K^{-1}$) means the influx of negantropy which increases the availability of the energy that allows all the phenomenology which occurs in the climate system. In other words, it allows the occurrence of irreversible processes within the climate system.

At the bottom of the atmosphere we estimate the values of the entropy fluxes associated with the downward flux of solar radiation (28.5 $mWm^{-2}K^{-1}$), with the upward sensible heat flux (83 $mWm^{-2}K^{-1}$), and with the upward radiation flux (249 $mWm^{-2}K^{-1}$).

When all fluxes of entropy into the atmosphere are considered we obtained a net outgoing flux of entropy of 588 $mWm^{-2}K^{-1}$. This means that for steady conditions this value must equal the mean rate of generation of entropy by internal processes.

From the meteorological data we computed directly the internal rate of production of entropy for the most important individual processes that occur within the atmosphere. The results are presented inside the boxes of figure 20.

Summing up all the evaluated generation processes we find a value of 578 $mWm^{-2}K^{-1}$ for the total mean generation rate of entropy. This value can be compared with the previous value of 588 $mWm^{-2}K^{-1}$ obtained from the global budget from the fluxes at the top of the atmosphere. The difference is only 10 $mWm^{-2}K^{-1}$ and can be attributed (apart from errors) to the approximations made and to some neglected effects (e.g. mixing of air masses).

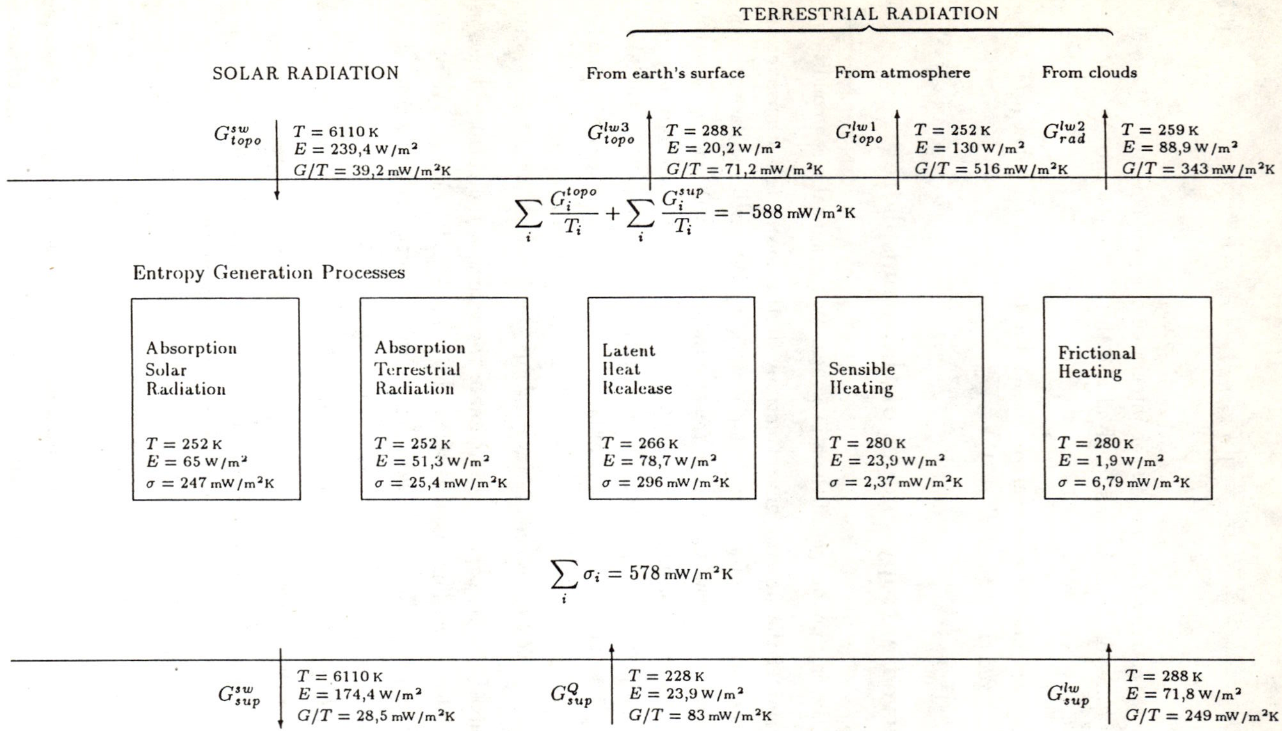

Fig. 20 :Entropy budget of the global atmosphere estimated for annual mean conditions. Shown are the equivalent temperature T^* in K, the energy flux G in W m^2 and the entropy flux G/T^* in units of mWm^{-2}K^{-1} for each energy component at the top of the atmosphere and at the earth's surface. The boxes in the middle of the figure contain for each component estimates of the atmospheric temperature where the absorption takes place in K, the rate where energy is absorbed E in Wm^{-2} and the rate of entropy production σ in units of mWm^{-2}K^{-1}.

The values presented inside the boxes of figure 20 show that the largest sources of entropy in the atmosphere are the absorption of solar radiation and the release of latent heat. The rate of production of entropy associated with the absorption of longwave radiation is an order of magnitude smaller, since the temperatures of infrared radiation emission and absorption are not very different. As it can be seen among the nonradiative processes the water phase transitions dominate the entropy generation. The generation of entropy associated with the latent heat release is essentially concentrated at low latitudes, as we will see, and at middle latitudes in the lower atmosphere.

The entropy balance just presented differs a little from the balance published by Peixoto *et al*, 1991 because the actual balance was done with a larger and more consistent set of data.

III.4 Vertical and horizontal distribution of the entropy generation within the atmosphere

As we have shown the most important mechanisms for the entropy production are the solar radiation absorption and the release of latent heat. Using the diabatic heating obtained from a 4th order general circulation model (Salstein, Program GARP) the geographical and vertical distribution of the entropy generation by absorption of solar radiation and by the release of latent heat were evaluated for a winter month (6 January to 4 of February of 1979).

The vertical distribution of the entropy generation due to the absorption of solar radiation is presented in figure 21, and the corresponding horizontal distribution is shown in figure 22. As expected, for this period of time the generation of entropy associated with the solar radiation absorption is more important in the southern hemisphere, with a maximum in the southern tropical region at 500 hPa.

The horizontal bidimensional distribution confirms the existence of the maximum of generation of entropy over the tropical region of the Southern hemisphere, decreasing both to the North and to the South. The decreasing of entropy generation towards the North is gradual, whereas towards the South it presents a strong meridional gradient over the subpolar region of the southern hemisphere.

The presence of the continents does not seem to be of primordial importance in the horizontal distribution of entropy generation by absorption of solar radiation excluding the regions where the elevation is relatively high as in the Andes and Hymalaias.

The vertical and the geographical distributions of entropy generation by condensation of water vapour are shown in figures 23 and 24, respectively. The inspection of these

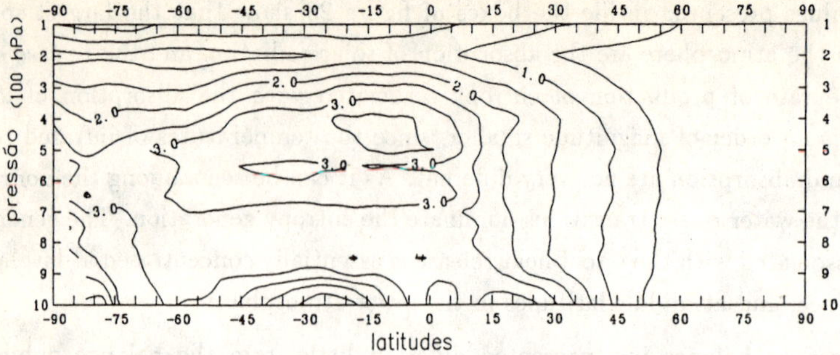

Fig. 21: Cross section of the generation of entropy by absorption of income solar radiation by the atmosphere in units of 10^{-5} W/kg K. The contour interval is 0.5×10^{-5} W/kg K.

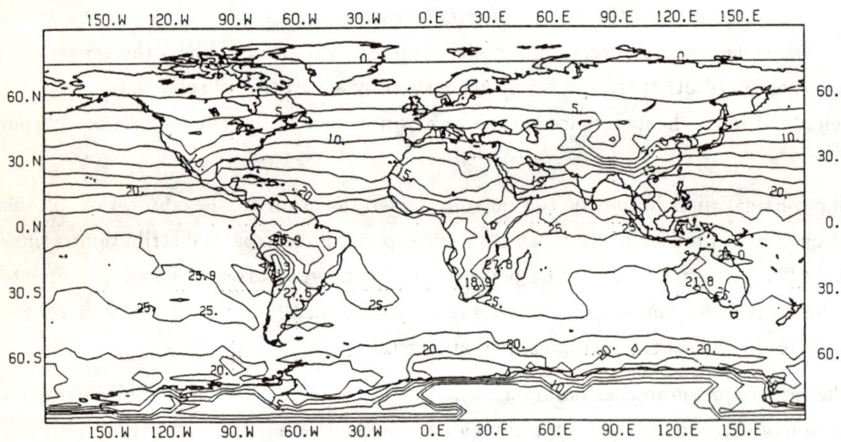

Fig. 22: Global distribution of the generation of entropy by the absorption of solar radiation by the atmosphere in units of 10^{-2} W/m^2 K. The contour interval is 2.5×10^{-2} W/m^2 K.

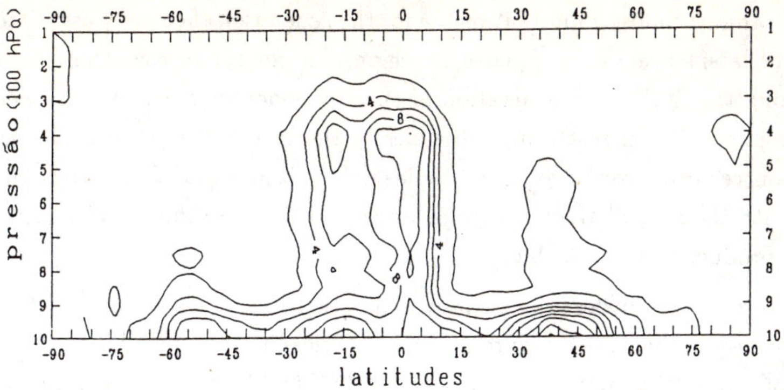

Fig. 23:Cross sections of the mean generation of entropy associated with the release of latent heat for 6 Jan–4 Fev (1979) period in units of 10^{-5} W/kg K. The contour interval is 2.0×10^{-5} W/kg K.

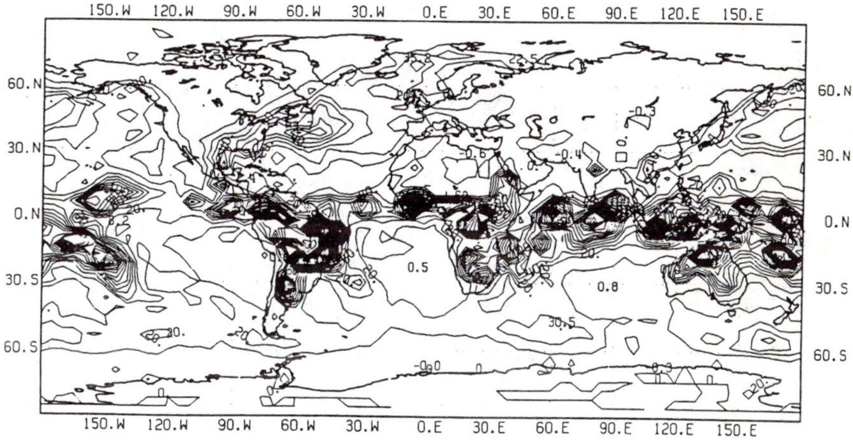

Fig. 24:Global distribution of the generation of entropy due to the release of latent heat in 10^{-2} W/m^2 K, for the 6 Jan–4 Fev, 1979 period mean conditions. The contour interval is 10×10^{-2} W/m^2 K.

figures shows that the most important region for the generation of entropy associated with the release of latent heat is the equatorial region, due mainly to the strong convection observed along the ITCZ. The generation of entropy is not confined to the lower layers of the atmosphere; it can reach high altitudes in the troposphere. There are also two secondary sources of entropy over the middle–high latitudes (one for each hemisphere) associated with the rainfall along the polar fronts. In this case the vertical extension of the entropy production is not as high.

The main regions of maxima of entropy generation by the release of latent heat are situated in the equatorial region (figure 24). Nevertheless it is also possible to identify secondary regions of maximum in middle latitudes (40^0N and 50^0S). For this generation the contrast continent-oceans seems to be important, because the main regions of maximum are situated in coastal areas.

Using a larger contour interval it was possible to get a better representation of the

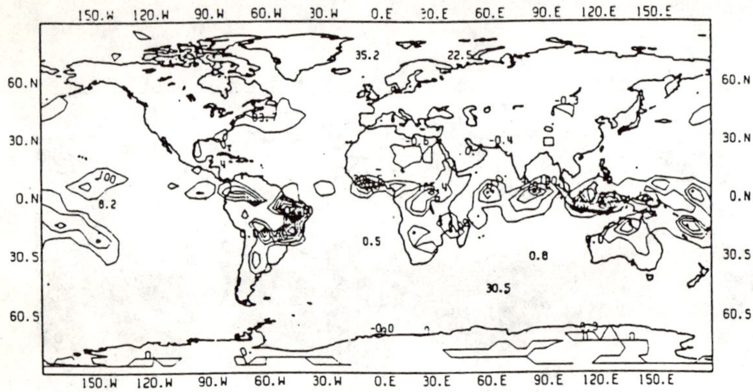

Fig. 25:Same as figure 24 but with an interval contour five times larger.

intense maxima of the equatorial region, losing, however, information in the middle latitudes (figure 25). The most relevant maximum of the equatorial region is situated over the North of Brazil. This maximum is due not only to the meridional transport of water by the Hadley cell, but also to the longitudinal transport of water vapour from the Atlantic Ocean by the Walker circulation (Tomé, 1992).

⋆

⋆ ⋆

Besides this contribution of the hydrological cycle for the generation of entropy within the atmosphere through condensation, there are other contributions in the climate system associated with other phase transitions. Among these we can mention the melting of snow and ice and the evaporation of liquid water at Earth's surface and the sublimation at the cryosphere.

We see now how important the contributions of the hydrological cycle are, as a whole, for the production of entropy and thus for the modulation of the climate.

IV. EPILOGUE: THE ATMOSPHERE AND THE HYDROLOGICAL CYCLE

The atmosphere plays a fundamental role to close and to keep moving the hydrological cycle. It is within the atmosphere that vertical ascending motions produce the adiabatic cooling that may lead to condensation and to cloud formation. The clouds are the generating factories of the precipitation, which, falling on the ground, triggers the terrestrial branch of the water cycle.

The atmosphere is also a reservoir with an adequate storage capacity, conditioned by the temperature, to receive and to keep the water vapour released from the Earth's surface. Furthermore, it is a laboratory that processes the water vapour, transforming it into liquid droplets and solid crystals which make the clouds that later will precipitate as rainfall, so that the water substance returns to its origins. Without this processing, the water vapour would reach higher and higher levels by convection and continuous diffusion in the atmosphere, where it would be destroyed by photodissociation. The atmosphere constitutes, then, an upper buffer that keeps, transforms, protects and preserves the water, as a substance, in the climatic system, preventing its molecular destruction. With the energy transformations associated with all this phase transition there is a strong production of entropy in the atmosphere, which latter will be exported to outerspace through the emission of longwave radiation

The general circulation of the atmosphere provides the ingredients for the cloud generation and the dynamics that models their shapes and characteristics: "clouds are the calligraphy of the skies, without spelling errors".

Another essential role of the atmosphere for the water cycle is the transport of water, not only in the condensed phases by the clouds, but, specially, in the vapour phase, generating the aerial branch of the hydrological cycle. Without this atmospheric runoff, the cycle would not be closed and no water recycling could ever take place.

The lack of adequate qualitative and quantitative representations has lead in the past to confusing ideas about the importance and nature of the role of the atmosphere in the hydrological cycle. For sometime hydrologists thought that a major part of the rainfall water over the continents came directly from the water vapour due to evaporation and evapotranspiration *in situ*. The transfer of water in the atmosphere was minimized and essentially was reduced to a compensation for the water runoff over the continents. The total transfer of the water vapour, however, constitutes a very important part of the generalized hydrological cycle equation.

In spite of having been popular in certain literature areas (Horton, 1943), the concept of *in situ* evaporation–rainfall correspondence can not be accepted anymore. The acceptance of this hypothesis would lead to the conclusion that to increase the rainfall *in situ* it would be sufficient to increase local evaporation and evapotranspiration using artificial lakes, irrigation, forest planting, and so on.

⋆

⋆ ⋆

It seems obvious that it is the atmosphere, through the dynamics of its general circulation that forces all the phases of the hydrological cycle. In fact, the general circulation provides all the necessary ingredients for the phase transitions of the water substance, through its vertical motion originating the adiabatic cooling, essential for the condensation. It is the resistance due to atmosphere that enables the formation of clouds preventing the fall of the ice crystals and water droplets. Finally, through subsequent microprocesses the clouds lead to the occurrence of precipitation. Furthermore, the general circulation together with the incoming solar radiation, play a crucial role in enhancing of evaporation of water at the earth's surface, modulating the sources and sinks of water vapour. Finally, it is through the aerial transport of water vapour by the general circulation that the precipitation processes are fed and the water cycle is closed and maintained.

The implications of these considerations are important when we study the possibility of enhancing precipitation by artificial processes over a certain area. Even if we were able to precipitate all the water contained in a cloud, this process could not be sustained without a strong water vapour flux to keep it moving, since the water content in a cloud in solid and liquid phases is very small. This flux of water vapour, however, is not conditioned by local or regional factors, but by the general circulation of the atmosphere.

⋆

⋆ ⋆

Without atmosphere there would be no clouds and so no water cycle. In fact, without clouds there would be no precipitation, runoff, infiltration or oceans and the energy budget of the Earth would be completely different. Let us suppose for a moment that the atmosphere disappeared and that oceans and the surface waters over the continents would remain. Of course the water would evaporate and be quickly photodissociated, so that soon everything on the Earth would dry out.

The atmosphere, with its aerial runoff, maintains the great routes that feed and control all the water economy on the Earth. It is through its own dynamics and associated mechanisms that the atmosphere determines, where, when and how the precipitation occurs.

The atmosphere plays then an essential role in the water cycle. It is a *sine qua non* condition for the existence of the hydrosphere (with its rivers, seas, oceans and lakes), of the biosphere (to reproduce, to grow and to blossom), and of the cryosphere (with its polar ice caps, glaciers, snow and ice). It is with water that the subsystems of the climate system can continue to model the ever changing climate of the Earth.

References

Baumgartner, A., Reichel, E., 1975. The world water balance. Elsevier, Amsterdam, 179pp.

Chen, T., 1985. Global water vapour flux and maintenance during FGGE. Monthly Weather Review, vol 113, pp1801-1819.

Henderson-Sellers,A., McGuffie,K., 1988. A Climate Modelling Primer. John Wiley & Sons.

Horton, R. E. 1943. Evaporation maps of the United States. Tans. Am. Geophys. Union. Vol 24.

Jaeger, L., 1976. Monthly precipitation maps for the entire earth (in German). Ber. Deutschen Wetterdienstes 18, 38pp.

Oort, A. H., 1983. Global atmospheric circulations statistics, 1958-1973 NOAA Prof. Pap.14 U.S. gov. printing office, Washington D.C., 180pp.

Peixoto, J.P., Kettani, M. A. 1973. The control of the water cycle. Sci. Am., 228, 46–61.

Peixoto, J.P., Oort, A. H. 1983. The atmospheric branch of the hidrological cycle and climate. In "Variations of the global water budget", Reidel, Londres, pp. 5–65.

Peixoto, J.P., Oort, A.H., Almeida, M., Tomé, A.R. 1991. Entropy Budget of the Atmosphere. J. Geophys. Res., 96, 981–988.

Peixoto, J.P., Oort, A.H., 1992. Physics of Climate. American Institute of Physics, 520pp.

Peixoto, J.P., Tomé, A.R. e Almeida M., 1994. Dinâmica do Ciclo Hidrológico; as fontes do vapor de água da atmosfera. Instituto Geofísico do Infante D. Luís, Lisboa.

Salstein, D.A., R.D. Rosen and J.P.Peixoto, 1983. Modes of variability in annual hemispheric water vapour and transport fields. J. Atmos. Sci. 40, 788-803.

Salstein, D.A., Y.C. Sud. Large–scale diabatic heating in a model with downdrafts. Tenth Conference on Numerical Weather Prediction. Portland, Oregon 1994.

Tomé, A.R., 1988 Contribuição para o estudo da dinâmica do ciclo hidrológico; importância das circulações celulares. Universidade de Lisboa. Lisboa, Portugal.

Tomé, A.R., 1992 Contribuição para o estudo da entropia da atmosfera. Universidade da Beira Interior. Covilhã, Portugal.

Whittaker, M.T., 1977. Automated streamline analysis. Monthly Weather Review, vol. 105, pp786-788.

Acknowledgements

We are grateful to Dr. António Tomé, from the University of Beira Interior, for his important contributions and excellent colaboration, to Dr. Mário Almeida, from the Geophysical Center of the University of Lisbon, for his interest and valuable sugestions and to Prof Abraham Oort from GFDL (Princeton University) for providing most of the data. This work was sponsored by a grant of the Geophysical Center of the University of Lisbon.

Laurentian Great Lakes Dynamics, Climate, and Response to Change

Thomas E. Croley II
Great Lakes Environmental Research Laboratory
National Oceanic and Atmospheric Administration
2205 Commonwealth Blvd.
Ann Arbor, Michigan 48105-1593
U. S. A.

1. Introduction

The Laurentian Great Lakes are one of North America's greatest water resources. Due to recent record high lake levels and to climate change issues, there is renewed interest in lake level trends and in factors affecting high water levels. Impacts on Great Lakes water supply components and basin storages of water and heat must be understood before lake level impacts can be assessed. The Great Lakes Environmental Research Laboratory (GLERL) developed conceptual simulation models for Great Lakes hydrology to address the impact questions. GLERL integrated the models to estimate lake levels, whole-lake heat storage, whole basin moisture storage, and water and energy balances for forecasts and for assessment of impacts associated with climate change. Because the Great Lakes possess tremendous water and heat storage capacities, they respond slowly to changed meteorologic inputs. This "memory" results in a filtering or dampening of most short-term meteorologic fluctuations and in a response to longer-period fluctuations characteristic of climate change. The large Great Lakes system, thus, is ideal for studying regional effects of climate changes.

This paper outlines large lake dynamics and climate, pertinent especially to the Laurentian Great Lakes, summarizes GLERL's Great Lakes hydrology models, presents results of recent climate change studies on the Great Lakes, and looks ahead to the next generation of interactively coupled models useful for assessing climate change.

2. Great Lakes Dynamics and Climate

There is a major tendency to think of Great Lakes water levels in terms of extremes rather than of normal conditions. Within recent memory we had the record low lake levels of 1964. This resulted in docks sitting out of the water, insufficient depths for navigation in many harbors and channels, and greatly reduced recreational opportunities. These low levels were followed in 1973 by record high lake levels with resultant flooding and shore damage and erosion. The lake levels remained high until 1989, whence they returned to near-average

NATO ASI Series, Vol. I 31
The Role of Water and the Hydrological Cycle
in Global Change
Edited by Howard R. Oliver and Sylvia A. Oliver

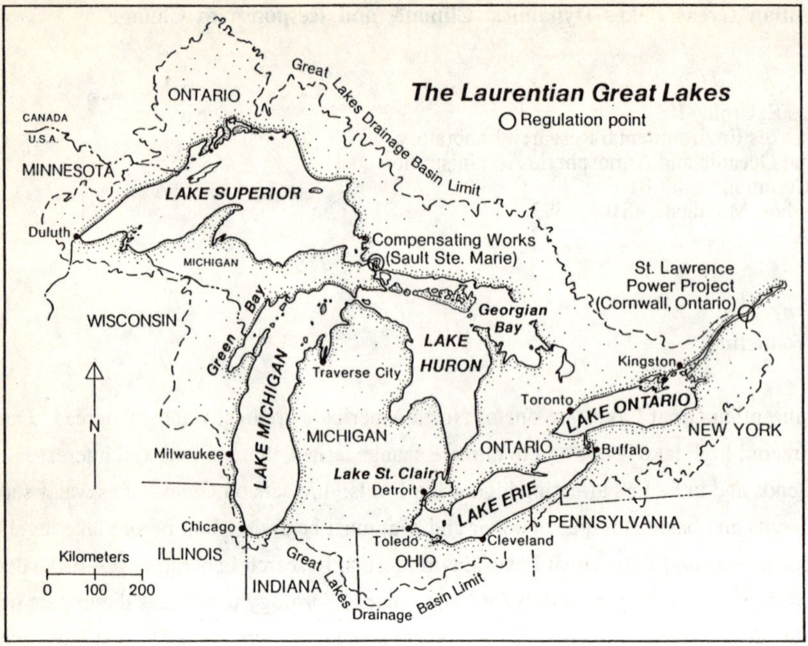

Figure 1. Great Lakes Basin.

conditions, and new record highs were once again set on Lakes Superior, Michigan-Huron, St. Clair, and Erie.

This section presents an overview of the physical characteristics of the Great Lakes from a water quantity perspective, outlines the basin and lake physical processes, summarizes the climatology of the Great Lakes, examines the types of natural lake level fluctuations and their causes, compares the natural fluctuations with existing diversions and regulation effects, describes current conditions, and concludes with a long-term perspective on lake levels.

2.1 Great Lakes Overview

The Great Lakes basin, shown in Figure 1, contains an area of approximately 770,000 km^2 (300,000 mi^2), about one-third of which is water surface. Cursory descriptions are given by *Freeman and Haras* (1978), the *U. S. Army Corps of Engineers* (1985), and the *Coordinating Committee on Great Lakes Basic Hydraulic and Hydrologic Data* (1977). The basin extends some 3,200 km (2,000 mi) from the western edge of Lake Superior to the Moses-Saunders Power Dam on the St. Lawrence River. The water surface drops in a cascade over this distance some 180 m (600 ft) to sea level. The most upstream, largest, and deepest lake, is Lake Superior. The lake has two interbasin diversions of water into the system from the Hudson Bay Basin: the Long Lac and Ogoki Diversions. Lake Superior waters flow through the lock and compensating works at Sault St. Marie and down the St. Marys River into Lake Huron where it is joined by water flowing from Lake Michigan. Lake Superior is completely regulated, to balance Lakes Superior, Michigan, and Huron water levels, according to Regulation Plan 1977, under the auspices of the International Joint Commission (*International Lake Superior Board of Control* 1981, 1982).

Lakes Michigan and Huron are considered to be one lake hydraulically because of their connection through the deep Straits of Mackinac. The second interbasin diversion takes place from Lake Michigan at Chicago. Here water is diverted from the Great Lakes to the Mississippi River Basin. The water flows from Lake Huron through the St. Clair River, Lake St. Clair, and Detroit River system into Lake Erie. The drop in water surface between Lakes Michigan-Huron and Lake Erie is only about 2 m (8 ft). This results in a large backwater effect between Lakes Erie, St. Clair, and Michigan-Huron; changes in Lakes St. Clair and Erie levels are transmitted upstream to Lakes Michigan and Huron. From Lake Erie the flow is through the Niagara River and Welland Diversion into Lake Ontario. The major drop over Niagara Falls precludes changes on Lake Ontario from being transmitted to the upstream lakes. The Welland Diversion is an intrabasin diversion bypassing Niagara Falls and is used for navigation and hydropower. There is also a small diversion into the New York State Barge Canal System which is ultimately discharged into Lake Ontario. Lake Ontario is completely regulated in accordance with Regulation Plan 1958D to balance damages upstream on Lake Ontario with those downstream on the St. Lawrence Seaway [estimated to have lowered Lake Ontario 0.75 m (2.5 ft) in 1986]. The outflows are controlled by the Moses-Saunders Power Dam between Massena, New York and Cornwall, Ontario. From Lake Ontario, the water flows through the St. Lawrence River to the Gulf of St. Lawrence and to the ocean.

Lakes Superior, Michigan, Huron, and Ontario are very deep, while Lakes Erie and St. Clair are very shallow. Table 1 contains pertinent gross statistics on the sizes of the Great Lakes, Lake St. Clair, and their basins.

2.2 Physical Processes

The behavior of the Laurentian Great Lakes system is governed by its huge storages of water and energy. There are three main conservation laws to consider relative to these huge storages: 1) mass balances in the basins, 2) mass balances in the lakes, and 3) energy balances in the

Table 1. Laurentian Great Lake Size Statistics.

Characteristic		Superior	Michigan	Huron	St Clair	Erie	Ontario
Basin area,	km^2	128,000	118,000	131,000	12,400	58,800	60,600
	mi^2	49,300	45,600	50,700	4,800	22,700	23,400
Surface area,	km^2	82,100	57,800	59,600	1,114	25,700	18,960
	mi^2	31,700	22,316	23,000	430	9,920	7,320
Volume,	km^3	12,100	4,920	3,540	3	484	1,640
	mi^3	2,900	1,180	850	1	116	393
Average depth,	m	147	85	59	3	19	86
	ft	482	280	190	10	62	280
Maximum depth,	m	405	281	229	6	64	244
	ft	1,330	923	750	21	210	802

lakes. There are also mass and energy balances to consider for the lakes' ice cover. The first conservation law (mass balance on the basins) comprises the primary process determining lake levels: the hydrologic cycle of the Great Lakes Basin (*Croley* 1983a). As shown in Figure 2, precipitation enters the snowpack, if present, and is then available as snow melt depending mainly on air temperature and solar radiation. Snow melt and rainfall partly infiltrate into the soil and partly run off directly to rivers, depending upon the moisture content of the soil. Infiltration is high if the soil is dry, and surface runoff is high if the soil is saturated. Soil moisture evaporates or is transpired by vegetation depending upon the types of vegetation, the season, solar radiation, air temperature, humidity, and wind speed. The remainder percolates into deeper basin storages which feed the rivers and lakes through interflows and groundwater flows. Generally, these river supplies are high if the soil and groundwater storages are large. Because of this buffering effect of the large snowpack and the large soil, groundwater, and surface storages, runoff from rivers into a lake can remain high for many months or years after high precipitation has stopped.

Mass conservation in the lake is the next major determinant of lake levels. Major sources of water into a lake include precipitation on the land basin which results in runoff into the lake, precipitation over the lake surface, inflow from upstream lakes, and diversions into the lake. Net groundwater flows directly to each of the Great Lakes are generally neglected (*DeCooke and Witherspoon* 1981). The outflows consist of evaporation from the lake surface, flow to downstream lakes, and diversions. The imbalance between the inflow and outflow results in the lake levels either rising if there is more inflow than outflow, represented by a positive change in storage, or falling if there is more outflow than inflow, represented by a negative change in storage. The large lake water storages provide a buffering of the input fluctuations with regard to output variations. The large surface areas of the lakes enable large storage changes with very small water level changes; hence, outputs (which are a function of water levels) change slowly.

Energy conservation in a lake actually must be considered together with a lake's mass balance. Lake heat storage is a function of the lake's size and shape and of its surface inputs of solar insolation and reflection (short wave exchanges), thermal emission and atmospheric emission (net long wave exchange), conduction to the atmosphere (sensible heat transfer), heat loss through evaporation (latent and some advection), other advection terms (precipitation, inflows, and outflows), and ice growth and melt. Evaporation is a function of surface temperature (heat storage), air temperature (atmospheric stability), humidity, and wind speed. Water surface temperatures generally peak in August (September for Superior) at 15-25 °C resulting in a stable summertime temperature stratification in the water column (high-density cool water at depth and low-density warm water at the surface). Surface temperatures drop during the fall and winter, and the water column in each lake "turns over" as temperatures drop through 4°C where water density is maximum (deep now lighter waters rise and mix with now-heavier surface layers). Turn over occurs again in the spring as surface temperatures rise to that of maximum density.

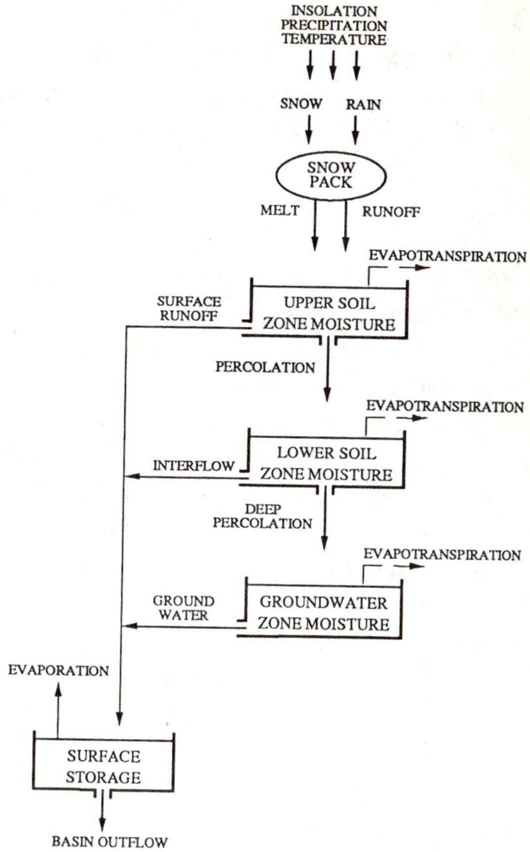

Figure 2. Runoff Hydrology Concepts.

There is also extensive ice cover on most of the lakes during most winters. Lake Superior averages about 75% ice-covered, Michigan is 45%, Huron is 68%, Erie is 45%, and Ontario is 24%. Ice formation and breakup is governed by additional mass and energy balances that take place simultaneously with those of the lakes' water bodies. The Great Lakes do not ordinarily freeze-over completely (*Assel et al.* 1983) because of the combination of their large heat storage capacity, large surface area, and their location in the mid-latitude winter storm track. Alternating periods of mild and cold air temperatures combine with episodic high and low wind stresses at the water surface to produce transitory ice conditions during the winter. Ice cover in mid-lake regions is often in motion. Lake Erie ice speeds have been observed to average 8 cm/s with a maximum speed of 46 cm/s (*Campbell et al.* 1987). Ice can form, melt, or be advected toward or from most mid-lake areas throughout the winter (*Rondy* 1976). When ice is advected into areas with existing ice cover, it can under- or over-ride the ice cover, forming rafted rubble 5-10 m thick. The normal seasonal progression of ice formation begins in the shallow shore areas of the Great Lakes in December and January. The deeper mid-lake areas normally do not form extensive ice cover until February and March. Ice is lost over all lake areas during the last half of March and during April.

Ice formation alters the surface thermodynamics of the lakes, changing subsequent ice formation, surface heating or cooling, lake evaporation, and lake responses to atmospheric changes. The large heat storages of the lakes provide a buffering; they forestall and reduce ice formation and shift the large evaporation response. Water temperatures lag air temperatures and evaporation lags surface heating (insolation). Evaporation peaks in October-November on Lake Erie and in November-December on Lake Superior.

The large basin and lake storages of water and ice and the large lake and ice storages of energy represent an "intrinsic memory" that allow scientists to forecast basin moisture storage and runoff (basin storage buffering) in the face of uncertain meteorology. It also allows prediction of evaporation (heat storage buffering) and lake levels (lake storage buffering) of up to about six months of low-frequency changes. It further enables estimation of ice formation amounts and timing as well as all secondary hydrological variables.

2.3 Climatology

Precipitation causes the major long-term variations in lake levels (*Quinn and Croley* 1981; *Quinn* 1985). Table 2 shows that annual precipitation ranges from about 82 cm (32 in) for Superior to 93 cm (37 in) for Ontario. Figure 3 depicts total annual precipitation over Lakes Michigan-Huron, St. Clair, and Erie for the 1900-79 period (*Quinn* 1981; *Quinn and Norton* 1982). From 1900 through 1939, a low precipitation regime predominated with the majority of the years falling below the mean. From about 1940 until recently, a high precipitation regime has existed. Of particular interest is the high precipitation in the early 1950s, the low precipitation in the early 1960s that led to the record lows, and a consistently very high pre-

Table 2. Partial Great Lakes Annual Water Balance (1951-88).

Component	Superior		Michigan		Huron		Erie		Ontario	
	(cm)	(in)	(cm)	(in)	(cm)	(in)	(cm)	(in)	(cm)	(in)
Lake Precip.[a]	82	32	83	32	87	34	91	36	93	37
Lake Runoff[a]	62	24	64	25	84	33	80	32	169	67
Lake Evap.[a]	56	22	65	25	63	25	90	35	67	26

[a]Equivalent depth over the lake area.

cipitation regime from the late 1960s through the late 1980s. Table 3 summarizes Great Lakes annual precipitation totals by basin for several periods. Of particular interest are the progressions of increasing precipitation for each basin. While the 1940-90 period is generally above normal (2-8% higher than the 1900-69 average and -2-6% higher than the 1900-90 average), the last 20 of those years are higher still (8-13% than the 1900-69 average and 2-11% higher than the 1900-90 average); 1985 set many new records with the highest precipitation to that date (8-40% higher than the 1900-69 average and 7-33% higher than the 1900-90 average).

Variations in air temperature also influence lake level fluctuations. At higher air temperatures, plants tend to use more water, resulting in more transpiration, and there are higher rates of evaporation from both the ground surface and the lake. This yields less runoff for the same amount of precipitation than would exist during a low temperature period when there is less evaporation and transpiration. Coupled with the higher lake evaporation, lake levels drop with increasing air temperature, all other things being equal. The annual mean air temperature around the perimeter of the Great Lakes since 1900, summarized in Figure 4, indicate three distinct temperature regimes; a low temperature regime from 1900-1929 to a higher temperature regime from about 1930-1959, and an additional low regime from 1960-present period. The difference between the previous and current regime is a drop of about 1°F.

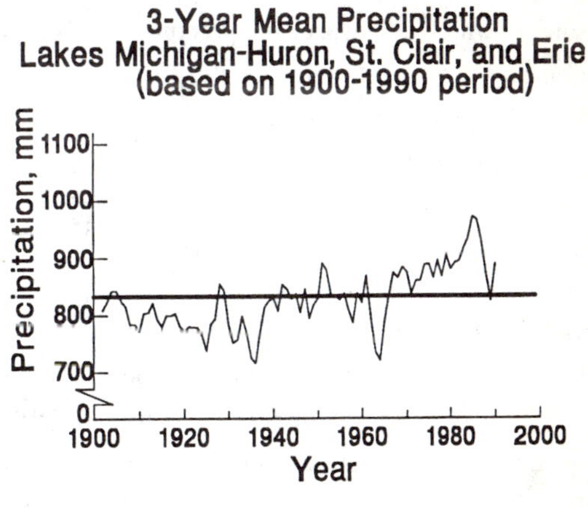

Figure 3. Historical Precipitation.

Table 3. Great Lakes Annual Precipitation Summary.

Period	Superior (cm)	Superior (in)	Michigan (cm)	Michigan (in)	Huron (cm)	Huron (in)	Erie (cm)	Erie (in)	Ontario (cm)	Ontario (in)
1900-39	72	29	78	31	77	31	85	34	86	34
1940-90	81	32	82	33	86	34	89	35	93	37
1970-90	84	33	86	34	89	35	94	37	98	39
1985[a]	105[b]	41[b]	97[b]	38[b]	106[b]	42[b]	107	42	94	37
1900-69[c]	75	30	79	31	80	32	87	34	87	34
1900-90[c]	79	31	84	33	84	33	89	35	88	35

[a]June-December 1985 provisional data from the U. S. Army Corps of Engineers.
[b]Record high for 1900-85.
[c]Long-term period averages are supplied for comparison.

The magnitude of the hydrologic variables vary with season, as shown in Figure 5 for Lake Erie (*Quinn* 1982; *Quinn and Kelley* 1983). The monthly precipitation is fairly uniformly distributed throughout the year, while the runoff has a peak during the spring which results primarily from the spring snow melt. The runoff is at a minimum in the late summer and early fall due to large evapotranspiration from the land basin. The lake evaporation reaches a minimum during the spring and gradually increases until it reaches a maximum in the late fall or early winter. The high evaporation period is due to very cold dry air passing over warm lake surfaces. The integration of these components is depicted in the net basin supply, which consists of the precipitation plus the runoff minus the evaporation. As seen from Table 2, these three components of net basin supply are all of the same order of magnitude for each lake. Annual runoff to the lake ranges from about 62 cm (24 in) for Superior to 169 cm (67 in) for Ontario, and annual lake evaporation ranges from about 56 cm (22 in) for Superior to 90 cm (35 in) for Erie. The net basin supply is seen in Figure 5 to reach a maximum in April

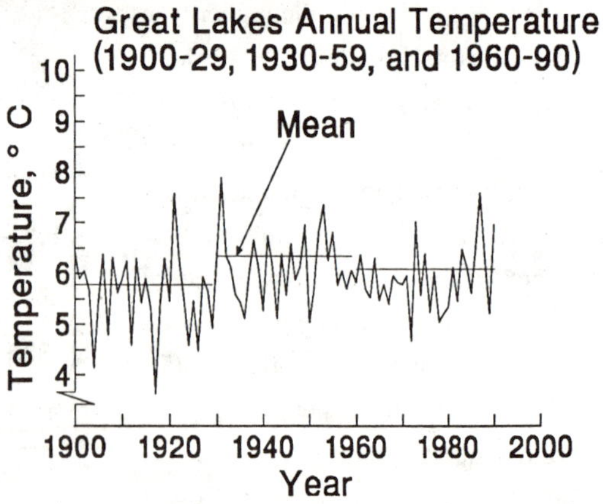

Figure 4. Historical Air Temperature.

and a minimum in the late fall. The negative values indicate that more water is leaving the lake through evaporation than is being provided by precipitation and runoff.

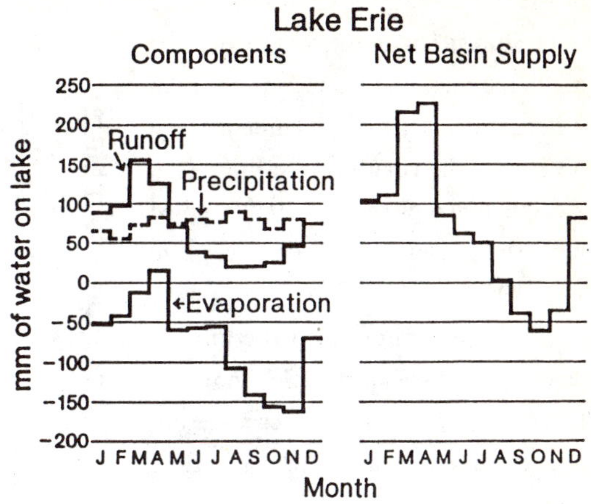

Figure 5. Seasonal Net Basin Supply.

2.4 Lake Level Fluctuation & Trends

There are three primary types of lake level fluctuations: long-term lake levels (represented on an annual basis), seasonal lake levels, and short-period lake level changes due to wind setup and storm surge. Annual fluctuations result in most of the variability leading to the record high and low lake levels. The annual lake levels are shown in Figure 6 from 1860 through the present to illustrate the long-term variability of the system. The record highs in 1952 and 1973 and record lows in 1935 and 1964 are readily apparent. There is an overall range of about 2 m (6 ft) in the annual levels. Of particular interest is the fall in the levels of Lakes Michigan and Huron occurring in the mid-1880's from which the lakes never recovered. This probably results from dredging for deeper draft navigation in the St. Clair River. Other changes in the St. Clair River include sand and gravel dredging between about 1908 and 1924, a 7.6 m (25 ft) navigational project in the mid-1930's, and an 8.2 m (27 ft) navigation project in the late 1950's and early 1960's. Without these changes, Lake Michigan-Huron would be approximately 0.5 m (1.5 ft) higher than it is today.

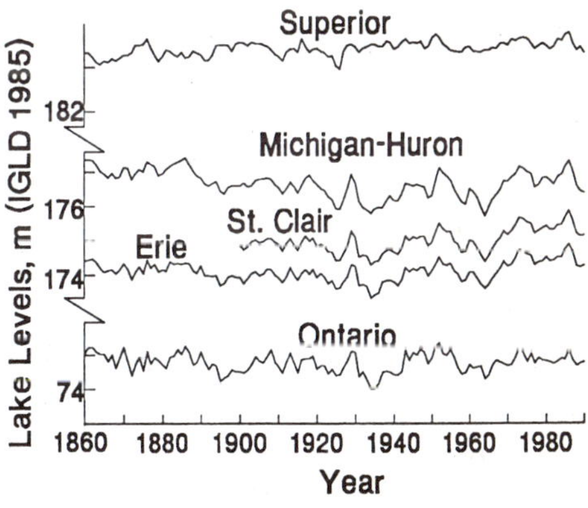

Figure 6. Historical Great Lake Levels.

The three-year precipitation mean in Figure 3 correlates very well with annual lake levels as observed by superimposing the annual precipitation on the annual Lake Erie water levels in Figure 7. The precipitation tends to lead the water levels by approximately one year, as shown here by the 1929 highs, the 1935 lows, the 1952 highs, and the 1963 lows. In particular, the last 15 years of high precipitation resulted in very high water levels. Thus, the continuing high levels are the result of the increased precipitation regime since 1940 coupled with the lower temperature regime since 1960.

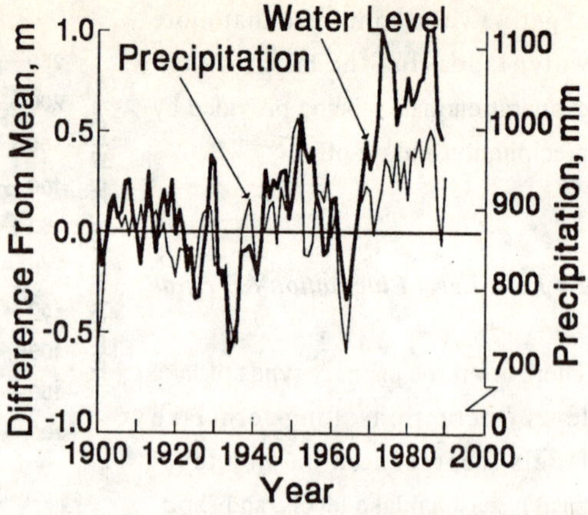

Figure 7. Annual Lake Erie Water Levels.

Superimposed on the annual levels are the seasonal cycles shown in Figure 8; each lake undergoes a seasonal cycle every year. The magnitude depends upon the individual water supplies. The range varies from about 30 cm (1 ft) on the upper lakes to about 38 cm (1.3 ft) on the lower lakes. In general, the seasonal cycles have a minimum in the winter, usually January or February. The levels then rise due to increasing water supplies from snow melt and spring precipitation until they reach a maximum in June for the smaller lakes, Erie and Ontario, and September in the case of Lake Superior. When the net water supplies diminish in the summer and fall, the lakes begin their seasonal decline.

The final type of fluctuation which is common along the shallower areas of the Great Lakes, particularly Lake Erie, Saginaw Bay, and in some cases on Green Bay, are storm surges and wind set-up. Under these conditions when the wind is blowing along the long axis of a shallow lake or bay, a rapid difference in levels can build up between one end of the lake and the other. This difference can be as large as 5 m (16 ft) for Lake Erie (storm of 2 December 1985). These storm conditions, when superimposed on high lake levels, cause most of the damage along the Great Lakes shoreline.

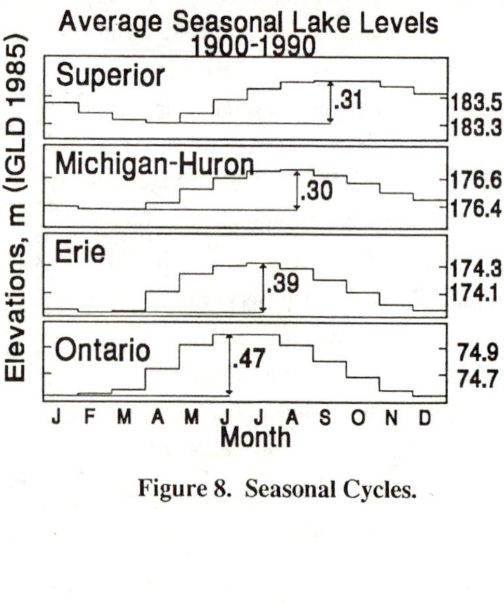

Figure 8. Seasonal Cycles.

Looking in more detail at the past trends in lake levels, along with the more recent conditions for Lake Erie, we see a steady progression of changes in the lake levels with time in Figure 9. These changes reflect the changes in precipitation, illustrated in Figure 3 and summarized in Table 3. At the bottom of Figure 9 are the record low lake levels for each month which were set primarily in 1964. Proceeding upwards we have the 40-year average from 1900-1939. From 1940-1979, the lake is at a still higher average level. Taking the 11-year period from 1970-1980, we see that the lake level average is higher yet, followed by the record highs set in 1985. Record levels for the month were set in April and May 1985 on Lakes Michigan-Huron, St. Clair, and Erie; they were set for November 1985 through April 1986 on Lakes Erie and St. Clair. Since that time, a record drought brought water levels back to their long-term normal values in the late 1980s and early 1990s.

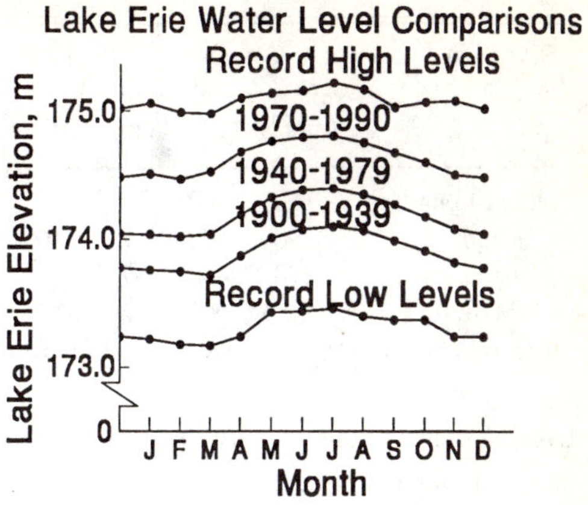

Figure 9. Lake Erie Level Comparisons.

2.5 Diversions

It is interesting to compare the impacts of the existing diversions on lake levels in Table 4 with natural lake-level fluctuations (*International Great Lakes Diversions and Consumptive Uses Study Board* 1985). This enables a comparison of man's impacts with natural fluctuations. The Long Lac and Ogoki Diversions average about 160 cms (5,600 cfs) and raise lake levels between 6 cm (0.21 ft) and 11 cm (0.37 ft). The Chicago Diversion averages about 90 cms (3,200 cfs) and lowers lake levels between 2 cm (0.07 ft) and 6 cm (0.21 ft). The Welland Canal, which bypasses Niagara Falls, averages about 270 cms (9,400 cfs) and lowers lake levels between 2 cm (0.06 ft) and 13 cm (0.44 ft) with no effect on Lake Ontario. The combined effect on the lakes ranges from a 2 cm (0.07 ft) rise for Lake Superior to a 10 cm (0.33 ft) drop for Lake Erie. The diversion effects are therefore small in comparison with the one or more meter (several foot) variation associated with short-term storm movements, the 30-38 cm (1-1.3 ft) seasonal cycle, and the 2 m (6 ft) range of annual variations.

The small effects of the diversions along with the long response time of the system illustrate why diversions are not suitable for lake regulation. Due to the large size of the Great

Table 4. Impact of Existing Diversions on Lake Levels.

Diversion	Amount		Superior		Mich-Hur		Erie		Ontario	
	(m^3s^{-1})	(cfs)	(cm)	(ft)	(cm)	(ft)	(cm)	(ft)	(cm)	(ft)
Ogoki-Long Lac	160	5600	+6	+0.21	+11	+0.37	+8	+0.25	+7	+0.22
Chicago	90	3200	-2	-0.07	-6	-0.21	-4	-0.14	-3	-0.10
Welland	270	9400	-2	-0.06	-5	-0.18	-13	-0.44	0	0
COMBINED			+2	+0.07	-1	-0.02	-10	-0.33	+2	+0.08

Lakes system, it responds very slowly to man-induced changes. This is illustrated in Figure 10 by the length of time it takes from the start of a hypothetical diversion on Lakes Michigan and Huron (of the magnitude of the Chicago diversion) until the ultimate effect of that diversion is reached on Lakes Michigan-Huron, and Erie. It takes approximately 3-3.5 years to achieve 50% of the ultimate effect and 12-15 years to get 99% of the effect. (These results depend somewhat on the lake levels at the beginning of the diversion.) Thus, regulation by diversion would not produce changes responsive to natural fluctuations. Recent studies at GLERL indicate that an increase of 10% in the Niagara River discharge from Lake Erie (and consequent increases in Lake Erie inflow) would lower it 27 cm (10.5 in) in about 11-12 years and lower Lakes Michigan and Huron 14 cm (5-6 in) in this same period. If Lake Erie inflows were held constant (not possible at the present time), then it would take 6 months to 1 year to achieve this lowering.

Additional interbasin diversions are a highly controversial issue at the present time around the Great Lakes. Possible uses of Great Lakes water outside the basin are flow augmentation for navigation, energy uses such as synthetic fuels or pipelines, agriculture and aquifer recharge, and municipal water supplies. A small pipeline project

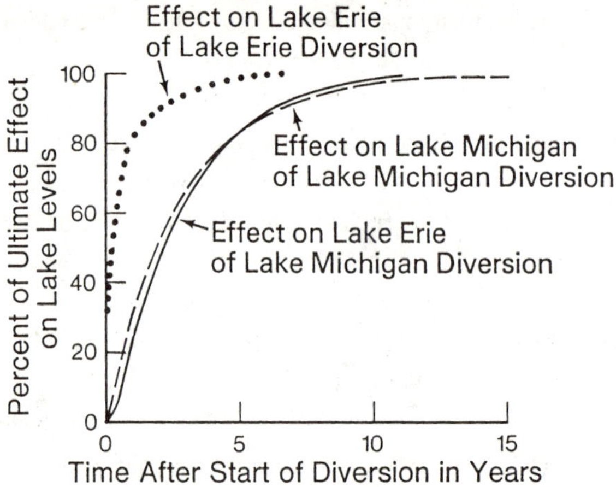

Figure 10. Response to Diversions.

such as the Powder River coal slurry pipeline would require 0.2 cms (5-8 cfs) of water and would have no measurable impact on lake levels. A synthetic fuels project, highly unlikely at this time, could require approximately 23 cms (800 cfs) and result in a lake level lowering of 1-2 cm (0.04-0.06 ft). A major agricultural or aquifer recharge project could require 300 cms (10,000 cfs) and would result in lake level decreases ranging from 12 cm (0.4 ft) on Lake Erie to 21 cm (0.7 ft) on Lake Michigan-Huron. It should be emphasized that these are hypothetical projections for illustration only.

2.6 Future

Water levels ordinarily do not change fast, as shown by the above consideration of diversions. Other studies at GLERL indicate that if normal meteorological conditions were realized ("normal" being the average conditions over 1900-69) instead of the record drought of the late 1980s, it would have taken about 6 years for Lake Michigan-Huron to return from its January 1986 level to its normal (1900-69) level. About 7 years would have been required for Lakes St. Clair and Erie to return to within 4 in of normal, and about 9 years would have been required for them to return to within 2 in of normal. Even supposing that we encountered a drought similar to the 1960-64 conditions, about 3.5 years would have been required for Lake Michigan-Huron and about 4 years would have been required for Lakes St. Clair and Erie.

A long-term perspective on Lake Michigan levels for 7,000 years was reconstructed through geologic and archaeologic evidence (*Larsen* 1985) under work sponsored by the Illinois Geologic Survey. Conditions several thousand years ago were not necessarily the same as today due to isostatic rebound and uplift during the intervening time. But, in general, this provides additional perspective on possible conditions we may experience in the future. Looking at just the last 2,500 years, during which time the Great Lakes were in their current state, there were major lake level fluctuations. During most of this time the levels were much higher and more variable than they have been during the last 120 years of record. If the past is any indication, lake levels in the future could go through a considerably larger range than we have experienced lately. Indeed, the period of record which makes up what many consider to be normal, the early 1900's through the 1960's, may be abnormal conditions.

2.7 Summary Comments on Great Lakes Dynamics

Huge storages of water in the basins and the lakes and of energy in the lakes give the Laurentian Great Lakes their characteristic behavior. They filter the variability of the meteorological inputs and enable hydrological predictions in the face of uncertain meteorology, if the storage amounts are known. Lake levels are most affected by temporal patterns of precipitation; air temperature patterns play a lesser but important role also. It is important to keep in perspective that while we have ranges in annual lake levels of 1-2 m (4-6 ft), and additional short term effects on the order of 2-3 m (7-8 ft), the effects of man on the system are relatively small, on the order of about 5 cm (0.2 ft). While the lakes are slow changing over the long term in the face of normal meteorology, past fluctuations have been very large. Future changes will depend mostly on future climate.

3. Laurentian Great Lakes Physical Process Models

GLERL developed, calibrated, and verified conceptual model-based techniques for simulating hydrological processes in the Great Lakes (including Georgian Bay and Lake St.Clair separately). GLERL integrated the models into a system to estimate lake levels, whole-lake heat storage, and water and energy balances for forecasts and for assessment of impacts associated with climate change (*Croley* 1990, 1993a,b; *Croley and Hartmann* 1987, 1989; *Croley and Lee* 1993; *Hartmann* 1990). These include models for rainfall-runoff [121 daily watershed models (*Croley* 1982, 1983a,b; *Croley and Hartmann* 1984)], over-lake precipitation (a daily estimation model), one-dimensional (depth) lake thermodynamics [7 daily models for lake surface flux, thermal structure, and heat storage (*Croley* 1989a,b, 1992a; *Croley and Assel* 1994)], channel routing [4 daily models for connecting channel flow and level, outlet works, and lake levels (*Hartmann* 1987, 1988; *Quinn* 1978)], lake regulation [a monthly plan balancing Lakes Superior, Michigan, and Huron (*International Lake Superior Board of Control* 1981, 1982) and a quarter-monthly plan balancing Lake Ontario and the St. Lawrence Seaway (*International St. Lawrence River Board of Control* 1963)], and diversions and consumption (*International Great Lakes Diversions and Consumptive Uses Study Board* 1981).

3.1 Runoff Modeling

The GLERL Large Basin Runoff Model (LBRM) consists of moisture storages arranged as a serial and parallel cascade of "tanks" (*Croley* 1983a,b); water flows from the snowpack to the upper soil zone tank, from the upper to the lower soil zone and surface storage tanks, from the lower to the groundwater and surface tanks, from the groundwater to the surface tank, and from the surface tank out of the watershed; see Figure 2. It makes use of physical concepts for snow melt and net supply to the watershed surface, infiltration, heat available for evapotranspiration, actual evapotranspiration, and mass conservation. As a conceptual model, the LBRM is useful not only for predicting basin runoff, but to facilitate understanding of watershed response to natural forces as well. The main mathematical feature of the LBRM is that it may be described by strictly continuous equations; none of the complexities associated with inter-tank flow rate dependence on partial filling are introduced. For a sufficiently large watershed, these nuances are not observed due to the spatial integration of rainfall, snow melt, and evapotranspiration processes.

Daily precipitation, temperature, and insolation (the latter available from climatological summaries as a function of location) may be used to determine snowpack accumulations and net surface supply based on degree-day determinations of snow melt. The net surface supply is divided into infiltration to the upper soil zone and surface runoff by taking infiltration proportional to the net surface supply rate and to the areal extent of the unsaturated portion of the upper soil zone. Outflow from each storage within the watershed is proportional to the moisture in storage. The evapotranspiration rate from the upper and lower soil zones is proportional to available moistures there and to the heat rate available for evapotranspiration; it also reduces the heat available for subsequent evapotranspiration. The total amount of heat in a day is split between that used for and that still available for evapotranspiration by empirical functions of air temperature based on a long-term heat balance. Mass continuity yields a first-order linear differential equation for each of the moisture storages (*Croley* 1982), which are tractable analytically; they are solved simultaneously to determine daily moisture storage, evapotranspiration, and basin runoff from daily data.

The Great Lakes basin is divided into 121 watersheds, each draining directly to a lake, grouped into the six lake basins. The meteorologic data from about 1,800 stations about and in the watersheds are combined through Thiessen weighting to produce areally-averaged daily time series of precipitation and maximum and minimum air temperatures for each watershed (*Croley and Hartmann* 1985b). Records for all "most-downstream" flow stations are combined by aggregating and extrapolating for ungauged areas to estimate the daily runoff to the lake from each watershed. The LBRM is calibrated to determine the set of parameters resulting in the smallest sum-of-squared-errors between model and actual daily flow volumes for the calibration period (*Croley* 1983b, *Croley and Hartmann* 1984, 1985a). After the LBRM is calibrated for each watershed, the model outflows are combined to represent each Great Lake basin; this distributed-parameter model integration filters individual sub-basin model errors. The LBRM calibration periods generally cover 1965-1982 depending upon flow data availability. Table 5 presents overall calibration results for the distributed-parameter applications. The LBRM was also used in forecasts of Lake Superior water levels (*Croley and Hartmann* 1987), and comparisons with climatic outlooks showed the runoff model was very close to actual runoff (monthly correlations of water supply were on the order of 0.99) for the period

Table 5. Large Basin Runoff Model Calibration Statistics[a].

Lake	Number of Sub-basins	Mean 1-day Flow (mm)[b]	Flow Std. Dev. (mm)[b]	Root Mean Square Error (mm)[b]	Correlation Calib.	Correlation Verif.
Superior	22	1.12	0.67	0.25	0.93	0.77
Michigan	29	0.89	0.47	0.18	0.93	0.86
Huron	27	1.06	0.69	0.26	0.92	0.69
St. Clair	7	0.90	1.36	0.62	0.89	0.87
Erie	21	1.01	1.28	0.54	0.91	0.90
Ontario	15	1.41	1.13	0.43	0.93	0.89

[a]Statistics and calibrations generally cover 1966-83; verification generally covers 1956-63.
[b]Equivalent depth over the land portion of the basin.

August 1982 - December 1984 which is outside of and wetter than the calibration period (*Croley and Hartmann* 1986). The model also was used to simulate flows for the time period 1956-63, outside of the period of calibration. The correlation of monthly flow volumes between the model and observed values during this verification period are also contained in Table 5. They are a little lower than the calibration correlations but quite good except for Lakes Superior and Huron (there were less than two-thirds as many flow gages available for 1956-63 as for the calibration period for these basins).

3.2 Over-lake Precipitation

The lack of over-lake precipitation measurements means that estimates typically depend on land-based measurements, and there may be differences between land and lake meteorology. Although gage exposures may significantly influence the results of lake-land precipitation studies (*Bolsenga* 1977, 1979), *Wilson* (1977) found that Lake Ontario precipitation estimates, based on only near-shore stations, averaged 5.6% more during the warm season and 2.1% less during the cold season than estimates based on stations situated in the lake. By using a network that also included stations somewhat removed from the Lake Ontario shoreline, *Bolsenga and Hagman* (1975) found that eliminating several gages not immediately in the vicinity of the shoreline increased over-lake precipitation estimates during the warm season and decreased them during the cold season. Thus, for the Great Lakes, where lake effects on near-shore meteorology are significant and the drainage basins have relatively low relief, the use here of all available meteorologic stations throughout the basin is probably less biased than the use of only near-shore stations.

3.3 Over-lake Evaporation

Current Great Lakes evaporation studies use mass transfer formulations that include atmospheric stability effects on the bulk transfer coefficients, applied to monthly data for water surface temperatures, wind speed, humidity, and air temperatures (*Quinn* 1979). The present study uses that approach applied to daily data but combined with models for over-water meteorology, ice cover, and lake heat storage and with a lumped representation of a lake's heat balance (*Croley* 1989a,b, 1992a); see Figure 11. As over-water data are not generally available, over-land data are used by adjusting for over-water conditions. *Phillips and Irbe's* (1978) regressions for over-water corrections are used directly by replacing the fetch (and derived quantities) with averages. Air temperatures and specific humidities over ice are used for over-ice evaporation calculations and over water for the over-water calculations; the two estimates are combined by weighting for the fraction of the surface covered in ice. Water and ice pack heat balances (*Croley and Assel* 1994) were used to relate ice cover extent to meteorology, heat storage, and surface fluxes between the atmosphere, the water body, and the ice pack.

Kraus and Turner's (1967) mixed-layer thermal structure concept is extended for the Great Lakes to allow the determination of a simple one-dimensional model for surface temperature increments or decrements from past heat additions or losses, respectively (*Croley* 1989a,b, 1992a). The effects of past additions or losses are superimposed to determine the surface temperature on any day as a function of heat in storage; each past addition or loss is parameterized by its age. Turnovers (convective mixing of deep lower-density waters with surface waters as surface temperature passes through that at maximum density) can occur as a fundamental behavior of this superposition model, and hysteresis between heat in storage and surface temperature, observed during the heating and cooling cycles on the lakes, is preserved.

Heat in storage in the lake at the end of each day is given by a simple conservation of energy by taking the change in storage equal to the sum of the fluxes integrated over the day. As summarized by *Gray et al.* (1973), short-wave radiation is interpolated from generalized maps of Canadian and northern U.S. mid-monthly clear-sky values and adjusted for cloud cover. Average short-wave reflection is taken simply as one-tenth of the incident or as a function of ice cover, and sensible heat transfers at the water or ice temperature (minimum of air temperature or freezing temperature) are computed directly from the same mass transfer formulation and assumption (that the bulk evaporation coefficient is equal to the sensible heat coefficient) that is used to derive evaporation. It is then added to evaporative advection and latent heat transfers. Evaporative heat transfers from ice include the heat of fusion as well. Net long-wave radiation exchange is derived from considerations of the water and atmosphere as gray bodies with correction for cloud cover only to atmospheric radiation (*Keijman* 1974). Net long-wave radiation exchange over ice is computed as for open water, ignoring the small effects of the ice surface on the exchange. Energy advected with precipitation is adjusted if the precipitation is snow, to account for the heat required in snow melt. Energy advected with precipitation onto the ice surface is uncorrected for melt since that is taken as occurring with ice melt, which is added to the budget when it happens. The energies advected into and out of the lake with other mass flows are relatively very small and are ignored. The equations representing evaporation, heat storage, and heat fluxes are solved simultaneously with daily data on over-land wind speed, air temperature, cloud cover, and humidity; details of an iterative solution technique are available elsewhere (*Croley* 1989a,b, 1992a).

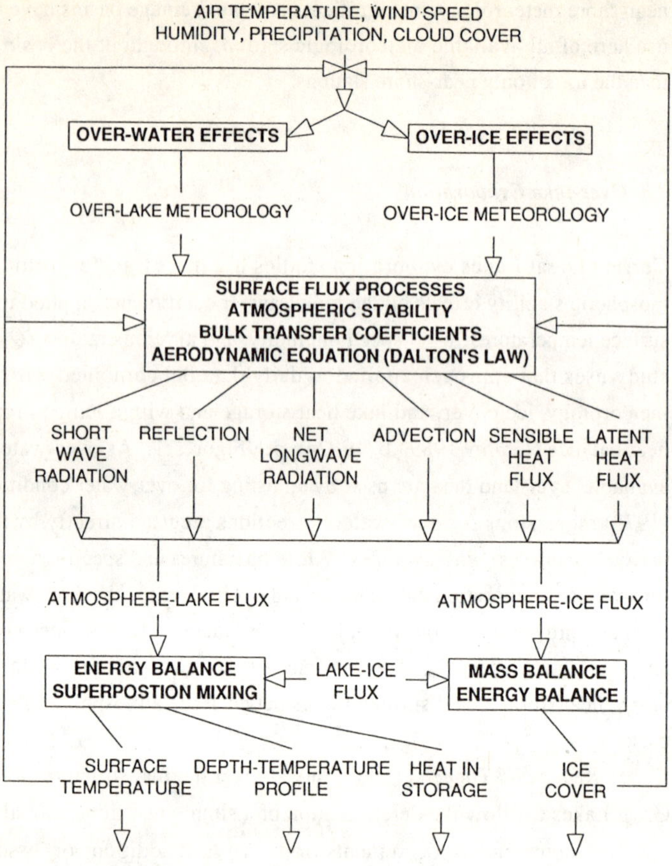

Figure 11. Evaporation Model Schematic

Unfortunately, there are no really good independent evaporation data to calibrate and verify evaporation models on the Great Lakes. Water balances are insufficient due to the large errors induced by subtracting nearly equal large inflows and outflows to each Great Lake, or due to errors in estimates of the water balance components. However, with the joint heat balance and evaporation model described here, it is possible to compare water surface temperatures with data, now available from the National Oceanic and Atmospheric Administration's Polar Orbiting Satellite Advanced Very High Resolution Radiometer (*Irbe et al.* 1982; *AES* 1988).

Daily meteorological over-land data at from five to seven near-shore stations about each Great Lake were assembled and averaged for correction to over-lake data. The heat balance model was calibrated to give the smallest sum-of-squared-errors between model and actual daily water surface temperatures observed by satellite during the calibration period of generally 1979-88; the results are summarized in Table 6. There is good agreement between the actual and calibrated-model water surface temperatures; the root mean square error is between 1.1-1.6°C on the large lakes [within 1.1-1.9°C for an independent verification period, 1966-78 (*Croley* 1989a,b, 1992a)].

3.4 Models Validity

Although GLERL uses a daily resolution of data with their models, basin-wide processes of runoff, over-lake precipitation, and lake evaporation (described with models here) respond discernibly to weekly changes at best, and monthly is usually adequate for net supply and lake level simulation (this ignores short-term fluctuations associated with storm movement which are not addressed in this study). Likewise, spatial resolution finer than about 1,000-5,000 km^2 (the present average resolution of GLERL's models and their applications) is unnecessary and much can be done in assessing hydrology changes at resolutions of 100,000 - 1,000,000 km^2 with lumped versions of the models. This coarse spatial resolution is still much finer than present general circulation model (GCM) grids.

The models were partially assessed by computing net basin supplies to the lakes with historical meteorological data for 1951-80 and comparing to historical net basin supplies. The absolute average annual difference ranged from 1.6% to 2.7% on the deep lakes while the Lake St. Clair and Lake Erie applications were 12.0% and 7.0% respectively; month-to-month differences showed more variation. These differences generally reflect poorer evaporation modeling on the shallow lakes and snow melt and evapotranspiration model discrepancies for the other lake basins. While monthly differences were generally small, a few were significant. The low annual residuals were felt to be acceptable to use these models in assessing changes from the current climate as they would be consistently applied to both a "present" and all budget term errors in the derived net basin supplies.

Table 6. Daily Lake Evaporation Model Calibration Results.

	Lake					
	Superior	Michigan	Huron	Georgian	Erie	Ontario
	CALIBRATION PERIOD STATISTICS					
	Water Surface Temperatures (1980-88)[a]					
Means Ratio[b]	1.00	1.01	0.98	1.01	1.03	0.99
Variances Ratio[c]	1.01	0.98	0.95	1.02	1.08	0.99
Correlation[d]	0.98	0.97	0.98	0.99	0.99	0.98
R. M. S. E.[e]	1.13	1.56	1.33	1.10	1.58	1.43
	Ice Concentrations (1960-1988)[f]					
Means Ratio[g]	0.92	0.72	0.70	0.98	1.15	0.39
Variances Ratio[h]	1.24	1.02	1.67	1.62	1.09	0.63
Correlation[i]	0.76	0.83	0.73	0.77	0.89	0.54
R. M. S. E.[j]	23.4	12.4	26.0	21.5	19.0	15.4
	VERIFICATION PERIOD STATISTICS					
	Water Surface Temperatures (1966-79)[k]					
Means Ratio[b]	0.96		1.03	0.98	1.05	0.94
Variances Ratio[c]	1.10		0.95	1.00	1.10	0.97
Correlation[d]	0.97		0.99	0.98	0.98	0.96
R. M. S. E.[e]	1.09		1.10	1.34	1.91	1.92

[a]Data between January 1, 1980 and August 31, 1988 for all lakes except Michigan and between January 1, 1981 and August 31, 1988 for Lake Michigan, with an initialization period for all lakes except Georgian Bay starting January 1, 1948 and January 1, 1953 for Georgian Bay.

[b]Ratio of mean model surface temperature to data mean.

[c]Ratio of variance of model surface temperature to data variance.

[d]Correlation between model and data surface temperature.

[e]Root-mean-square error between model and data surface temperatures in degrees C.

[f]Data between January 1, 1960 and August 31, 1988 for all Great Lakes except Superior and between March 1, 1963 and August 31, 1988 for Lake Superior, with an initialization period for all lakes starting January 1, 1958.

[g]Ratio of mean model ice concentration to data mean.

[h]Ratio of variance of model ice concentration to data variance.

[i]Correlation between model and data ice concentration.

[j]Root-mean-square error between model and data ice concentrations in %.

[k]Data between January 1, 1966 and December 31, 1979 for all lakes except Michigan with an initialization period for all lakes except Georgian Bay starting January 1, 1948 and January 1, 1953 for Georgian Bay.

a "changed" climate. Further assessment of model deficiencies with comparisons to historical net basin supplies is difficult since the latter are derived from water budgets which incorporate

There is some indication of model applicability outside of the time periods over which the models were calibrated as indicated above and in Tables 5 and 6. To assess the applicability of the process models to a climate warmer than the one under which they were calibrated and verified requires access to meteorologic data and process outputs for the warmer climate which unfortunately do not exist. Warm periods early in this century are not sufficiently documented for the Great Lakes. In particular, data are lacking on watershed runoff to the lakes, water surface temperatures, wind speed, humidity, cloud cover, and solar insolation.

It is entirely possible that the models are tied somewhat to the present climate; empiricism is employed in the evapotranspiration component of the LBRM and in some of the heat flux terms in the heat balance and lake evaporation model. Coefficients were determined or selected in accordance with the present climate. The models are all based on physical concepts that should be good under any climate; however, the assumption is made that they represent processes under a changed climate that are the same as the present ones. These include linear reservoir moisture storages, partial-area infiltration, lake heat-storage relations with surface temperature, and gray-body radiation. However, the calibration and verification periods for the component process models include a range of air temperatures, precipitation, and other meteorological variables that encompass much of the changes in these variables predicted for a changed climate. Even though the changes are transitory in the calibration and verification period data sets, the models appear to work well under these conditions.

4. Laurentian Great Lakes Climate Change Response

Climatic change will impact many aspects of the hydrologic cycle with interrelated consequences for mankind. A doubling of atmospheric CO_2 will impact Great Lakes water supply components and basin storages of water and heat that must be understood before lake level impacts can be assessed. Because the Laurentian Great Lakes possess tremendous water and heat storage capacities, they respond slowly to changed meteorologic inputs. This "memory" results in a filtering or dampening of most short-term meteorologic fluctuations and in a response to longer-period fluctuations characteristic of climate change. The large Great Lakes system thus is ideal for studying regional effects of climate changes.

Preliminary estimates of the impact of climatic warming on Great Lakes water resources are summarized elsewhere (*Croley and Hartmann* 1989; *Croley* 1990). The Environmental Protection Agency (EPA) coordinated several regional studies of various impacts of a doubling of atmospheric CO_2 at the direction of the U.S. Congress. As part of that study,

GLERL assessed steady-state and transient changes in Great Lakes hydrology consequent with simulated atmospheric scenarios from three GCMs. Those studies, in part, and the high water levels of the late 1980s prompted the International Joint Commission to reassess climate change impacts on Great Lakes hydrology and lake thermal structure.

The methodology established in the EPA studies was adopted with slight modifications for use in the IJC studies. The methodology integrates hydrology and lake heat storage models (*Croley* 1991, 1992b) to consider climate scenarios supplied by the Canadian Climate Centre (CCC) from its GCM (*Louie* 1991; *McFarlane* 1991). *Cohen* (1991) discusses the problems with this approach. The CCC provided a "present-climate" meteorology simulation (1xCO_2) and a changed-climate scenario (2xCO_2) developed from their atmospheric global circulation model. GLERL abstracted differences between the CCC-generated 1xCO_2 and 2xCO_2 atmospheres, made these changes to historical data, and observed the impact of the changed data in the hydrological outputs of their models.

The EPA studies included partial assessments of large-lake heat storage associated with climate change on Lakes Michigan (*McCormick* 1989) and Erie (*Blumberg and DiToro* 1989). The IJC study looked in less detail but more breadth at large-lake thermodynamics in that while only lake-wide effects were considered, all lakes were assessed. This section presents the methodology of linkage between regional hydrological models and the GCMs, describes their limitations, and presents and interprets the IJC studies of hydrological changes predicted through use of the Canadian Climate Centre's GCM.

4.1 Methodology

GLERL constructed a master computer procedure to integrate the Large Basin Runoff Model, over-lake precipitation estimates, and the lake evaporation models for all lakes to provide a net water supply model for the entire Great Lakes system. They developed it specifically to look at the impact of changed climate by doing simulations with changed meteorology that represent scenarios of changed climate and comparing with simulations based on historical meteorology (representing an unchanged climate). Inputs are areal-average daily precipitation and maximum and minimum air temperatures for each of the 121 watersheds about the Great Lakes and areal-average daily air temperature, cloud cover, humidity, and wind speed for each of the five Great Lakes and Lake St. Clair.

GLERL's general procedure for the investigation of steady-state behavior under a changed climate is similar to that used for the EPA, as detailed elsewhere (*Croley* 1991; *Louie* 1991); it required that GLERL first simulate 38 years of "present" hydrology by using historical daily maximum and minimum air temperatures, precipitation, wind speed, humidity, and cloud cover data for the 1951-88 period; this is called the "base case" or "1xCO_2" scenario. The initial conditions were arbitrarily set, but an initialization simulation period of 1 January 1948 through 31 December 1950 was used to allow the models to converge to conditions (basin moisture storages, water surface temperatures, and lake heat storages) initial to the 1 January 1951 through 31 December 1988 period. GLERL then attempted to estimate "steady-state" conditions, but there were problems.

The procedure to estimate "steady-state" conditions is to repeat the 41-year simulation with initial conditions (basin moisture storages, lake heat storages, and surface temperatures) set equal to their values at the end of the simulation period, until they are unchanging. This procedure requires many iterations for a few sub-basins with very slow groundwater storages and suggests very different initial groundwater storages than were used in calibrations. Actually, the original calibrations of the models used arbitrary (but fixed) initial conditions. GLERL should have determined initial conditions in the calibrations, but that was unfeasible; there is little confidence in calibrated parameter sets that suggest very slow groundwater storages (half-lives on the order of several hundred years in some cases) since only 10-20 years were used in the calibrations. Therefore, the best estimate of "present" hydrology is to use calibrated parameters with initial conditions on "the same order" as those assumed for the calibrations. GLERL did the latter and then conducted simulations with adjusted data sets.

Average monthly absolute air temperatures, specific humidities, cloud cover, precipitation, and wind speed were supplied for each month of the year by the Canadian Climate Centre as resulting from their second-generation global circulation model; see *McFarlane* (1991). While available at grid points spaced 3.75 degrees latitude by 3.75 degrees longitude, *Louie* (1991) interpolated monthly averaged data to a grid of 1 degree latitude by 1 degree longitude for both the "present" and "future" atmospheres (with one and two times the CO_2 content of the "present" atmosphere). He weighted values at surrounding grid points inversely to the square of the distance to each point. GLERL computed ratios of "future" (2xCO_2) to "present" (1xCO_2) monthly average absolute air temperatures, specific humidities, cloud cover, and precipitation and monthly average differences of 2xCO_2 to 1xCO_2 wind speeds at each of these grid points. They then used these ratios and differences with the historical data to estimate the 41-year sequences (1948-88) of atmospheric conditions associated with a changed climate, referred to as the "2xCO_2" scenario.

GLERL inspected each of the 770,000 square kilometers within the Great Lakes Basin to see which grid point it is closest to and applied the monthly adjustment at that grid point to data representing that square kilometer. By combining all square kilometers representing a watershed or the lake surface, GLERL derived an areally-averaged adjustment to apply to their areally-averaged data sets for the watershed or lake surface, respectively. They then used the 2xCO_2 scenario in simulations similar to the base case scenario. They repeated the

41-year simulation with initial conditions set equal to their values at the end of the simulation period, until they were unchanging to estimate "steady-state" future conditions. They then interpreted differences between the $2xCO_2$ scenario and the base case scenario, for the 1951-88 period, as resulting from the changed climate.

Transfer of information between the GCMs and GLERL's hydrologic models in the manner described involves several assumptions. Solar insolation at the top of and through the atmosphere on a clear day are assumed to be unchanged under the changed climate, modified only by cloud cover changes. Over-water corrections are made in the same way, albeit with changed meteorology, which presumes that over-water/over-land atmospheric relationships are unchanged. GLERL's procedure for transferring information from the GCM grid to their spatial data is an objective approach but simple in concept. It ignores interdependencies in the various meteorologic variables as all are averaged in the same manner. Of secondary importance, the spatial averaging of meteorologic values over a box centered on the GCM grid point (implicit in the use of the nearest grid point to each square kilometer of interest) filters all variability that exists in the GCM output over that box. If GCM output was interpolated between these point values, then at least some of the spatial variability might be preserved. The interpolation performed by *Louie* (1991) from the original GCM grid to a finer grid reduced this problem, but it still exists in the use of the finer grid with the hydrology models. Of course, little is known about the validity of various spatial interpolation schemes and, for highly variable spatial data, they may be inappropriate. However, the same is true for the spatial averaging that was used to supply the GCM results for this study.

Steady-state behavior, in all aspects of the hydrological cycle, are exemplified here in figures for the Lake Superior basin and summarized for all lakes and all climate-change scenarios for the entire period in tables.

4.2 Basin Meteorology

The annual cycles, of all meteorologic variables, were averaged over the 1951-88 period and inspected. The $2xCO_2$ climate air temperatures are higher throughout the annual cycle than the $1xCO_2$ climate (base case); the difference is smallest during the late fall to early winter and largest during the late winter to early spring for all lakes; see Figure 12. The difference is smallest and largest (most variability in the seasonal cycle) for the southern-most lakes. The average annual air temperatures are 4.4-6.1°C higher, depending on the basin; see Table 7. The $2xCO_2$ climate precipitation is generally higher during the spring and lower during the fall and winter than the $1xCO_2$ climate precipitation over all of the Great Lake basins, although generally lower to the south; see Figure 12. The average steady-state annual precipitation is 8% higher over the Superior basin to 10% lower over the Erie basin with a fairly smooth change with longitude; see Table 7. Precipitation changes and air temperature changes are both fairly consistent with longitude as illustrated in Table 7.

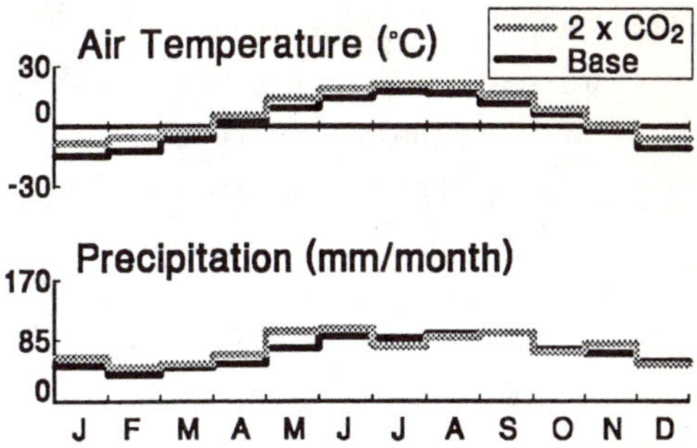

Figure 12. Over-Basin Meteorology Changes.

4.3 Basin Hydrology

The resulting average annual steady-state evapotranspiration from the land portion of the basins is higher for the $2xCO_2$ climate in all lake basins, with a fairly smooth change with longitude from 26% higher over the Superior

Table 7. Average Annual Steady-State Basin Hydrology Differences.

Basin	Air Temperatures and Absolute Differences			Precipitation[a] and Relative Changes		
	$1xCO_2$	$2xCO_2$	Diff.	$1xCO_2$	$2xCO_2$	Diff.
Superior	2.4°	6.8°	4.4°	817 mm	880 mm	8 %
Michigan	7.2°	12.8°	5.6°	825 mm	797 mm	-3 %
Huron	5.4°	10.4°	5.0°	870 mm	852 mm	-2 %
St. Clair	8.3°	14.4°	6.1°	849 mm	772 mm	-9 %
Erie	9.1°	15.1°	6.0°	905 mm	817 mm	-10 %
Ontario	7.2°	12.2°	5.0°	930 mm	879 mm	-6 %

Basin	Evapotranspiration[a] and Relative Changes			Runoff[a] and Relative Changes		
	$1xCO_2$	$2xCO_2$	Diff.	$1xCO_2$	$2xCO_2$	Diff.
Superior	423 mm	534 mm	26 %	394 mm	346 mm	-12 %
Michigan	507 mm	600 mm	18 %	317 mm	196 mm	38 %
Huron	493 mm	608 mm	24 %	377 mm	243 mm	-36 %
St. Clair	535 mm	632 mm	18 %	315 mm	140 mm	-56 %
Erie	565 mm	659 mm	17 %	341 mm	158 mm	-54 %
Ontario	472 mm	575 mm	22 %	459 mm	304 mm	-34 %

[a]Expressed as a depth over the land portion of the basin.

basin to 17% higher over the Erie basin; see Table 7. However, over the seasonal cycle, $2xCO_2$ evapotranspiration exceeds the base case most often in the late spring to early summer (late on Lake Superior basin) and is actually smaller in the early fall, see Figure 13. Runoff from the land portion of the basin is reduced by the $2xCO_2$ climate in all basins, changing from only 12% lower over the Superior basin to 56% lower over the St. Clair basin in a fairly smooth variation with latitude; see Table 7. The average annual cycle of runoff, depicted in Figure 13, has changed as well; runoff peaks slightly earlier and with smaller magnitude under the $2xCO_2$ climate than under the $1xCO_2$ climate. This results largely from big changes in snowpack accumulation and ablation, and in other moisture storages.

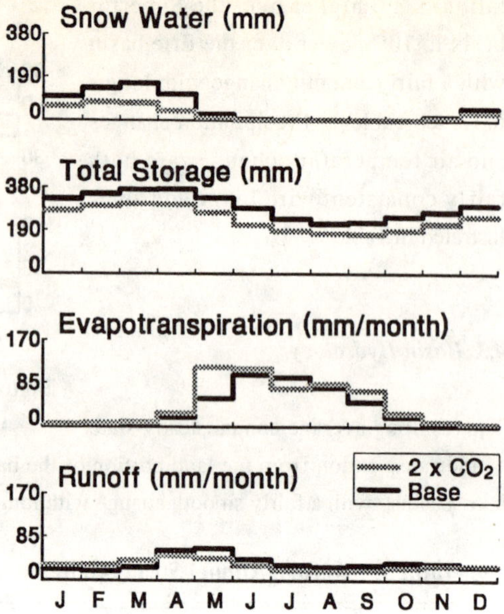

Figure 13. Over-Basin Hydrology Changes.

On the Superior basin, the average steady-state snowpack storage is reduced by more than half; on the other basins, more to the south, the snowpack is almost entirely absent under the $2xCO_2$ climate; see Figure 13 and Table 8. This reduction in snowpack accumulation results from the higher air temperatures, especially during the winter, that accompany the changed climate. The snow season is shortened more than one month. The effects on the snowpack are felt throughout the basin in terms of the derived moisture storages in the soil zone, groundwater, and surface zones. Figure 13 illustrates the general impact on all Great Lake basins of generally lower moisture storages that peak earlier in the $2xCO_2$ climate than in the $1xCO_2$ climate scenarios. This general lowering of moisture in storage in each of the basins is summarized in Table 8 and in some cases represents greater than a 50% reduction in available moisture (see "Total Basin Storage" column).

4.4 Over-Water Meteorology

The over-lake air temperature, humidity, and wind speed differs from over-land since the lower atmospheric layer is affected by the water surface over which it lies. The model correc-

Table 8. Average Annual Steady-State Basin Storage Differences.

Basin	Snow Water Equivalent[a] and Relative Changes			Soil Moisture[a] and Relative Changes		
	$1xCO_2$	$2xCO_2$	Diff.	$1xCO_2$	$2xCO_2$	Diff.
Superior	50 mm	24 mm	-51 %	42 mm	36 mm	-14 %
Michigan	12 mm	2 mm	-87 %	35 mm	22 mm	-37 %
Huron	28 mm	6 mm	-79 %	54 mm	40 mm	-26 %
St. Clair	9 mm	1 mm	-91 %	6 mm	2 mm	-67 %
Erie	6 mm	1 mm	-90 %	7 mm	3 mm	-63 %
Ontario	16 mm	2 mm	-85 %	21 mm	14 mm	-31 %

Basin	Groundwater Moisture[a] and Relative Changes			Total Basin Storage[a] and Relative Changes		
	$1xCO_2$	$2xCO_2$	Diff.	$1xCO_2$	$2xCO_2$	Diff.
Superior	146 mm	124 mm	-15 %	295 mm	237 mm	-20 %
Michigan	61 mm	41 mm	-33 %	114 mm	68 mm	-40 %
Huron	8 mm	5 mm	-39 %	99 mm	57 mm	-43 %
St. Clair	10 mm	5 mm	-51 %	28 mm	9 mm	-67 %
Erie	9 mm	4 mm	-52 %	24 mm	8 mm	-65 %
Ontario	11 mm	7 mm	-36 %	61 mm	33 mm	-46 %

[a]Expressed as a depth over the land portion of the basin.

tions to over-land meteorology observations for over-water conditions depend heavily on the water surface temperature which in turn is a function of the over-lake meteorology and heat balance at the surface of the lake. Figure 14 illustrates the $2xCO_2$ and the base case annual cycles for Superior over-lake meteorology (air temperature, absolute humidity, and wind speed) and Figure 15 illustrates water temperature, evaporation, and net basin supply. In general, the synergistic relationship that exists between air and water temperature in the $2xCO_2$ scenario yields a general increase in both that follows the $1xCO_2$ climate patterns, similar to over-land behavior in Figures 14 and 15. The most pronounced increase in both occurs in the summer for Lake Superior. Table 9 shows that the average steady-state air temperature difference between the $2xCO_2$ and base cases varies from 5.3°C on Lakes Superior and Huron to 5.9°C on Lake Michigan. Variations in the impact with latitude or longitude or size of the lake are not pronounced, in terms of volume or heat capacity. Relative humidity over the lakes is increased, probably due to the increased lake evaporation, and cloud cover generally has decreased slightly for the $2xCO_2$ climate; see Figure 14 and Table 9. Again, the difference is most pronounced in the summer. Over-water wind speed is not greatly affected after correction of over-land values for over-water conditions at increased water temperatures; Figure 14 and Table 9 show only slight decreases for each lake in average steady-state wind speed.

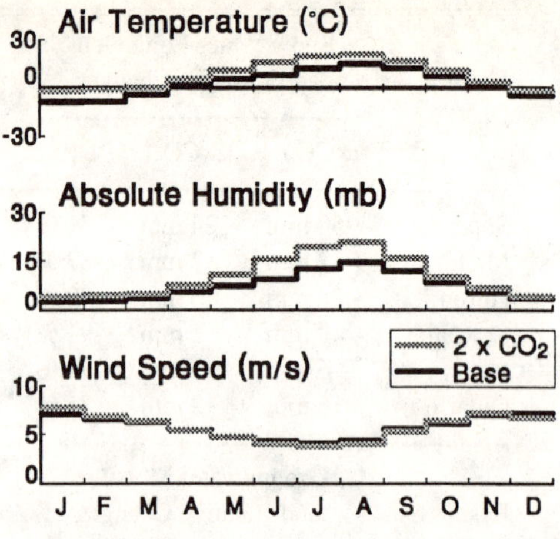

Figure 14. Over-Lake Meteorology Changes.

4.5 Lake Heat Balance

The heat budget gives rise to increased water surface temperatures as seen in Figure 15 and summarized in Table 10. The average steady-state increase in water surface temperatures for the $2xCO_2$ scenario range from 4.8°C on Lake St. Clair to 5.6°C on Lake Michigan. The heat storage capacity of a lake influences the increase in water surface temperatures that can almost be seen in Figure 15. Water surface temperatures are seen to peak earlier on deep lakes under the $2xCO_2$ climate than under the $1xCO_2$ climate. Large amounts of heat now reside in the deep lakes throughout the year, increasing latent and sensible transfers to the atmosphere; see Figure 15. The increased heat in storage also means that ice formation will be greatly reduced over winter on the deep Great Lakes. The higher water surface temperatures under the $2xCO_2$ climate result in increased annual lake evaporation of about 31-33%; see Table 10. The increased heat storage also changes the temperature-depth profiles within the lakes.

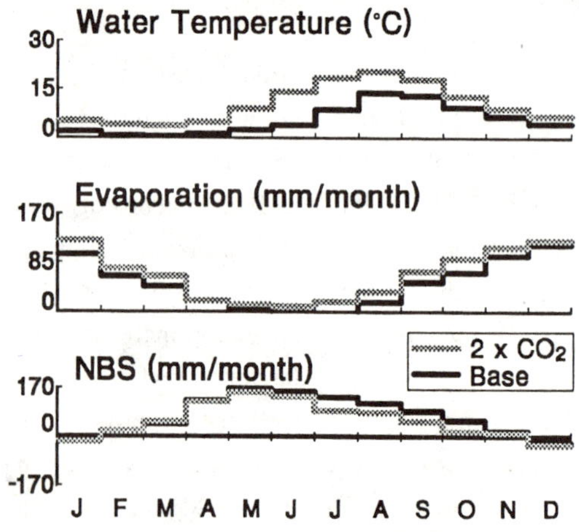

Figure 15. Over-Lake Hydrology Changes.

Some of the deep lakes (Michigan, Huron, and Ontario) show water surface temperatures that stay above 3.98°C throughout the average annual

Table 9. Average Annual Steady-State Over-Lake Meteorology Differences.

Basin	Air Temperatures and Absolute Differences			Absolute Humidity and Absolute Differences		
	$1xCO_2$	$2xCO_2$	Diff.	$1xCO_2$	$2xCO_2$	Diff.
Superior	3.1°	8.4°	5.3°	7.2 mb	10.3 mb	3.1 mb
Michigan	7.7°	13.6°	5.9°	9.7 mb	13.3 mb	3.6 mb
Huron	6.5°	11.8°	5.3°	9.0 mb	12.4 mb	3.4 mb
St. Clair	10.3°	16.0°	5.7°	11.2 mb	15.1 mb	3.9 mb
Erie	9.7°	15.3°	5.6°	10.9 mb	14.7 mb	3.8 mb
Ontario	8.0°	13.4°	5.4°	9.8 mb	13.6 mb	3.8 mb
Basin	**Cloud Cover and Relative Changes**			**Wind Speed and Relative Changes**		
	$1xCO_2$	$2xCO_2$	Diff.	$1xCO_2$	$2xCO_2$	Diff.
Superior	0.57	0.58	2 %	5.7 m/s	5.7 m/s	0 %
Michigan	0.43	0.42	-2 %	6.1 m/s	5.9 m/s	-3 %
Huron	0.55	0.55	-1 %	6.0 m/s	5.9 m/s	-2 %
St. Clair	0.50	0.47	-4 %	5.7 m/s	5.6 m/s	-2 %
Erie	0.63	0.61	-3 %	6.3 m/s	6.1 m/s	-3 %
Ontario	0.59	0.58	-1 %	6.1 m/s	6.0 m/s	-2 %

Table 10. Selected Average Annual Steady-State Hydrology Differences.

Basin	Water Temperature and Absolute Differences			Over-Lake Evaporation[a] and Relative Changes		
	$1xCO_2$	$2xCO_2$	Diff.	$1xCO_2$	$2xCO_2$	Diff.
Superior	5.4°	10.5°	5.1°	561 mm	736 mm	31 %
Michigan	8.5°	14.1°	5.6°	647 mm	854 mm	32 %
Huron	8.0°	13.0°	5.0°	627 mm	829 mm	32 %
St. Clair	1.0°	15.8°	4.8°	936 mm	1234 mm	32 %
Erie	10.9°	15.8°	4.9°	898 mm	1197 mm	33 %
Ontario	9.0°	14.4°	5.4°	665 mm	874 mm	31 %

[a]Expressed as depths over the lake.

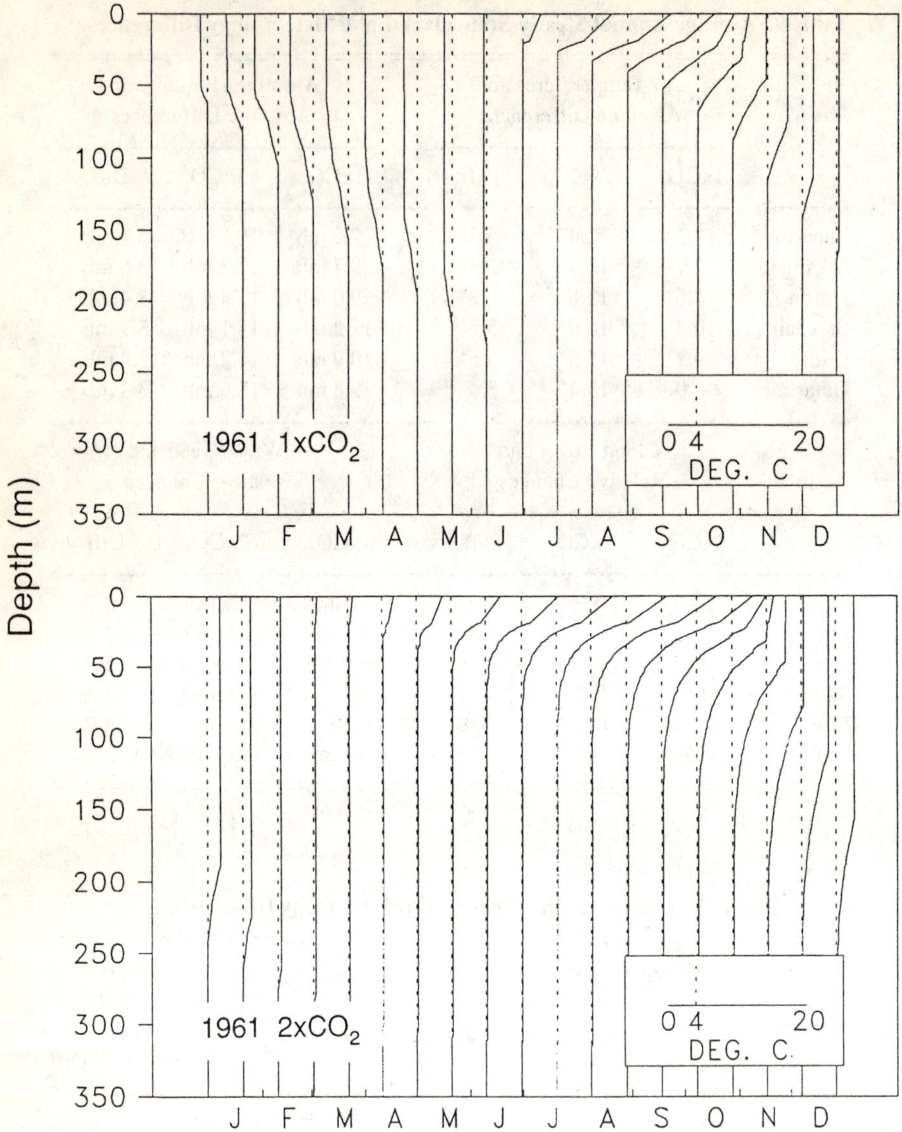

Figure 16. Steady-State Lake Superior Temperature-Depth Profiles.

cycle. Figure 16 illustrates this for 1961 on Lake Superior. This means that buoyancy-driven turnovers of the water column do not occur in the same way as they do at present.

In some years, the large lakes are changed from dimictic lakes (turnovers occur twice a year as water temperatures pass through the point of maximum density, 3.98°C) to monomictic lakes (maximum turnover occurs at the temperature "reversal" where temperatures stop declining and start rising again and the minimum temperature is greater than 3.98°). Figure 16 shows that the 1xCO_2 temperature profile for Lake Superior passed through 3.98°C in June 1961 and approached, in December, the January 1962 transition. Under the 2xCO_2 scenario, temperatures remain above 3.98°C but approach a vertical profile most in March. This represents a change from dimictic to monomictic.

Table 11 shows that the large lakes remain dimictic under the 2xCO_2 climate only between 2% and 76% of the time. The largest change is associated with Lake Ontario which is the furthest south of the deep lakes. Least effected are Lakes Erie and St. Clair which are very shallow and have relatively little heat storage. As the lakes move to one reversal per year in some years, instead of two turnovers per year, the interarrival times of the maximum mixing extent increase. Table 11 illustrates that the average interarrival time grows to nearly a full year on Lake Ontario since only 2% of the years have dimictic behavior. Table 11 also illustrates the monomictic reversal temperature is, of course, well above the point of maximum water density.

The timing of maximum turnovers or temperature reversals shifts. Table 12 shows the time between the spring turnover and the fall turnover (for dimictic behavior) increases. The spring turnover occurs earlier and the fall turnover occurs later in the annual cycle. For monomictic behavior, the single maximum turnover occurs even earlier in the year than the dimictic turnovers. These are consequences of greater heat storage in, and heat inputs to, the lakes.

Temperature-depth profiles for every day of a single model year can be combined and depicted as depth-time plots of temperature isolines; see Figure 17 for an example on Lake Superior. Then, not only are the turnover timing changes depicted between 1xCO_2 and 2xCO_2 climates, but depth changes are more apparent as well. Table 12 also summarizes the maximum depths at turnover in the lakes. Dimictic spring turnovers exhibit shallower average depths under 2xCO_2 conditions and fall turnovers are deeper, where not limited by the depth of the lake. Monomictic turnovers are generally even deeper.

There is a normal hysteresis observed in graphs of lake heat plotted with surface temperature, such as in Figure 18. This reflects the mixing of heat at depth. Surface temperatures rise quickly and heat storage follows after the spring turnover. When surface temperatures then begin to drop in the fall, stored heat does not initially. Then heat storage drops more slowly. Similar behavior occurs after the fall turnover, and both result in the characteristic double "loop" in the plot. Under the warmer climate change scenario, temperatures sometimes never drop below that at maximum density (3.98°C). This results in only one hysteresis loop, but it is much larger.

Table 11. Average Characteristics of Turnovers/Reversals

	Fraction Dimictic		Interarrival Times		$2xCO_2$ Monomictic Reversal
	$1xCO_2$	$2xCO_2$	$1xCO_2$	$2xCO_2$	Water Temperature
Superior	100 %	67 %	182 d	211 d	4.5 °C
Michigan	100 %	15 %	182 d	318 d	4.9 °C
Huron	100 %	24 %	182 d	292 d	4.9 °C
St. Clair	100 %	100 %	183 d	185 d	
Erie	100 %	76 %	183 d	206 d	4.9 °C
Ontario	100 %	2 %	182 d	356 d	5.9 °C

Table 12. Average Dates and Depths of Maximum Turnover or Temperature Reversal.

Basin	$1xCO_2$		$2xCO_2$		
	Dimictic		Dimictic		Monomictic
	Spring	Fall	Spring	Fall	
			DATES		
Superior	02 Jul	23 Dec	24 Apr	11 Feb	24 Mar
Michigan	27 May	01 Jan	27 Mar	28 Jan	22 Feb
Huron	26 May	10 Jan	26 Mar	11 Feb	10 Mar
St. Clair	30 Apr	20 Nov	04 Mar	28 Nov	
Erie	30 Apr	24 Dec	03 Mar	06 Jan	01 Feb
Ontario	20 May	18 Jan	10 Mar	29 Jan	03 Mar
			DEPTHS		
Superior	234 m	162 m	127 m	257 m	297 m
Michigan	132 m	111 m	52 m	199 m	231 m
Huron	229 m[a]	229 m[a]	140 m	229 m[a]	229 m[a]
St. Clair	6 m[a]	6 m[a]	6 m[a]	6 m[a]	
Erie	64 m[a]	64 m[a]	48 m	64 m[a]	53 m
Ontario	232 m	242 m	70 m	244 m[a]	244 m[a]

[a]Maximum average depth of the lake.

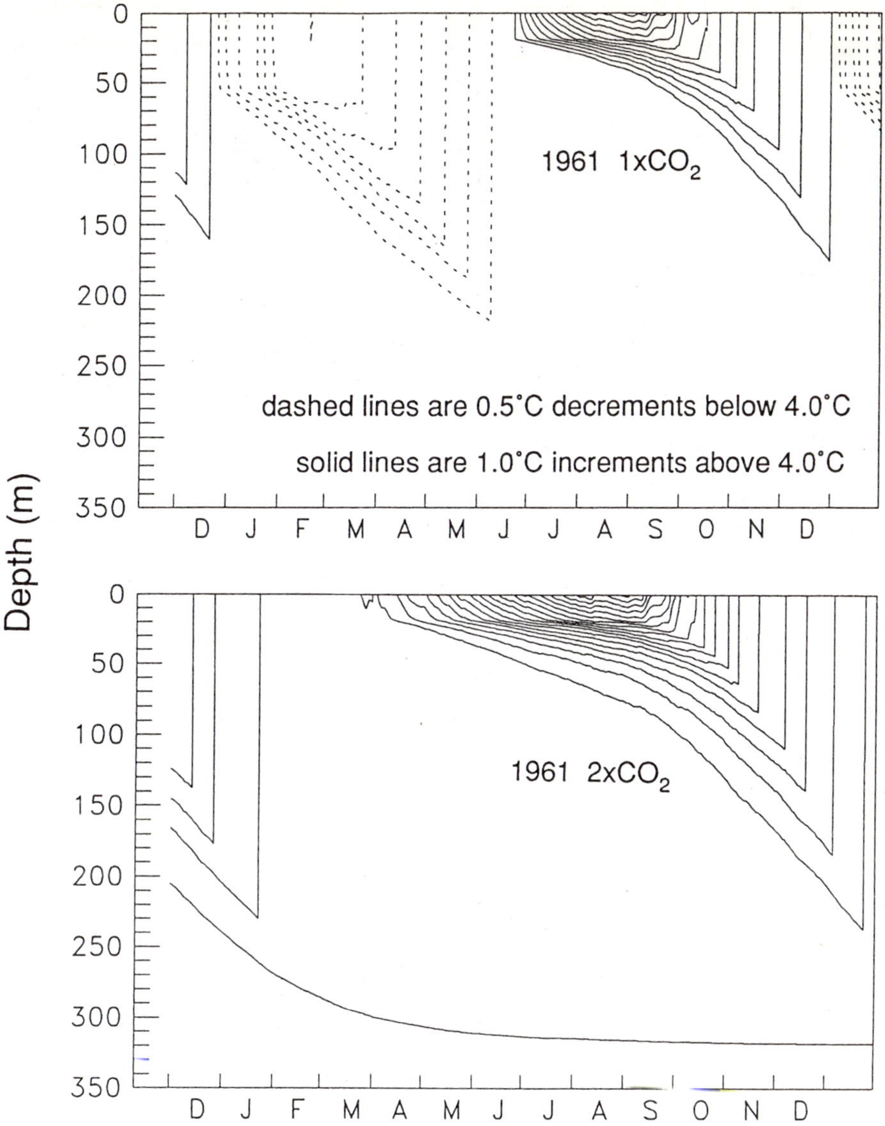

Figure 17. Steady-State Lake Superior Depth-Time Temperature Isolines.

4.6 Net Supply Components

Over-lake precipitation, runoff, and lake evaporation sum algebraically as the net basin supply and are presented again in Table 13 for convenience. Since over-lake precipitation is taken

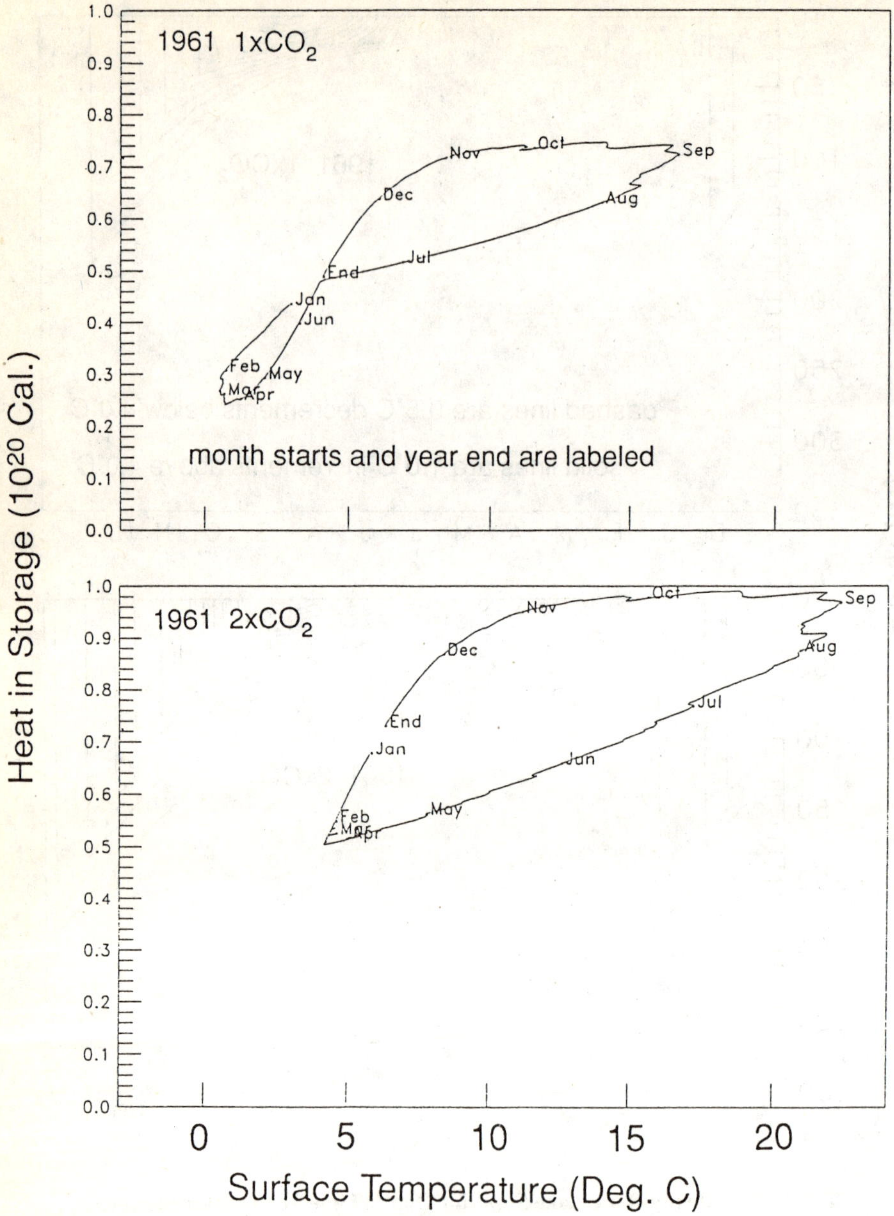

Fig. 18. Steady-State Lake Superior Heat-Temperature Hysteresis.

Table 13. Average Annual Steady-State Net Supply Components Differences.

Basin	Over-Lake Precipitation[a] and Absolute Differences			Basin Runoff[a] and Absolute Differences		
	$1xCO_2$	$2xCO_2$	Diff.	$1xCO_2$	$2xCO_2$	Diff.
Superior	817 mm	880 mm	63 mm	615 mm	539 mm	-75 mm
Michigan	825 mm	797 mm	-28 mm	635 mm	393 mm	-242 mm
Huron	869 mm	852 mm	-18 mm	836 mm	539 mm	-297 mm
St. Clair	849 mm	772 mm	-77 mm	4453 mm	1980 mm	-2474 mm
Erie	905 mm	817 mm	-88 mm	803 mm	372 mm	-431 mm
Ontario	930 mm	879 mm	-51 mm	1694 mm	1123 mm	-571 mm
Basin	**Over-Lake Evaporation[a] and Absolute Differences**			**Net Basin Supply[a] and Relative Changes**		
	$1xCO_2$	$2xCO_2$	Diff.	$1xCO_2$	$2xCO_2$	Diff.
Superior	561 mm	736 mm	175 mm	871 mm	684 mm	-21 %
Michigan	647 mm	854 mm	207 mm	813 mm	336 mm	-59 %
Huron	627 mm	829 mm	202 mm	1079 mm	562 mm	-48 %
St. Clair	936 mm	1234 mm	298 mm	4367 mm	1517 mm	-65 %
Erie	898 mm	1197 mm	299 mm	810 mm	-8 mm	-101 %
Ontario	665 mm	874 mm	209 mm	1959 mm	1127 mm	-42 %

[a]Expressed as depths over the lake.

here as the same as over-land, Table 13 shows the same relations for $2xCO_2$ vs. base case precipitation as does Table 8. Net basin supply in Table 13 is less under the $2xCO_2$ climate than under the $1xCO_2$ climate on all lakes. This is true throughout the year for Lake Erie and nearly true on the other lakes (see Figure 15). Net basin supplies are lower throughout the annual cycle except for February and March on Lake Superior, February on Lakes Michigan and Huron, January on Lake St. Clair, and January and February on Lake Ontario. Table 14 summarizes the changes in the hydrologic and net basin supply components for the entire Great Lakes basin; they were computed by converting the equivalent depths of Table 13 to annual flow rates on each lake and adding them over all the lakes. The changes from the base case are also expressed relatively in Table 14. Also expressed relatively are changes from other studies that used other GCMs (*Croley* 1990); they are provided for comparison. Net basin supplies to all Great Lakes are seen to drop to about one half under the CCC GCM; this corresponds to the GCM from the Geophysical Fluid Dynamics Laboratory (GFDL) in the earlier studies. This drop in net basin supply seems to result from the increases in lake evapo-

Table 14. Average Annual Steady-State Great Lakes Basin Hydrology and Net Basin Supply Components.

Scenario	Over Land Precipitation (cms)	Evapotranspiration (cms)	Basin Runoff (cms)	Over Lake Precipitation (cms)	Over Lake Evaporation (cms)	Net Basin Supply (cms)
$1xCO_2$	13815	7825	5987	6604	4992	7783
$2xCO_2$[a]	13598	9518	4077	6578	6587	4180
CCC[a]	-2%	22%	-32%	0%	32%	-46%
GISS[b]	2%	21%	-24%	4%	27%	-37%
GFDL[c]	1%	19%	-23%	0%	44%	-51%
OSU[d]	6%	19%	-11%	6%	26%	-23%

[a]Canadian Climate Centre, prepared by *Louie* (1991).
[b]Goddard Institute for Space Studies GCM, used by *Croley* (1990).
[c]Geophysical Fluid Dynamics Laboratory GCM, used by *Croley* (1990).
[d]Oregon State University GCM, used by *Croley* (1990).

ration and overland evapotranspiration (reducing subsequent runoff to the lakes) observed in the $2xCO_2$ scenario from the GCM.

4.7 Sensitivities

Without temperatures below freezing, the snowpack is insensitive to precipitation. Although the steady-state scenarios on different lakes show different estimates of precipitation change, each shows increases in air temperatures that significantly reduce the snowpack, especially in the southern basins. Thus, even if precipitation increases more than suggested by the GCM, the snowpack will be much reduced under warmer winters. Similarly, regardless of actual changes in precipitation, the Great Lakes basin will experience reduced soil moisture storage and runoff. Both peak shortly after snow melt and then drop throughout the summer and fall due to high evapotranspiration; each climate scenario produces earlier snow melt and a longer period of evapotranspiration. Soil moisture and runoff are most sensitive to precipitation in midsummer when at annual minimums. Thus, within the limits of precipitation produced by the GCMs, soil moisture and runoff scenarios are relatively insensitive to precipitation.

Of the meteorological variables that affect lake evaporation (air temperature, humidity, cloud cover, and wind speed) wind speed is probably the most critical, although air temperature and humidity are also important. Across all lakes and scenarios, daily evaporation was reduced (compared to the base case) only when the scenarios showed reduced wind speeds. If wind speeds remain near historical levels, evaporation will still increase somewhat, however, due simply to the increase in air temperatures that then increase water temperatures. Thus, within the range of other meteorologic variables shown by the GCMs, only if wind speeds are less than historical levels by about 0.5% will lake evaporation not increase.

Because net basin supplies are a sum of lake evaporation, runoff, and precipitation, they are equally sensitive to changes in any of the components. Thus, as long as wind speeds are not much less than historical levels, regardless of precipitation changes (unless precipitation increases are much larger than any shown by the GCMs), net basin supplies are likely to drop due to higher air temperatures that increase evaporation and decrease runoff.

4.8 Summary Comments on Great Lakes Response to Climate Change

The study results should be received with caution as they are, of course, dependent on the GCM outputs, which have large uncertainties. The linkage method used here does not recognize interdependencies between meteorological variables. It also simply changes the magnitude of meteorological time series without affecting their temporal structures. Therefore, changes in variabilities that would take place under a changed climate are not addressed. Seasonal timing differences in the GCM for the changed climate are not reproduced with this method of coupling. Instead, while seasonal meteorology patterns are preserved in the 2xCO_2 scenario as they exist in the 1xCO_2 historical data, one still can observe seasonal changes induced by storage effects. Water temperatures increase and peak earlier; heat resident in the deep lakes increases throughout the year. Mixing of the water column diminishes, as most of the lakes become mostly monomictic, and lake evaporation increases. Changes in annual variability are less clear, again as a result of using the same historical time structure for both the base case and the changed climate scenarios.

The higher air temperatures under the 2xCO_2 scenarios lead to higher over-land evapotranspiration and lower runoff to the lakes with earlier runoff peaks since snowpack is reduced up to 100% and the snow season is shortened more than one month. This also results in a reduction in available soil moisture. Water surface temperatures peak earlier and are higher, with larger amounts of heat resident in the deep lakes throughout the year. Also, buoyancy-driven turnovers of the water column do not occur as often on all lakes except St. Clair. Without biannual turnovers, hypolimnion chemistry may be altered; oxygen may be depleted, releasing nutrients and metals from lake sediments. The lakes may experience more than a single winter turnover if temperature gradients are small and winds are strong enough to induce mixing (*Hutchinson* 1957). Ice formation will be greatly reduced over winter on the deep Great Lakes, and lake evaporation will increase; average steady-state net supplies drop.

5. Coupled Hydrosphere Atmosphere Research Model (CHARM)

The linkage between a GCM and hydrology models allows no feedbacks between these independent models. While the GCMs have crude hydrologic process models, they represent inappropriately large scales and use very simplified conceptualizations. The regional hydrologic impact models may do a much better job of representing the hydrology of an area. However, their use with GCM outputs does not allow the GCM simulations to benefit from these refined processes. Feedback from land and lake surfaces' hydrometeorological properties cannot exist without incorporating regional hydrology models into atmospheric models.

Modelers are turning their attention to mesoscale atmospheric models to enable better assessment of local to regional effects. The leading approach now is to embed mesoscale atmospheric models within GCMs for a region of interest and to couple relevant surface hydrology models to the mesoscale atmospheric model (*Dickinson et al.* 1989; *Giorgi* 1990; *Hostetler et al.* 1993). This allows both the use of more relevant scales for regional impact estimation and the consideration of dynamic linkages between the atmosphere and the surface, now recognized as essential in describing the hydrology and meteorology of an area. This approach has generally been limited in the past to 50-km grids or larger because of the complexity of the modeling system that is required and because of the computer power that was required. The science panel of the GEWEX Continental-Scale International Project and the WMO-CAS Working Group on Numerical Experimentation launched their joint Project for Intercomparison of Land-Surface Parameterization Schemes (PILPS). The National Center for Atmospheric Research (NCAR) is exploring possibilities of operating their atmospheric, hydrologic, and lake flux models embedded in their GCM at scales finer than 50 km.

To estimate impacts associated with both large and fine scales of parameter changes, the Great Lakes research community can address these scales separately. This offers the advantage that we can begin <u>now</u> to look at large-scale parameter changes (such as lake levels, lake-wide heat storage, and annual and monthly water and energy balances) by combining *existing* process models appropriate to these scales. This can be underway while fine-scale parameter changes are investigated. They will require more development and integration of process models. Thus, we have two components to physical modeling of climate-change impacts over the Great Lakes. The first is the integration and use of existing Great Lakes hydrologic process models (lumped-parameter, applying to irregular-shaped areas over spatial scales of 30-100 km for the land surface and 100-300 km for the lake surface and temporal scales of 1-30 days). The second is the development and integration of fine-scale second-generation (gridded) surface hydrologic process models (at scales from 1-30 km or for simulations of many years) with atmospheric mesoscale models.

5.1 Large-Scale Parameter Changes

We must explore linkages to atmospheric models for existing large-scale irregular-area surface models that already represent excellent portrayals of the hydrology and lake thermodynamics in the Great Lakes. Since hydrological models now exist for large-scale parameter change estimates, large-scale couplings will be useful in beginning derivative studies (such as socio-economic, food-web dynamics, and other secondary impacts identified as dependent on large-scale parameter changes). They also will prove useful as a starting point for subsequent second-generation joint atmospheric-hydrological parameterizations and in the verification of same and of like developments by other investigators. They also will be useful as a base-line for comparing multiple approaches in modeling the atmosphere and hydrology.

GLERL, in cooperation with the Air Resources Laboratory (ARL), is now linking their hydrology models with the Regional Atmospheric Modeling System or "RAMS" (*Pielke et al.* 1990; *Lyons et al.* 1990, 1991a,b). The combination will be used for large-scale parameter investigations, requiring assessment of the temporal and spatial incompatibilities that exist between mesoscale meteorological and regional hydrology models. A modest target is to arrange for coupled modeling by using a 40-km grid, with time steps of 90 seconds in the atmospheric component, coupled to some components of the surface models defined over irregular areas on 12- to 24-hour intervals. RAMS-predicted atmospheric momentum, temperature, moisture, and precipitation fields will be input to the large-scale hydrological models which will use these fields to update sea surface temperature, soil moisture, and snowpack variables. These hydrological parameters will then be input into RAMS to drive the surface energy fluxes over both land and water. Since there is some overlap in function between parts of the atmospheric model and the surface models, decisions are required about which model should be used for some purposes; this is discussed elsewhere (*Croley and Lofgren* 1994).

5.2 Second-Generation Fine-Scale Atmospheric-Hydrologic Integrations

Only when sufficiently fine grids become available for surface hydrology models will surface runoff at points into the lakes be directly estimable from purely gridded models. These fine grids will be approached in the next few years. Likewise, lake heat storage models for the

Great Lakes exist at several levels, from one-dimensional superposition models to three-dimensional circulation models. Again, researchers are approaching fine grids that are usable in long continuous simulations.

Two fine-scale approaches are possible now. The first uses developing atmospheric-hydrologic mesoscale models to estimate joint meteorology and hydrology for surface areas of interest in the Great Lakes and then refines the hydrological estimates through use of the better-calibrated GLERL hydrology models for the Great Lakes. This approach is similar to that taken in linking hydrology models to GCM outputs, described previously. Again, there is no dynamic interaction between the final hydrology models and the atmospheric model. Outputs from the joint atmospheric-hydrologic mesoscale model are inputs to the hydrology models. However, better agreement should be possible since the scales of both sets of models are closer than was true in the GCM-hydrology model studies.

The second fine-scale approach consists of developing second-generation fine-scale Great Lake hydrologic and lake thermodynamic models on finer grids to interface directly with atmospheric models applied at ever-finer resolution and of assessing the importance of two-way runoff-atmospheric interactions unique to CHARM. These will complement similar efforts elsewhere (NCAR) that use alternate models. The matching of spatial and temporal scales between models will proceed at different levels. Linkage will begin with coarse irregular spatial and temporal scales, where existing hydrological models are established over large areas in the Great Lakes (as in the above section), and proceed to finer scales as hydrological models are redeveloped in atmospheric-hydrologic studies. Comparisons will be made between scales to see what is resolved and which process refinements make no difference with regard to different uses (water level estimation, sea breeze predictions, and so forth). Both the atmospheric and hydrological models will be run in three dimensions on the same grid. The grid spacings will be reduced from 30 km to 15, 10, 5, and 1 km scales. For the smaller scales, non-hydrostatic physics and explicit cloud microphysics will be employed. To start out, interactions will be performed at the time step of the atmospheric model (between 5-90 seconds depending on the horizontal resolution of the grids). Sensitivity experiments will be performed to determine an optimum update frequency between the atmospheric and hydrologic models since it may not be necessary to interact the models every time step.

5.3 Summary Comments on CHARM

Earlier assessments used atmospheric GCM outputs as meteorologic scenarios to drive process models for generating hydrologic scenarios. Climate change effects were inferred by comparing process model outputs for a base case with the changed climate scenario. As the linkage methods of these assessments constrained spatial and temporal meteorologic variabilities to those present in the historical records, impact assessments began with the transference of existing climates to the Great Lakes. Lack of feedback between surface process models and atmospheric models is still a problem.

Researchers are now developing and verifying multi-scale hydrologic models, with appropriate links to mesoscale atmospheric models, using spatially extensive observations based upon satellite and *in situ* measurements and supported by field experiments. These linked models are slated to be embedded in GCM or other boundary condition simulations to assess climate change effects. GLERL is working with ARL to investigate alternative CHARM possibilities. Now underway are a large-scale coupling, that employs GLERL's existing irregular-area surface models, and a series of finer-scale couplings where surface models are defined over the same (surface) grid as used in the mesoscale atmospheric models. GLERL also plans to work with existing and developing coupled atmospheric-hydrologic mesoscale models over the Great Lakes by refining hydrologic estimates with more-detailed hydrologic and lake surface flux models.

6. Acknowledgments

This paper is GLERL contribution no. 884.

7. References

Assel RA, Quinn FH, Leshkevich GA, Bolsenga SJ (1983) *NOAA Great Lakes Ice Atlas*. Great Lakes Environmental Research Laboratory, Ann Arbor

Atmospheric Environment Service (1988) Monthly and annual evaporation from the Great Lakes bordering Canada. Environment Canada, Atmospheric Environmental Service, Downsview

Blumberg AF, DiToro DM (1989) The effects of climate warming on Lake Erie water quality. In *The Potential Effects of Global Climate Change on the United States*, JB Smith and DA Tirpak (eds), U. S. Environmental Protection Agency report no. EPA-230-05-89-051, 7-1 - 7-28

Bolsenga SJ (1977) Lake-land precipitation relationships using Northern Lake Michigan data. *Journal of Applied Meteorology*, **16**(11):1158-1164

Bolsenga SJ (1979) Determining overwater precipitation from overland data: the methodological controversy analyzed. *Journal of Great Lakes Research*, **5**(3-4):301-311

Bolsenga SJ, Hagman JC (1975) On the selection of representative stations for Thiessen polygon networks to estimate Lake Ontario overwater precipitation. *International Field Year for the Great Lakes Bulletin 16*, National Oceanic and Atmospheric Administration, Rockville, 57-62

Campbell JE, Clites AH, Green GM (1987) Measurements of ice motion in Lake Erie using satellite-tracked drifter buoys. *NOAA Data Report ERL GLERL-30*, Great Lakes Environmental Research Laboratory, Ann Arbor

Cohen SJ (1991) Caveats in applying GCM output to regional impact assessments. *Proceedings of A Symposium on Climate Scenarios*, The Water Network, University of Waterloo, Waterloo, 16-32

Coordinating Committee on Great Lakes Basic Hydraulic and Hydrologic Data (1977) *Coordinated Great Lakes physical data*. U. S. Army Corps of Engineers, Detroit

Croley TE II (1982) Great Lakes basins runoff modeling. *NOAA Technical Memorandum ERL GLERL-39*, National Technical Information Servicer, Springfield

Croley TE II (1983a) Great Lake basins (U.S.A.-Canada) runoff modeling. *Journal of Hydrology*, **64**:135-158

Croley TE II (1983b) Lake Ontario basin (U.S.A.-Canada) runoff modeling. *Journal of Hydrology*, **66**:101-121

Croley TE II (1989a) Lumped modeling of Laurentian Great Lakes evaporation, heat storage, and energy fluxes for forecasting and simulation. *NOAA Technical Memorandum ERL GLERL-70*, Great Lakes Environmental Research Laboratory, Ann Arbor

Croley TE II (1989b) Verifiable evaporation modeling on the Laurentian Great Lakes. *Water Resources Research*, **25**(5):781-792

Croley TE II (1990) Laurentian Great Lakes double-CO_2 climate change hydrological impacts. *Climate Change*, **17**:27-47

Croley TE II (1991) Use of general circulation model simulations in regional hydrological climate change impact studies. *Proceedings of A Symposium on Climate Scenarios*, The Water Network, University of Waterloo, Waterloo, 33-48

Croley TE II (1992a) Long-term heat storage in the Great Lakes. *Water Resources Research*, **28**(1):69-81

Croley TE II (1992b) Climate change impacts on Great Lakes water supplies. *Proceedings of the Symposium on Managing Water Resources During Global Change*, Twenty-Eighth American Water Resources Association Conference, 241-250

Croley TE II (1993a) CCC GCM 2xCO_2 hydrological impacts on the Great Lakes. In: *Climate, Climate Change, Water Level Forecasting and Frequency Analysis, Supporting Documents, Volume 1 - Water Supply Scenarios*, International Joint Commission, Washington

Croley TE II (1993b) Probabilistic Great Lakes hydrology outlooks. *Water Resources Bulletin*, **29**(5):741-753

Croley TE II, Assel RA (1994) A one-dimensional ice thermodynamics model for the Laurentian Great Lakes. *Water Resources Research*, (in press)

Croley TE II, Hartmann HC (1984) Lake Superior basin runoff modeling. *NOAA Technical Memorandum ERL GLERL-50*, National Technical Information Service, Springfield

Croley TE II, Hartmann HC (1985a) Lake Champlain water supply forecasting. GLERL Open File Report, Contribution No. 450, Great Lakes Environmental Research Laboratory, Ann Arbor

Croley TE II, Hartmann HC (1985b) Resolving Thiessen polygons. *Journal of Hydrology*, **76**:363-379

Croley TE II, Hartmann HC (1986) Near-real-time forecasting of large-lake water supplies; a user's manual. *NOAA Technical Memorandum ERL GLERL-61*, Environmental Research Laboratories, Boulder

Croley TE II, Hartmann HC (1987) Near real-time forecasting of large lake supplies. *Journal of the Water Resources Planning and Management Division*, ASCE, **113**(6):810-823

Croley TE II, Hartmann HC (1989) Effects of climate changes on the Laurentian Great Lakes levels. In *The Potential Effects of Global Climate Change on the United States, Appendix A: Water Resources*, JB Smith and DA Tirpak (eds), U. S. Environmental Protection Agency report no. EPA-230-05-89-051, 4-1 - 4-34

Croley TE II, Lee DH (1993) Evaluation of Great Lakes net basin supply forecasts. *Water Resources Bulletin*, **29**(2):267-282

Croley TE II, Lofgren BM (1994) Coupled hydrologic atmospheric research model (CHARM). *Proceedings of the Symposium on Effects of Human-Induced Changes on Hydrologic Systems*, American Water Resources Association Summer Symposium, Jackson Hole, Wyoming, 26-29 June (in press)

DeCooke BG, Witherspoon DF (1981) Terrestrial water balance. In *IFYGL - The International Field Year for the Great Lakes*, EJ Aubert and TL Richards (eds), Great Lakes Environmental Research Laboratory, Ann Arbor, 210-219

Dickinson RE, Errico RM, Giorgi F, Bates GT (1989) A regional climate model for the western United States. *Climatic Change*, **15**:383-422

Freeman NG, Haras WS (1978) *What you always wanted to know about Great Lakes levels and didn't know whom to ask.* Environment Canada, Ottawa

Giorgi F (1990) On the simulation of regional climate using a limited area model nested in a general circulation model. *J. Climate*, **3**:941-963

Gray DM, McKay GA, Wigham JM (1973) Energy, evaporation, and evapotranspiration. In *Handbook on the Principles of Hydrology*, DM Gray (ed), Water Information Center, New York, 3.1-3.66

Hartmann HC (1987) An evaluation of Great Lakes hydraulic routing models. *NOAA Technical Memorandum ERL GLERL-66*, Great Lakes Environmental Research Laboratory, Ann Arbor

Hartmann HC (1988) Historical basis for limit on Lake Superior water level regulations. *Journal of Great Lakes Research*, **14**(3):316-324

Hartmann HC (1990) Climate change impacts on Laurentian Great Lakes levels. *Climate Change*, **17**:49-67

Hostetler SW, Bates GT, Giorgi F (1993) Interactive coupling of a lake thermal model with a regional climate model. *J. Geophys. Res.*, **98**:5045-5057

Hutchinson GE (1957) *A Treatise on Limnology, Volume I, Part 1 - Geography and Physics of Lakes*, John Wiley and Sons, New York

International Great Lakes Diversions and Consumptives Uses Study Board (1981) *Great Lakes Diversions and Consumptive Uses.* International Joint Commission, Washington

International Great Lakes Diversions and Consumptive Uses Study Board (1985) *Great Lakes Diversions and Consumptive Uses.* International Joint Commission, Washington

International Lake Superior Board of Control (1981) *Regulation of Lake Superior: Plan 1977 Development, Description, and Testing.* International Joint Commission, Washington

International Lake Superior Board of Control (1982) *Regulation of Lake Superior: Operational Guides for Plan 1977.* International Joint Commission, Washington

International St. Lawrence River Board of Control (1963) *Regulation of Lake Ontario: Plan 1958-D.* International Joint Commission, Washington

Irbe JG, Cross RK, Saulesleja, A (1982) Remote sensing of surface water temperature of the Great Lakes and off the Canadian east coast. Northwest Atlantic Fisheries Organization Scientific Council Studies No. 4, Special Session on Remote Sensing, Dartmouth, Canada, 31-39

Keijman JQ (1974) The estimation of the energy balance of a lake from simple weather data. *Boundary-Layer Meteorology* **7**:399-407

Kraus EB, Turner JS (1967) A one-dimensional model of the seasonal thermocline II; the general theory and its consequences. *Tellus*, **19**:98-105

Larsen CE (1985) A stratigraphic study of beach features on the southwestern shore of Lake Michigan: new evidence of holocene lake level fluctuations. Environmental Geology Notes 112, Illinois State Geological Survey, Champaign

Louie PYT (1991) Proposed GCM scenarios for the phase 2 study. *Proceedings of A Symposium on Climate Scenarios*, The Water Network, University of Waterloo, Waterloo, 67-78

Lyons WA, Pielke RA (1990) Predicting 3-D wind flows at Cape Canaveral Air Force Station using a mesoscale numerical model. Final Report NASA SBIR Phase I, 325 pp

Lyons WA, Pielke RA, Cotton WR, Keen CS, Moon DA, Lincoln NR (1991a) Some considerations of the role of the land/lake breezes in causing elevated ozone levels in the southern Lake Michigan region. *Environmental Modelling*, P Melli and P Zannetti (eds), Computational Mechanics Publications, Elsevier Applied Science, 151-170

Lyons WA, Keen CS, Pielke RA, Cotton WR, Moon DA, Lincoln NR (1991b) A prototype real-time forecasting system for mesoscale dispersion in land/lake breezes. *Proc. Third Topical Meeting on Emergency Preparedness and Response*. Chicago, Illinois, 99-102

McCormick MJ (1989) Potential climate changes to the Lake Michigan thermal structure. In: *The Potential Effects of Global Climate Change on the United States*, JB Smith and DA Tirpak (eds), U. S. Environmental Protection Agency report no. EPA-230-05-89-051, 6-1 - 6-26

McFarlane NA (1991) The Canadian Climate Centre second-generation global circulation model. *Proceedings of A Symposium on Climate Scenarios*, The Water Network, University of Waterloo, Waterloo, 3-15

Phillips WD, Irbe JG (1978). Land-to-lake comparison of wind, temperature, and humidity on Lake Ontario during the International Field Year for the Great Lakes (IFYGL). *Report CLI-2-77*, Environment Canada, Atmospheric Environment Service, Downsview

Pielke RA, Lyons WA, McNider RT, Moran MD, Moon DA, Stocker RA, Walko RL, Uliaz M (1990) Regional and mesoscale modeling as applied to air quality models. *Proc. 18th NATO/CCMS International Tech. Meeting on Air Pollution Dispersion Modeling and its Application*, Vancouver, British Columbia, 30 pp

Quinn FH (1978) Hydrologic response model of the North American Great Lakes. *Journal of Hydrology*, **37**:295-307

Quinn FH (1979) An improved aerodynamic evaporation technique for large lakes with application to the International Field Year for the Great Lakes. *Water Resources Research*, **15**(4):935-940

Quinn FH (1981) Secular changes in annual and seasonal Great Lakes precipitation, 1854-1979, and their implications for Great Lakes water resource studies. *Water Resources Research*, **17**:1619-1624

Quinn FH (1982) Trends and extremes of Lake Erie water supplies. *Proceedings International Symposium on Hydrometeorology*, American Water Resources Association, Minneapolis, 267-270

Quinn FH (1985) Great Lakes water levels. In *Great Lakes Water Levels*, Briefing of Senators and Representatives from the Great Lakes Basin July 19, International Joint Commission, Washington, 1-24

Quinn FH, Croley TE II (1981) The role of precipitation climatology in Hydrologic Design and Planning on the Laurentian Great Lakes. *Proceedings Fourth Conference on Hydrometeorology*, American Meteorological Society, Boston, 218-223

Quinn FH, Kelley RN (1983) Great Lakes monthly hydrologic data. *NOAA Technical Memorandum ERL GLERL-39*, National Technical Information Service, Springfield

Quinn FH, Norton DC (1982) Great Lakes precipitation by months, 1900-80. *NOAA Technical Memorandum ERL GLERL-20*, National Technical Information Service, Springfield

Rondy DR (1976) Great Lakes ice cover, Appendix 4, *Great Lakes Basin Framework Study*, Great Lakes Basin Commission, Ann Arbor

U. S. Army Corps of Engineers (1985) *Great Lakes water level facts*. U. S. Government Printing Office, Washington

Wilson JW (1977) Effect of Lake Ontario on precipitation. *Monthly Weather Review* **105**:207-214

MODELING OF RUNOFF AND STREAMFLOW AT REGIONAL TO GLOBAL SCALES

Dennis P. Lettenmaier
Department of Civil Engineering FX-10
University of Washington
Seattle, WA 98195
USA

ABSTRACT

Global change problems place a new set of demands on hydrologic models. The primary purpose for representation of land surface hydrology in the context of coupled land-atmosphere models is to partition downward solar and longwave radiation into latent, sensible, and ground heat fluxes, and upward longwave radiation, rather than to predict streamflow. Nonetheless, past work in the development and application of conceptual streamflow simulation models for operational applications, such as forecasting, can provide valuable lessons, especially with respect to model parameter parsimony, for the development and application of land surface parameterizations for coupled land-atmosphere models. Some important issues in model development and application are illustrated in the context of the Variable Infiltration Capacity two-layer (VIC-2L) model. Application of the VIC-2L model to FIFE (central Kansas Grassland) and ABRACOS (cleared Amazonia tropical forest) field data are described. A version of VIC-2L that incorporates streamflow routing is described, along with some results of its application off-line (climatalogical forcing) to the Columbia River basin. Finally, an approach for regional estimation of the parameters of VIC-2L is described, along with preliminary results for an application to the Columbia River basin (drainage area approximately 615,000 km^2) at a one degree by one degree spatial resolution.

1.0 INTRODUCTION

This paper focuses on dynamic hydrologic modeling at regional to global scales. By dynamic I mean time scales of not more than a day, and by regional I mean spatial domains that include at least several grid cells of a numerical weather prediction or climate model, that is, length scales of at least several hundred km. Both the threshold time and space scales are recognized to be arbitrary. The paper consists of four parts: 1) a review of the relevance of past work in operational hydrology in the context of streamflow simulation at regional to global scales; 2) some key issues and approaches to land surface parameterizations for large scale atmospheric models; 3) an outline of an approach to modeling large North American rivers with preliminary results for the Colorado River; and 4) some ideas and early results from the use of regionalization methods to estimate the parameters of a coupled land-atmosphere model.

NATO ASI Series, Vol. I 31
The Role of Water and the Hydrological Cycle
in Global Change
Edited by Howard R. Oliver and Sylvia A. Oliver

2.0 HYDROLOGIC MODELS IN THE CONTEXT OF CLIMATE CHANGE

There is an extensive experience summarized in the hydrologic literature with so-called conceptual streamflow simulation models, which have been applied at scales from a few km^2 to tens of thousands of km^2, for operational purposes such as flood and water supply forecasting, water resources project design, and planning. The grandfather of such models is the Stanford Watershed Model (Crawford and Linsley, 1964). An outgrowth of the Stanford model, the Sacramento model (Burnash et al., 1973) remains the mainstay of the U.S. National Weather Service River Forecast System, and has been widely used throughout the world for flood and water supply forecasting, and other operational purposes. Such conceptual simulation models are characterized by spatial lumping of catchment attributes, generally to the level of gauged catchments, often of area hundreds of km^2 and up. Recently, with the widespread availability of GIS data layers representing land surface characteristics, variations of the older conceptual simulation models have appeared. The most promising operational use of GIS data in the context of conceptual simulation models has been to identify and segregate land cover classes, which effectively amounts to subdivision of a catchment into a small number of similar land cover classes (e.g., Kite and Kouwen, 1992; Kite, 1993). While such models might be termed quasi-distributed, their functional design remains quite similar to that of the older, spatially lumped models.

The conceptualizations chosen in operational hydrological models are based primarily on a desire to maximize accuracy of streamflow prediction, rather than to represent the underlying physical processes. The driving variables characterize surface meteorological conditions, and usually including precipitation and surface air temperature, and sometimes downward solar radiation, surface wind, and surface humidity or dew point, from which potential evapotranspiration is estimated. In many cases, where the latter variables are not available, potential evapotranspiration is estimated using simplified methods, such as Priestley-Taylor (Shuttleworth, 1993).

Conceptual hydrologic models have been quite successful for their designated purposes, and a general experience is that more sophisticated physically based models are unable to do better, and often do worse, in terms of streamflow prediction accuracy. However, global change problems place a completely different set of demands on hydrologic models, which conceptual simulation models are unable to meet. The primary purpose for representation of land surface hydrology in the context of coupled land-atmosphere models is to partition downward solar and longwave radiation into latent, sensible, and ground heat fluxes, and upward longwave radiation. In short, it is the coupled surface moisture and energy balance that is of interest, with emphasis on the latter. Although streamflow is a byproduct of the computation, it is not the primary, or necessarily even a major, concern.

Many of the simplifications made by conceptual hydrologic models were motivated by the absence of accurate, site-specific meteorological data, and experience that physically based models that use, for instance, PET based on the combination equation, yielded no improvement, and perhaps reduced, streamflow prediction accuracy. When a land surface model is embedded within an atmospheric model, the requisite forcings to the land surface are internal variables (e.g., fluxes at the lowest atmospheric level) within the coupled model. It is not only desirable, but essential from the point of view of self-consistency, that land surface schemes in coupled land-atmosphere models explicitly utilize this information to predict the surface energy and water budget. These requirements have lead to a new generation of hydrologic models. Another key difference between land surface parameterizations used within atmospheric models and conceptual hydrological models is the focus on grid-based, rather than catchment-based, geometry, although the importance of this difference for large scale applications has perhaps been exaggerated. In any event, the extensive experience in hydrology in parameter estimation, model parsimony, model transferrability, and related issues has not generally been considered in the development of the new models.

The question then, is what, if anything, can be transferred from procedures developed for operational hydrologic models to the implementation of GCM land surface parameterizations? The answer, I believe, is that it is possible to reduce the parameter space somewhat, and to increase the confidence in surface flux predictions, through local estimation of those parameters that most strongly affect streamflow generation. In section 5, such an approach is outlined, with preliminary results.

3.0 GCM LAND SURFACE PARAMETERIZATION

Rather than attempting to review the plethora of GCM land surface schemes, or even to classify them, I will focus on one such scheme, the two-layer VIC model (VIC-2L, Liang et al., 1994). The aim will be to call attention to evolving issues in the development and implementation of such schemes.

3.1 VIC-2L Model

One of the few operational hydrologic models that has found use in coupled land-atmosphere models is the Nanjing model (Zhao, 1977), variations of which have also have been termed the Arno model by Francini and Pacciani (1991), and the variable infiltration model, or VIC, by Wood et al. (1992). Early uses of the Nanjing model were for flood forecasting, but the model has found favor in coupled land-atmosphere models because it has an explicit structure to represent the subgrid spatial variability of soil moisture storage, which is ignored by most other models. Variations of the model have been used for GCM land surface parameterization by Dumenil and Todini (1992) in the Max Planck Institute model, by Stamm et al. (1994) in the GFDL model, and by Laval et al. (1994) in the LMD model. Interest in the model is attributable in large part to its parameterization of subgrid variability in soil moisture capacity, which can alternatively be considered to represent, in an approximate way, saturation excess runoff generation. In particular, the model assumes that the variation in infiltration capacity (which can be defined as the amount of "active" moisture that can be stored in the soil column; i.e., the difference between porosity and permanent wilting point) is given by

$$i = i_m[1 - (1 - A)^{1/B}] \qquad 1)$$

where A ($0 \leq A \leq 1$) represents the fraction of the grid cell (or catchment) for which the infiltration capacity is less than i, and i_m is the maximum infiltration capacity.

Liang et al. (1994) extended the model (to a version termed VIC-2L) by inclusion of a second soil moisture layer, and explicit representation of transpiration via a resistance formulation, that represents the full surface energy and moisture budget. The so-called Arno drainage scheme described by Francini and Pacciani (1991), which can be considered to be a nonlinear baseflow representation, is applied in VIC-2L to the lower zone (second soil layer), which is treated as spatially lumped. The Arno drainage formulation is

$$Q_b = \alpha D_m \frac{W}{W_c} \qquad W < Y \qquad 2a)$$

$$Q_b = \alpha D_m \frac{W}{W_c} + (1-\alpha) D_m \frac{W-Y}{W_c - Y} \qquad W \geq Y \qquad 2b)$$

where α, D_m, Y, and W_c are model parameters, and α is constrained to be between zero and one. W_c is the moisture storage capacity of the lower layer, and the remaining three parameters describe a nonlinear recession (linear for $W < Y$).

Another feature of the model is that it incorporates a relatively simple energy balance snowmelt model described by Wigmosta et al. (1994). The structure of VIC-2L is shown schematically in Figure 1.

3.2 Model Validation

One of the major difficulties in validating models such as VIC-2L is the absence of coincident surface meteorological, streamflow, and land surface energy flux data. The

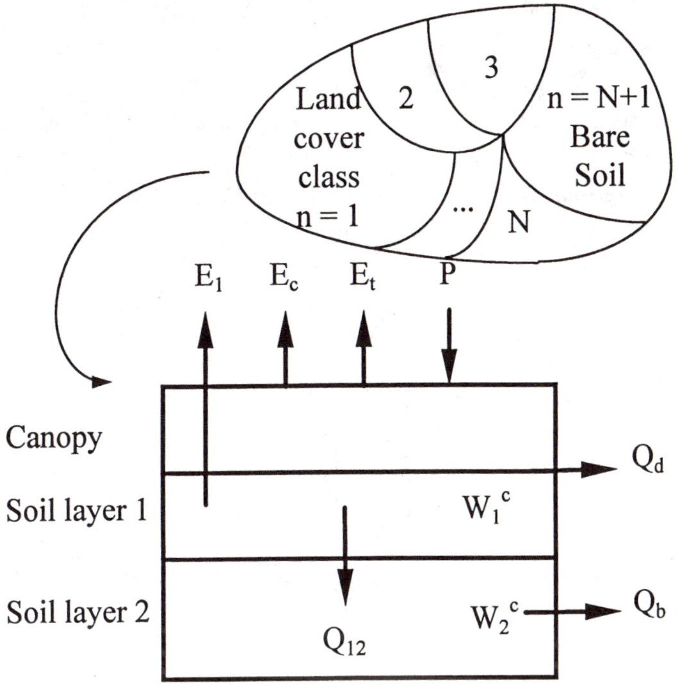

Figure 1: Schematic representation of the two-layer VIC model (from Liang, 1994).

latter are most difficult to obtain, and are mostly available only from field experiments, such as HAPEX, FIFE, ABRACOS, and BOREAS, that have been specifically designed for this purpose. The design of these experiments usually consists of a relatively small number of heavily instrumented surface flux towers, integrated with aircraft flux measurements and satellite remote sensing. Most such experiments have incorporated either high density rain gauge networks, or precipitation radar, to estimate the spatial distribution of rainfall. They are designed with the intent of characterizing land surface fluxes from a particular vegetation type or types. For instance, in the case of FIFE, a remote sensing field experiment conducted in central Kansas, USA during summer, 1987, the land cover was (relatively) homogeneous tall grass prairie, while in the case of ABRACOS, it was short grass pasture.

Validation of the VIC-2L model has been carried out using FIFE surface flux data (see Sellers, et al., 1992 for a description of the experiment), along with historical daily streamflow records for Kings Creek, Kansas, a 11.7 km^2 catchment lying completely within the FIFE study area, and daily precipitation for Manhattan, KS, about 11 km from the centroid of the Kings Creek catchment. Daily streamflow and precipitation data for water years 1983 and 1984 were used to estimate the baseflow (drainage) parameters of the VIC-2L model. Water years 1986 and 1987 were used for verification. The remaining parameters (upper and lower zone moisture capacity, soil wilting and critical points, residual moisture content, and saturated hydraulic conductivity) were estimated based on physical measurements at the site (Liang et al., 1994). Figure 2 shows observed and predicted Kings Creek streamflow for the calibration and validation periods. Model performance was deemed acceptable, given

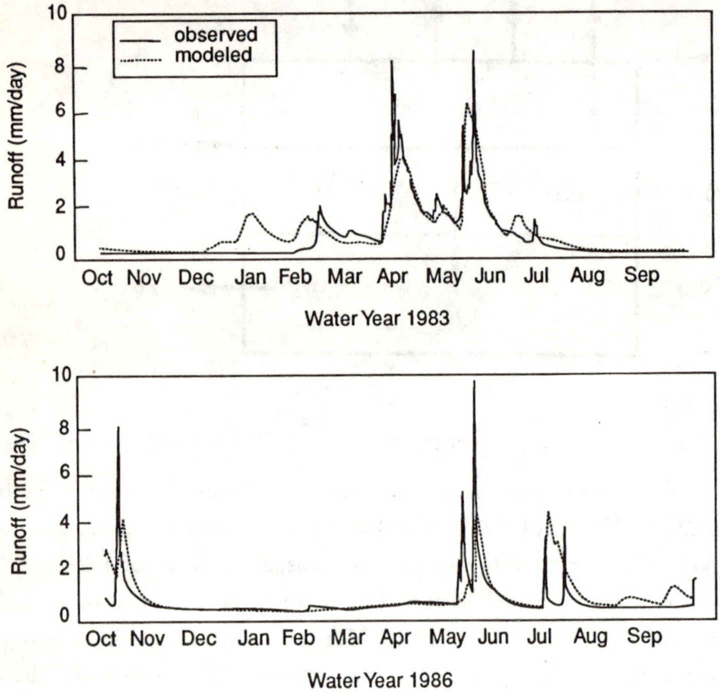

Figure 2: Predicted and observed streamflow for Kings Creek for calibration year 1983 and verification year 1986 (from Liang, 1994).

the discrepancy between the response time of the catchment (several hours at most) and the daily reporting interval for the precipitation and streamflow data. During these long-term simulations, potential evapotranspiration was estimated using the method of Hamon (1961), which requires only surface air temperature. For the period of the intensive field campaigns (IFCs) in summer, 1987, Penman-Monteith PET was also computed, and an adjustment ratio based on the average aerodynamic resistance was developed to convert Hamon into equivalent Penman-Monteith PET.

Figure 3 compares computed and observed surface fluxes during FIFE IFC 2, from June 30 to July 11, 1987. Agreement is quite good, with the exception that the diurnal peak latent heat flux is underestimated, and sensible heat overestimated, during the last few days of the IFC, which was a windy, warm period following a precipitation event on July 7. Soil moisture stress during the entire IFC 2 period was modest, which is evidenced by small magnitude of the sensible relative to latent heat.

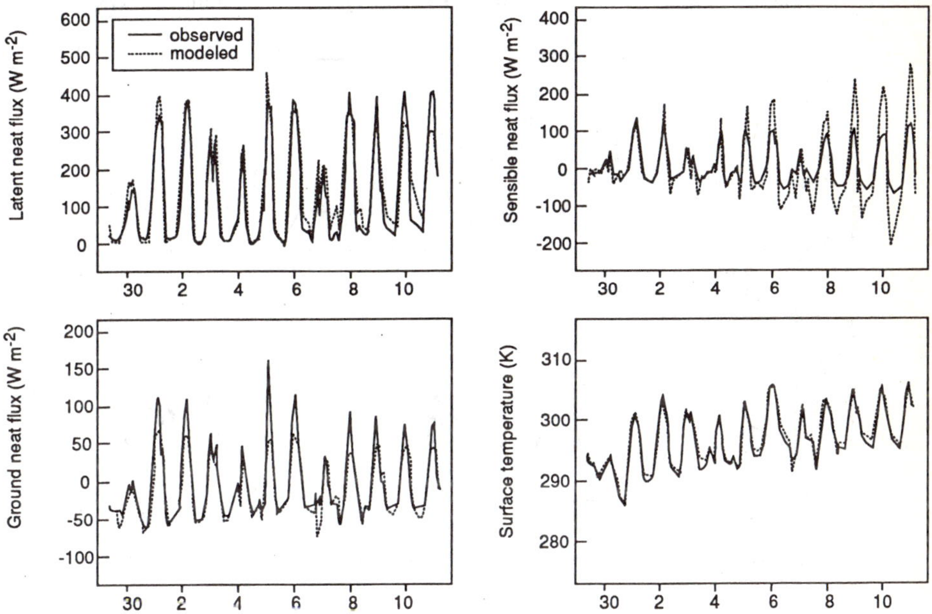

Figure 3: Predicted and observed surface fluxes and surface temperature at the FIFE site for IFC 2, Junc 30 July 11, 1987 (from Liang, 1994).

Model validation of VIC-2L has also been carried out at the ABRACOS (Anglo Brazilian Amazonian Climate Observation Study) described by Shuttleworth et al., (1991). The ABRACOS ranchland site, located in northwestern Brazil, was selected to characterize the local and large-scale effects of Amazonian deforestation. The site is part of a large cattle ranch that was created by felling and burning the forest and sowing the clay soil with hardy pasture grasses in the early 1980s. It was instrumented with a surface flux tower, a cluster of ground heat flux plates, upward and downward-looking short and longwave radiometers, and an automated weather station. All of the instrumentation was located in a fairly small area (within a few tens of meters). Soil moisture was measured at multiple levels in the soil column using a cluster of neutron probe tubes. However, there was no well-defined surface catchment, so no stream discharge data were available.

In the absence of stream discharge data, soil properties, specifically, porosity, permanent wilting point, field capacity, saturated hydraulic conductivity, and pore size distribution index measured and reported by Wright et al. (1992) and Hodnett et al. (1993) were used to the extent possible to estimate the VIC model parameters. The measured properties were sufficient to estimate all of the VIC-2L model parameters except for the infiltration shape parameter (B in Eq. 1), and the nonlinear recession

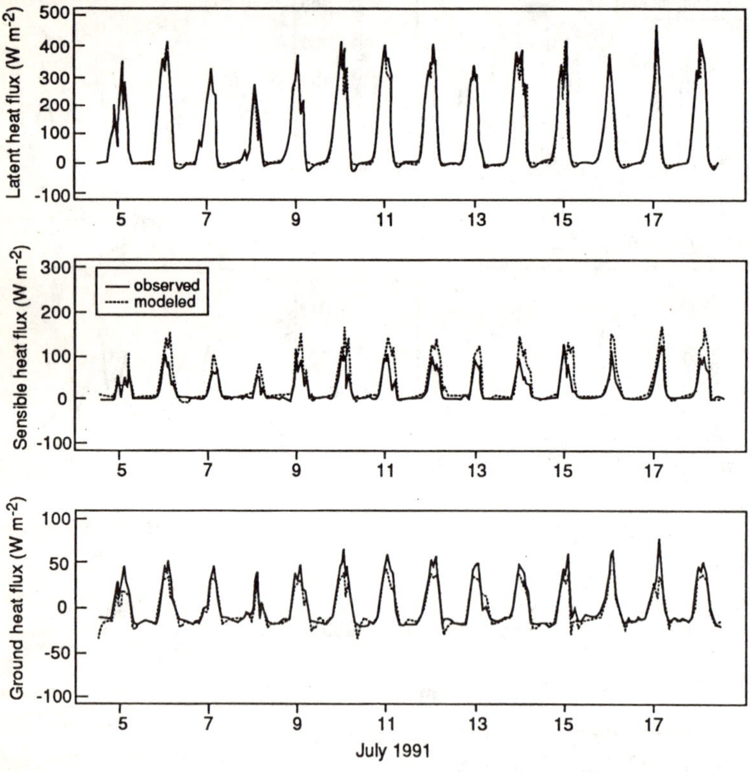

Figure 4: Comparison of predicted (dotted) and observed (solid) surface fluxes at the ABRACOS site for July 5-18, 1991 portion of Mission 2 (from Liang, 1994).

parameters (see Liang, 1994 for details). These parameters were assigned values based on ranges suggested by Dumenil and Todini (1992).

Results for part of one of two ABRACOS periods of intense observation, Mission 2, are shown in Figure 4. This was a relatively wet period, during which the model predicted latent and ground heat fluxes quite accurately. There is some overprediction of diurnal peak sensible heat fluxes. Similar results were obtained in other periods, although the results are not quite as good during dry periods, when there is a transition from atmospheric to soil limitation of latent heat fluxes.

Given the absence of stream discharge data, and hence the inability to estimate directly some of the parameters that most strongly affect the water balance, the ABRACOS results are surprisingly good. The reason for the apparent insensitivity of the surface fluxes to key hydrological parameters was further explored by Liang (1994) using the factorial design method of Box et al. (1978) previously applied to land surface models by Henderson-Sellers (1992; 1993). The factorial design method considers not only the sensitivity of model predictions to parameters in a classical "one at a time" sense, but also to interactions within groups of parameters. Liang (1994) applied the method to the VIC-2L model using model-generated inputs being used as part of the Project for Intercomparison of Land Surface Parameterizations (PILPS; see Pitman et al., 1994 for details) for two conditions: one a tropical forest climate typical of Amazonia, and the other a northern hemisphere grassland climate typical of parts of the UK. As part of PILPS, hourly forcings (surface meteorology; downward solar and longwave radiation) were provided for a one-year period, and each model (including VIC-2L) was run to equilibrium using a long multiyear sequence of back-to-back repetitions of the one year of data.

One of the key results of the Liang et al. analysis was that surface temperature at both sites was almost completely insensitive to the infiltration shape parameter, B, and that the grassland site (which was almost always atmospherically limited) was almost completely insensitive to the infiltration shape parameter. There was modest sensitivity to B for latent and sensible heat fluxes at the forest site during the period of most severe moisture stress. There was, however, appreciable sensitivity of monthly runoff to B. For the drainage parameters (α, D_m and Y, expressed in a slightly different form by Liang et al.), the results were even more striking. At both sites, the surface energy fluxes were almost completely insensitive to these parameters. However, the pattern of annual runoff was strongly sensitive to Ds and Ws. These results suggest that under some conditions (notably, where moisture stress is low) the energy and water budgets are somewhat decoupled, even though evaporation is a common term. In this respect, although the climates for the PILPS grassland and forest sites are strikingly different, evaporation at both sites is dominantly atmospherically limited. The interactions between the moisture and energy budget can be expected to be much stronger when moisture stress plays a more important role (e.g., at the FIFE site).

4.0 STREAMFLOW SIMULATION IN COUPLED LAND-ATMOSPHERE MODELS

Although there has been considerable recent attention to modeling land surface hydrology at the large scale, this attention has been directed almost wholly to description of runoff production, and other land surface moisture and energy fluxes, within one GCM grid cell. In the context of coupled land-atmosphere models, a distinction must be made between runoff and streamflow. Runoff is the excess of precipitation over evaporation and storage changes within a grid cell represented by the model. In the short term, it incorporates not only streamflow, but change in groundwater storage, which is not represented by any of the current generation of GCMs. Of equal importance is that streamflow is a directional quantity that flows from one grid cell to another; GCM runoff, on the other hand, essentially vanishes (unlike evaporation, which is accounted for in the water budget of the atmosphere).

In principle, it is relatively straightforward to develop algorithms to route runoff in such a way as to simulate the streamflow of large continental rivers. This can be done either by prescribing the form of the channel network, or by developing algorithms to determine grid cell connectivity based on topographic information. The latter approach, applied at the four degree latitude by five degree longitude grid scale, was used by Miller and Russell (1992) to estimate the long-term discharge of the largest world rivers as predicted by the Goddard Institute for Space Studies GCM. Because they focused on mean annual runoff, time of travel was ignored. However, more general models, which predict large river runoff at shorter time scales, must account for the time lag between production of runoff locally (at the subgrid scale), to the appearance of streamflow at the grid cell exit point. This is relevant in particular for applications such as numerical weather prediction models, where comparison of model predictions with observed flows provides a check on the model's water balance.

Wetzel (1994) has developed an approach to route runoff produced by grid-based land-atmosphere models to predict streamflow for large continental rivers. She tested her approach off-line, by subdividing the Missouri River basin into grid cells of size one degree latitude by one degree longitude. Within each grid cell, runoff is predicted using the VIC-2L model. The drainage patterns were defined using digital elevation data. The grid routing model accounts for the time of travel within a grid cell for runoff generated in the grid cell, and travel time and hydrograph diffusion effects for runoff entering the grid cell from an adjacent cell. Both forms of routing are plausibly represented by unit hydrographs, the ordinates of which can be estimated, to first order, using digital topographic data.

The VIC-2L based gridded runoff scheme has also been run in stand-alone mode for the Columbia River basin (see Figure 5), forced with station precipitation and temperature data, and long-term average Hamon (1961) potential evapotranspiration.

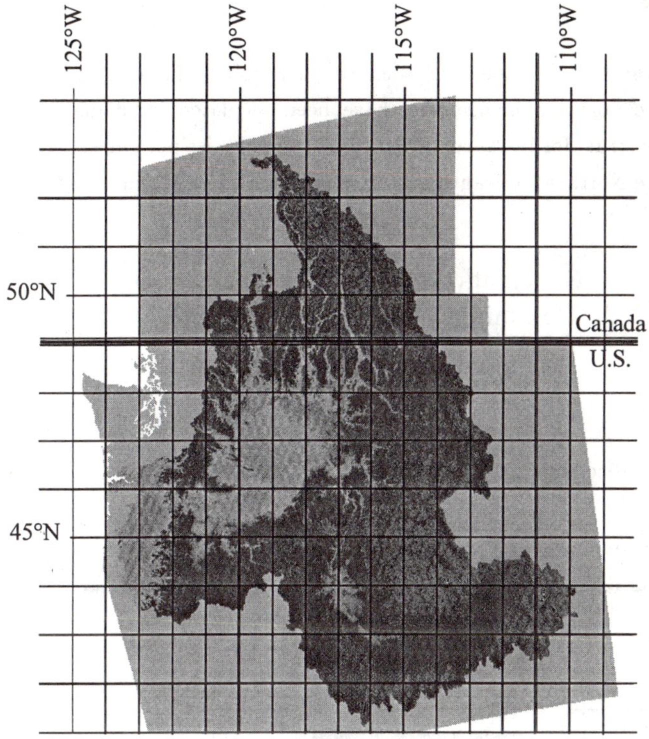

Figure 5: Relief map of Columbia River basin with one degree grid.

Figure 6 shows the one degree by one degree schematization of the basin, and the flow directions as represented in the model, along with locations at which naturalized flows (reservoir effects removed) were available. There is a strong gradient in precipitation within the basin from south to north, and from the dry interior to the mountainous western and eastern extremes. Most of the runoff at the Dalles, the most downstream location modeled, is derived from spring and summer snowmelt. Therefore, a key aspect of the simulation is the proper representation of the gradient of precipitation with elevation in the headwater grid cells. Some results for the Dalles are shown in Figure 7, which indicates that the timing of runoff is reasonably well represented, on average.. The results at the other nodes of the river system were comparable to those at the Dalles, with the exception of the upper Snake River basin (southeastern portion of the basin), where groundwater-surface water interactions, which are not well represented in the VIC-2L model, strongly affect streamflow. Comparison of simulated and observed coefficients of variation of monthly and annual flows showed that the model provides reasonable estimates of the year-to-year and inter-seasonal variability of runoff.

One major complication in modeling the runoff of large rivers is that many, especially in the northern hemisphere, have been regulated by dams and diversions. This is especially true, for instance, in the case of the Mississippi and its tributaries, and many other large North American rivers. For purposes of validating the Columbia

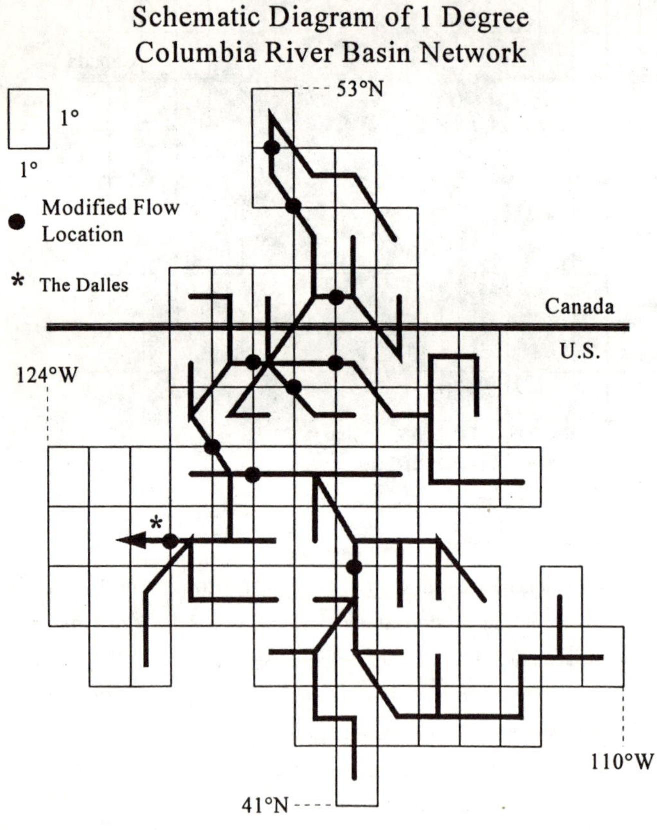

Figure 6: Schematic of Columbia River network at one degree resolution, with location of system nodes with long-term naturalized flows.

River application, we have used so-called naturalized flows, which have had the effectsof reservoir storage change and diversions removed. However, in practice, especially for application to numerical weather prediction models in near-real time, what is available for comparison is the regulated (measured) flow. Therefore, a capability is included in the model to represent storage reservoirs, and consumptive withdrawals. This affords the possibility of using predicted streamflow in near-real time as a diagnostic in numerical weather prediction models, and in a distributional sense, as a diagnostic for

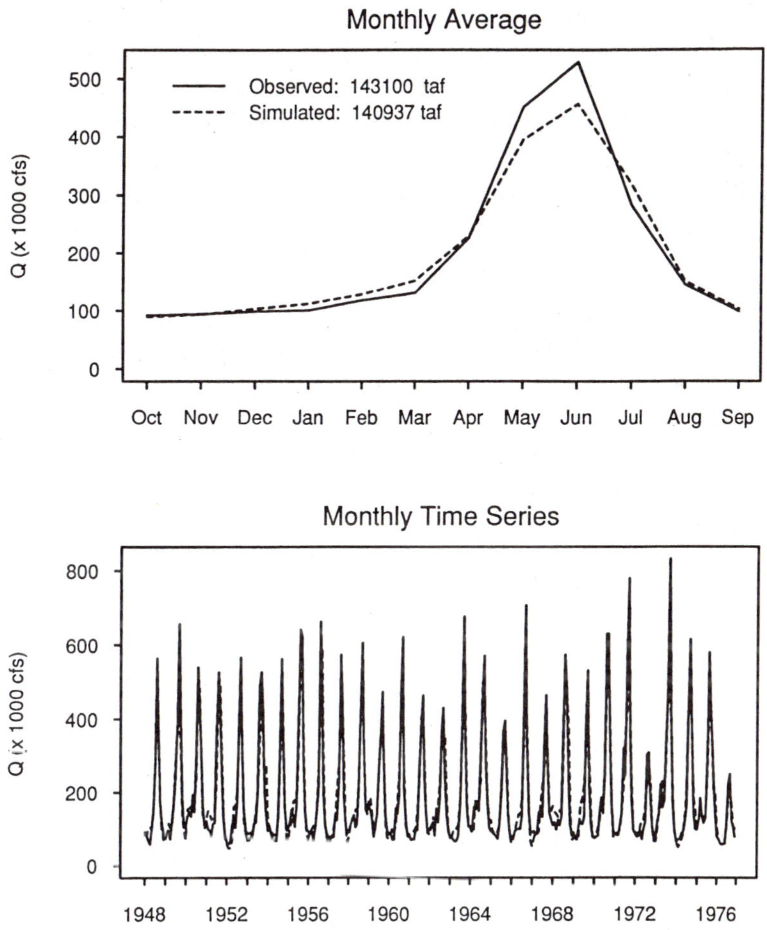

Figure 7: Simulation results at the Dalles for VIC-2L model applied off-line to Columbia River basin.

climate simulations. Ultimately, such methods offer the possibility for examining theeffects of water management on regional weather and climate.

5.0 PARAMETER ESTIMATION

The general approach to parameter estimation for land surface schemes, such as VIC-2L, has been either to fix parameters globally at "reasonable" values, or to use "literature values" for the appropriate land cover. Better methods of estimating parameters are clearly needed. We have initiated work in the Arkansas-Red River basins in the U.S. Southern Great Plains, some early results of which, taken from Abdulla and Lettenmaier (1994) are described here. The approach utilizes parameter search techniques to identify the optimal VIC-2L infiltration and drainage parameters based on hydrograph reproduction. Parameters have been estimated for 26 streams with drainage areas ranging from 179 to 5280 km^2, which are nominally unaffected by upstream regulation and diversions; the method will ultimately be extended to about 100 such streams. Regional equations for parameter prediction, based on readily estimated soil, topographic, and meteorological parameters, have been found to explain about 60 percent of the variance of most parameters. A key element in the success of the regionalization procedures has been reparameterization of the model to reduce parameter interdependence. Eqs. 2a and 2b can be replaced by the alternative form

$$Q_b = K_1 \frac{W}{W_c} \qquad W < K_2 \qquad \text{3a)}$$

$$Q_b = K_1 \frac{W}{W_c} + K_3 \frac{W-K_3}{W_c-K_3} \qquad W \geq K_2 \qquad \text{3b)}$$

which reduces the number of parameters in the base flow formulation by one. The reparameterization also has the important effect of reducing parameter interdependence, which reduces the diagonal terms (variances) in the parameter covariance matrix, and improves the performance of the regional parameter estimation equations.

We have also implemented a two-stage procedure to estimate the baseflow (drainage) parameters using recession-period streamflow data only. The remaining parameters are then estimated conditionally (given the drainage parameters), using a variation of the Shuffled Complex Evolution (SCE) method of Duan et al. (1992), where the objective function is the sum of squares of the difference between observed and simulated daily streamflow. The estimated parameters for each catchment are then regressed on a series of meteorological, soils, and geomorphic catchment characteristics.

Some preliminary results for a single layer version of the VIC model similar to that described by Wood et al. (1992) are given in Table 1, which summarizes the form of the regional equations for two parameters, B, the infiltration parameter, and $W_c = i_m/(1+B)$, the water holding capacity based on stepwise regression. In the case of both W_c and B a maximum of six variables were found to be significant. However, because stepwise regression assumes that the explanatory variables are independent, an assumption that cannot be met strictly in this case, we show the results only for the first four explanatory variables. The percentage of the variance explained is given in parenthesis. Table 1 shows, for instance, that if an equation with a single explanatory variable is used, that variable is AI, if two variables are used, the are AI and P1 (which enter the regression in that order), and so on. The nine explanatory variables listed are only those that appear in at least one equation, and were selected from a list of 25 candidates.

One interesting result of the regionalization is that most of the variables in the equations for both W_c and B are meteorological, e.g., mean annual precipitation and mean storm intensities. This occurred even though the 25 candidate variables included soil characteristics such as water holding capacity which would be expected to enter into the equation for W_c. We believe that the reason the soil variables, in particular, did not enter into the regional equation is that they were taken from a county summary data base, which represents selected characteristics to a fixed soil depth, which in turn resulted in an absence of clear discrimination between catchments. On the other hand, there is a strong climatic gradient over the Arkansas-Red basin. The hydrologic response of the 26 catchments follows the climatic gradient to some extent, which explains the predominance of climatic factors in the explanatory equations. We are presently extracting soils data for each of the catchments from the more detailed U.S. Soil Conservation Service State Geographic Data Base (STATSGO). Preliminary results suggest that the more detailed data do result in inclusion of soil characteristics as explanatory variables in the regional equations for W_c and B.

Figure 8 shows long-term mean historical streamflow for four of the 26 catchments as well as the mean simulated streamflow for the same period based on optimized parameters of the one-layer VIC model (based on the best fit to the data during a separate four-year calibration period), and simulated streamflow for the same period using parameters based on the regional equations. In the case of the South Fourche River, the most humid of the four catchments, the regional and locally estimated parameters perform almost the same, although there is a considerable bias between the model and observed results. For Little Cypress Creek, the simulation based on the regional parameters is slightly better than that using the locally optimized parameters. The worst performance is for the most arid catchment, the Whitewater

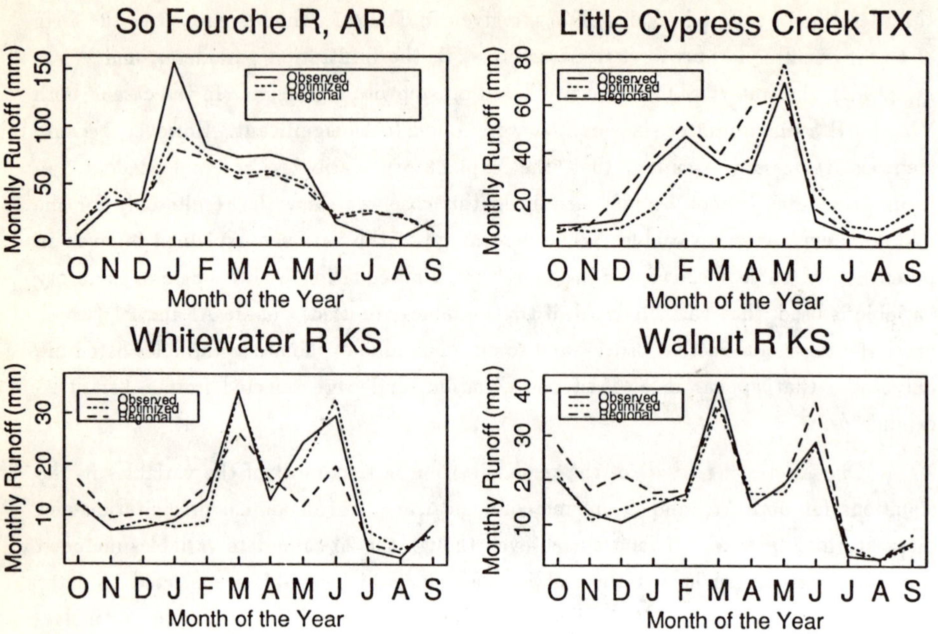

Figure 8: Observed and simulated streamflows using locally estimated and regional parameters for four catchments within the Arkansas-Red River basin.

River. In general, these preliminary results are promising, and are expected to improve as better land surface attribute data become available. Ultimately, the hope is that regional equations applicable at the continental to global scale can be derived that will allow specification of the parameters of the VIC-2L model in data sparse regions.

5.0 CONCLUSIONS

Global change problems place a new set of demands on hydrologic models. While there are important lessons to be learned from past experience with operational hydrologic models, it must be recognized that the primary purpose for representation of land surface hydrology in the context of coupled land-atmosphere models is to partition downward solar and longwave radiation into latent, sensible, and ground heat fluxes, and upward longwave radiation, rather than to predict streamflow. Some of the issues in development and validation of land surface schemes are illustrated with the VIC-2L model, and in particular, through applications of the model to FIFE 1987 field data (central North American grassland) and ABRACOS 1991 data (Amazonia cleared tropical forest). The potential for such a macroscale hydrologic model to simulate the streamflow of large continental rivers is illustrated through application of a network-based version of VIC-2L to the Columbia River basin. On a long-term average basis, the model performs quite well for prediction of flows at the basin outlet (The Dalles, OR) as well as intermediate system nodes. For applications at continental and global scales, estimation of the parameters of land surface schemes such as VIC-2L is problematic, as traditional hydrologic estimation schemes, which are based on fitting of predicted to observed streamflow during a calibration period, are inappropriate due to data scarcity and the presence of anthropogenic effects, such as reservoirs and diversions. Preliminary results of research in progress in the Arkansas-Red River basin show that regional estimation of the parameters of the VIC-2L model, using long-term climatalogical, physiographic, and geomorphic data appears feasible.

Table 1: Parameters in Regional Prediction Equations for Arkansas-Red River Catchments

	Wc				B			
Explanatory	n[a]				n[a]			
Variable[b]	1	2	3	4	1	2	3	4
	(27)	(58)	(63)	(73)	(19)	(27)	(32)	(50)
Pr				1				
PR				2			2	
L			2			2		4
Re								2
S							1	1
AI	1	1	1	3				
P1		2	3	4				
P2								
P3					1	1	3	3

[a]number of explanatory variables in equation
[b]independent variable definitions:

Pr = average permeability
PR = average permeability x relief ratio
L = length of the main channel
Re = maximum relief
S = average watershed slope
AI = mean annual precipitation
P1 = mean storm depth, Nov-May
P2 = mean storm depth, Jun-Aug
P3 = mean storm depth, Sep-Oct

ACKNOWLEDGMENTS

Dr. Xu Liang (Department of Civil Engineering and Operations Research, Princeton University) provided the simulation results for FIFE and ABRACOS. Mr Bart Nijssen (Department of Civil Engineering, University of Washington) performed the Columbia River simulations, and Mr. Fayez Abdulla (Department of Civil Engineering, University of Washington) performed the Arkansas-Red River regional parameter estimation. The research on which this paper is based was funded by the U.S. Department of Energy Western Center for Global Environmental Change, by the NOAA Office of Global Programs, and the National Science Foundation Hydrologic Science Program.

REFERENCES

Abdulla, F., and D.P. Lettenmaier, "Development of regional equations for parameterization of land surface schemes", paper presented at American Geophysical Union Fall Meeting, San Francisco, December, 1994.

Box, G.E.P., W.G. Hunter, and J.S. Hunter, Statistics for experiments: An introduction to design, data analysis, and model building, 653 pp., John Wiley and Sons, New York, 1978.

Burnash, R.J.C., R.L. Ferral, and R.A. McGuire, "A generalized streamflow simulation system -- Conceptual modeling for digital computers", U.S. National Weather Service, Sacramento River Forecast Center, 1973.

Crawford, N.J., and R.K. Linsley, "Digital simulation in hydrology: Stanford Watershed Model IV", Stanford University, Department of Civil Engineering Technical Report 39, 1966.

Duan, Q., S. Sorooshian, and V. Gupta, "Effective and efficient global optimization for conceptual rainfall-runoff models", Water Resources Research 28(4):1015-1032, 1992.

Dumenil, L., and E. Todini, A rainfall-runoff scheme for use in the Hamburg climate model, pp. 129-157 in Advances in theoretical hydrology, A tribute to James Dooge, European Geophysical Society Series on Hydrological Sciences 1, P.E. O'Kane, ed., Elsevier, New York, 1992.

Francini, M., and M. Pacciani, Comparative analysis of several conceptual rainfall-runoff models, Journal of Hydrology 122:161-219, 1991.

Hamon, R.W., "Estimating potential evapotranspiration", Journal of the Hydraulics Division, ASCE 87:107-120, 1961.

Henderson-Sellers, A., "Assessing the sensitivity of a land-surface scheme to parameters used in tropical-deforestation experiments", Quarterly Journal of the Royal Meteorological Society 118:1101-1116, 1992.

Henderson-Sellers, A., "A factorial assessment of the sensitivity of the BATS land-surface parameterization scheme", Journal of Climate 6:227-247, 1993.

Hodnett, M.G., L.P. da Silva, H.R. Rocha, and R. Cruz Senna, "Seasonal soil water storage changes beneath central Amazonian pasture land and rainforest", unpublished manuscript, 1993.

Kite, G.W., "Application of a land class hydrological model to climatic change", Water Resources Research 29(7):2377-2384, Jul., 1993.

Kite, G.W., and N. Kouwen, "Watershed modeling using land classifications", Water Resources Research, 28(12):3193-3200, 1992.

Laval, K., A. Ducharne, and J. Polcher, "Sensitivity of the LMD-GCM to the subgrid scale variability of field capacity", paper presented at EGS General Assembly, Grenoble, April, 1994.

Liang, X., D.P. Lettenmaier, E.F. Wood, and S.J. Burges, "A simple hydrologically based model of land surface water and energy fluxes for general circulation models", Journal of Geophysical Research 99(D3):14,415-14,428, Jul., 1994.

Liang, X., "A two-layer variable infiltration capacity land surface representation for general circulation models", Water Resources Series Technical Report No. 140, Department of Civil Engineering, University of Washington, May, 1994.

Miller, J.R., and G.L. Russell, "The impact of global warming on river runoff", Journal of Geophysical Research 97(D3):2757-2764, Feb. 28, 1992.

Pitman, A.J. et al., "Results from the off-line control simulation phase of the project for intercomparison of land surface parameterization schemes (PILPS)", in press, Bulletin of the American Meteorological Society, 1994.

Sellers, P.J., F.G. Hall, G. Asrar, D.E. Strebel, and R.E. Murphy, "An overview of the First International Satellite Land Surface Climatology Project (ISLSCP) Field Experiment", Journal of Geophysical Research 97(D17):18,345-18,371, 1992.

Shuttleworth, W.J., J.H.C. Gash, J.M. Roberts, C.A. Nobre, L.C.B. Molion, and M. de Nazare Goes Ribeiro, "Post-deforestation Amazonian climate: Anglo-Brazilian research to improve prediction", Journal of Hydrology 129:71-85, 1991.

Shuttleworth, W.J., "Evaporation", Chapter 4 in Handbook of Hydrology, D.R. Maidment, ed., McGraw Hill, New York, 1993.

Stamm, J.F., E.F. Wood, and D.P. Lettenmaier, "Sensitivity of a GCM simulation of global climate to the representation of land surface hydrology", Journal of Climate, 7(8):1218-1239, Aug., 1994.

Wetzel, S., "A hydrological model for predicting the effects of climate change", B.S. Thesis, Department of Civil Engineering and Operations Research, Princeton University, Apr., 1994.

Wigmosta, M., L. Vail, and D.P. Lettenmaier, "A distributed hydrology-vegetation model for complex terrain" Water Resources Research 30(6):1665-1679, Jun., 1994.

Wood, E.F, D.P. Lettenmaier, and V.G. Zartarian, A land-surface hydrology parameterization with subgrid variability for general circulation models, Journal of Geophysical Research 97(D3), 2717-2728, 1992.

Wright, I.R., J.H.C. Gash, H.R. da Rocha, W.J. Shuttleworth, C.A. Nobre, G.T. Maitelli, C.A.G.P. Zamparoni, and P.R.A. Carvalho, "Dry season micrometeorology of central Amazonian ranchland", Quarterly Journal of the Royal Meteorological Society 118:1083-1099, 1992.

Zhao, R.J., Flood forecasting method for humid regions of China, East China College of Hydraulic Engineering, 1977.

New Trends in Modelling Soil Processes from Hillslope to GCM Scales.

Ezio TODINI
University of Bologna
Institute for Hydraulic Construction
Viale del Risorgimento, 2
40136 Bologna - Italy

SUMMARY

The paper deals with the need for new parameterizations of the soil processes in order to improve the representation of the soil atmosphere exchanges at large horizontal spatial scales, typical of catchments and also applicable to the General Circulation Models (GCMs) used for climatic studies. The principal aim of this paper is to underline the relevance that the horizontal flow, particularly in the unsaturated zone, may have to the overall exchanges between soil and atmosphere. The basic physical hypotheses imbedded into two widely applied hydrological models (the ARNO and the TOPMODEL) are here reviewed, together with their pitfalls, in order to underline the main positive aspects to be retained in a new model formulation. In order to substantiate the basic hypotheses, allowing for lumping at catchment or at GCM scales, several numerical experiments were performed and are briefly reported in the paper. Finally the early stage of a new model formulation, TOPKAPI, is presented, as a logical consequence of the requirements and of the underlying physical hypotheses analysed and discussed in the paper.

INTRODUCTION

In the last decade new trends in hydrology were originated by the increasing interest of meteorologists in resolving the interactions

NATO ASI Series, Vol. I 31
The Role of Water and the Hydrological Cycle in Global Change
Edited by Howard R. Oliver and Sylvia A. Oliver

between soil and atmosphere at scales far beyond the size of the conventional hydrological scales.

Traditionally meteorologists are concerned with the vertical fluxes of water and heat, which are computed according to soil and atmosphere "column models", while paying little attention to the horizontal fluxes of water leaving the soil; at the same time hydrologists tend to give more emphasis to the horizontal flows as they do not really need to model the evapo-transpiration processes since their main objective is the estimation of river discharges.

Until recent years, hydrologists have always treated precipitation and runoff as *measured* quantities; models relating precipitation to runoff accounted for infiltration, percolation and evapo-transpiration as *losses*, which were indirectly derived from the water balance.

A simple evaluation of the orders of magnitude reveals how wrong this approach can be: under normal conditions, for timescales of the order of a year, if one takes the precipitation equal to 100%, evapo-transpiration generally lies in the range of 40-60%, while the combined horizontal fluxes (measured as river discharges) range between 20% and 40%.

How is it possible to correctly represent the physical nature of processes if one neglects one component, such as the evapo-transpiration, which has the same order of magnitude as the other two?

A similar criticism may be levelled at meteorologists, since they tend to neglect horizontal fluxes and often fail to take into account the available river discharge measurements for their model validation. River discharges are extremely important in that they incorporate, in lumped form, all the information relevant to the filling and depletion mechanisms taking place in the upper soil layer under the action of precipitation, gravity forces and solar radiation.

Global atmospheric models, known as General Circulation Models (GCMs), generally used by meteorologists, are based upon spatial discretization meshes of one or two hundred kilometres in size. At this scale, a wide variety of dynamical phenomena cannot be explicitly resolved, leaving an average overall effect to be accounted for ("parametrized").

This is the main justification invoked by most GCM modelers for their use of column models, generally based on the direct application of Richards equation at scales far beyond the infinitesimal dimension of the vertical column for which it was originally derived.

Unfortunately, the strong and interdependent non-linearities present in the land phase of the hydrological cycle and the extreme variability of land and soil properties do not allow for this direct extension of the column-type models to the GCM scales; this therefore calls for new lumped parameterizations of the soil response to atmospheric forcing, such as rainfall, solar radiation and wind-driven evaporation, at catchment or larger scales.

In order to derive such new lumped parameterization, after discussing the dominant mechanisms in the generation of streamflows, namely surface runoff, saturated and unsaturated subsurface flow and deep aquifer flow (Freeze, 1972a,b; Smith and Hebbert, 1983), two well-known hydrological models (the ARNO and the TOPMODEL) are presented and analysed in terms of their conceptualisation of the phenomena involved.

In view of their possible inclusion in GCMs, the problems related to the performance of these hydrological models and the physical meaning of their variables and parameters will be discussed, in order to derive an item list which would form the basis of a new conceptualisation. In order to substantiate a new model formulation, the results of a number of numerical tests on the possibility of lumping parameters from the point to the catchment scale, will then be presented. Finally a brief description of the newly proposed approach, the TOPKAPI, based upon the conclusions drawn from the experimental evidence, will be provided.

BRIEF DESCRIPTION OF RAINFALL-RUNOFF PROCESSES

It is exceptional for as much as 5% of catchment rainfall to fall directly on rivers and lakes. The remainder reaches rivers by the hillslopes or is lost upwards through evapotranspiration, or downwards to deep aquifers. An accurate understanding of the detailed processes acting within a uniform element of soil remains to date incomplete. The cost and time involved in first deriving and then applying such detailed knowledge for even a small catchment is likely to remain prohibitive for most applications. Hydrology has therefore been dominated by attempts to simplify those processes, using a variety of assumptions, until it becomes feasible to construct a model. The main components into which precipitation may be partitioned are evapotranspiration, surface runoff and saturated and unsaturated subsurface flow.

Surface runoff may occur because of soil saturation from above (Hortonian excess), because of saturation from below (saturation excess) or when subsurface flow is forced up to the surface by the soil or slope configuration (return flow). Hortonian surface runoff occurs on soils exhibiting low infiltration capabilities such as on unvegetated surfaces and in desert and urban areas. It is however very rare for it to occur in steeply sloping catchments in humid vegetated areas. Surface runoff due to saturation from below is in these catchments the primary mechanism at work (Freeze, 1972b; Sloan and Moore, 1984).

The saturation excess mechanism, firstly proposed by Horton (1933, 1939), takes place when rainfall intensity exceeds the hydraulic conductivity of the soil. The moisture content at the surface increases as a function of time and, at some stage, the surface becomes saturated and an inverted zone of saturation begins to propagate downward into the soil. At this time the infiltration rate drops below the rainfall rate and surface runoff is produced. As originally presented, Horton's infiltration excess mechanism inferred that most rainfall events exceed infiltration capacities and that surface runoff is common and spatially widespread. But in general only a part of a catchment contributes to surface runoff. A great heterogeneity in soil types over the catchment area and the very irregular patterns of rainfall in time and space create a very complex hydrologic response at the land surface. The infiltration excess mechanism (Horton) tends to dominate runoff production in most desert or semiarid regions, where the absence of vegetation and other organic matter prevents the development of a porous soil structure, through which water can move easily. On vegetated surfaces in humid regions however, Hortonian surface runoff occurs rarely (Dunne et al., 1975).

The second mechanism of surface runoff production occurs where the soil becomes saturated to the surface from below due to both rainfall inputs and downslope subsurface flow. This saturation excess (Dunne) mechanism is most common on near-channel wetlands. The area of saturation will expand and contract during and between storms. The Dunne mechanism can result in surface runoff production in humid and vegetated areas with shallow water tables, even where infiltration capacities of the soil surface are high relative to normal rainfall intensities. The main controls on the saturated contributing areas are the topographic and hydrogeologic characteristics of the hillslopes.

The literature also reports a third mechanism of surface runoff which is called return flow (Kirkby, 1985). This process can even occur after rainfall has ceased where subsurface flow is forced up to the surface by the soil or slope configuration. The return flow occurs usually on concave slope profiles and areas of flow convergence. For the sake of simplicity of many rainfall-runoff models, this process of surface runoff is mostly neglected.

Subsurface flow is likely to be significant in catchments with soils having high hydraulic conductivity and an impermeable or semi-permeable layer at shallow depth that can support a perched water table. Such conditions often occur in humid vegetated catchments where the organic litter protects the mineral soil and maintains high surface permeability. This upper soil profile is often interlaced with macropores caused by roots, decayed root holes, worm holes and other structural channels. When percolating water moving vertically in such a medium reaches an impermeable layer, lateral subsurface flow is generated (Beven and Germann, 1982). Under such conditions water movement occurs in two domains: through the micropores (i.e. the saturated soil matrix) and through the interconnected macropores of the soil system (unsaturated conditions). Where a significant portion of the total flow takes place in the macropores, the response time of subsurface flow to rainfall approaches that of surface runoff, giving rise to a high perceived hydraulic conductivity for the soil profile as a whole (Whipkey, 1965; Mosley, 1979). The response time of flow within the soil matrix is much slower, and except under special conditions such flow may provide only a small contribution to the total storm flow response of a catchment (Sloan and Moore, 1984).

The last mentioned deep aquifer flows give no significant response to storm events. The response time caused by the vertical transport of water through the thick soil above this aquifer is so large that we can speak of an almost constant horizontal flow in the aquifer with no significant response on one specific storm event in a catchment.

The literature contains many works that summarise the level of understanding of the physics of the complex problem of rainfall-runoff transformation (Dunne, 1978, Freeze, 1980). Many efforts have been made to schematise the whole process in order to develop mathematical models (Dooge, 1957, 1973, Amorocho and Hurt, 1964, Freeze and Harlan, 1969, Todini, 1989). This ranges from the simple calculation of design discharge to the two-dimensional representation of the various processes based on suitably and reciprocally conditioned mass balance, energy and momentum equations (SHE, Abbott et al. 1986a,b, Bathurst, 1986, IHDM, Beven et al, 1987), and to the three-dimensional representation of all the exchanges (Binley et al., 1989). Taken together, these latter kinds of model comprise the broad category of distributed differential models (Todini, 1988); they are frequently referred to as "physical models" to highlight the fact that their respective parameters are (or should be) reflected in the field measurements (Beven, 1989). Given their nature, they are mainly used in investigations and research as a mathematical support for the interpretation of physical reality.

Under the pressure of GCM modelers in the last years there has been a demand for models that are simple in terms of calibration and validation and with only a few parameters, corresponding with measurable quantities, to describe the physical process at catchment scale.

Two components can be identified in all lumped conceptual rainfall-runoff models; the first represents the water balance at soil level and the second the transfer to the basin outlet. The water balance at the soil level is the component that characterises the model and constitutes the most important part. Cordova and Rodriguez-Iturbe (1983) summarise this concept most succinctly, saying "... the problem is not *how* to route but *what* to route".

Possible solutions to the routing part of the models can be found in Todini (1988) and Naden (1992), where the surface runoff and river discharge are approximated by linear parabolic equations, the analytical solution of which is widely available. This component is absolutely necessary in GCMs when the actual river discharges are taken into account, in order to correctly evaluate the travel times to the river mouths.

The rest of this paper will concentrate on the water balance in the soil component, given its major relevance to the soil-atmosphere interactions.

THE ARNO MODEL

The ARNO model derives from the Xinanjiang model developed by Zhao (1977, 1984), who expressed the spatial distribution of the soil moisture capacity in the form of a probability distribution function, similar to that advocated by Moore and Clarke (1981) and Moore (1985). Successively, in order to more effectively account for soil depletion due to drainage, the original Xinanjiang model scheme was modified by Todini (1988), who originated the ARNO model within the framework of the hydrological study of the river Arno.

The basic concepts expressed in the ARNO model are:

- The precipitation is considered uniform over the catchment (or sub-catchment) area;
- All the precipitation falling over the soil infiltrates unless the soil has already reached saturation at that particular location, thus providing a mechanism for dynamically varying in time the surface runoff contributing areas;
- The catchment is composed of an infinite number of elementary areas (each with a different soil moisture capacity and a different soil moisture content) for each of which the continuity of mass can be written and updated in time;
- The total runoff is the spatial integral of the infinitesimal contributions deriving from the different elementary areas;

- The soil moisture capacity is related to the total catchment area by means of a spatial probability distribution: therefore the position of the elementary area inside the catchment is no longer important;

- The soil moisture storage is depleted by the evapo-transpiration as well as by the sub-surface flow towards the drainage network and the percolation to deeper layers;

- Both drainage and percolation are expressed by simple empirical expressions.

The ARNO model has been extensively used in hydrological practice: in particular it has become the kernel of a real-time operational system developed on behalf of the Commission of the European Communities (the European Flood Forecasting Operational Real-Time System - EFFORTS, ET&P, 1992) and already operational on several rivers in many countries: the Fuchun in China; the Danube in Germany; the Po, the Arno, the Tiber, the Adda, the Oglio in Italy. Recently it has been tested by a number of meteorologists (Rowntree et al., 1994) and despite the simplicity of its formulation it was successfully included in the Hamburg climate model (Dümenil and Todini, 1992).

The major advantage of the ARNO model is the fact that it is entirely driven by the total soil moisture storage, which is functionally related, by means of simple analytical expressions, to the directly contributing areas, to the drainage and to the percolation amounts: this makes the model extremely useful in evaluating the total amount of soil moisture available for evapotranspiration, which is one of the major requirements for inclusion in GCMs.

The major disadvantage is the lack of physical grounds for establishing some of the parameters, which have to be estimated on the basis of the available precipitation and runoff data, which is not particularly critical in hydrological applications but it becomes so when the model is used in GCMs, where direct input output observations are not really available.

THE TOPMODEL

The TOPMODEL (Beven and Kirkby, 1979; Sivapalan et al., 1987) introduces the use of topography in hydrological modelling. TOPMODEL is a variable contributing area conceptual model in which the predominant factors determining the formation of runoff are represented by the topography of the basin and a negative exponential law linking the transmissivity of the soil with the distance of the saturated zone below ground level (Franchini et al, 1993). Although synthetic, this model is described by Sivapalan et al. (1987) as a "simple physically based conceptual model" in the sense that its parameters can be measured directly in situ. This definition is somewhat optimistic in view of the considerable simplification inherent in the structure of the model and the doubts and uncertainties encountered in defining the parameters of "physical models" themselves. But the inclusion of the effects of variability of topography on contributing area dynamics represents a major advance over previous models based on "point" hydrological responses assumed to apply at catchment scale.

It is not simple to summarise the characteristics of the TOPMODEL since as Beven usually points out it is more a philosophical approach rather than a specific model. In the perspective of using it for GCM modelling, and without loss of generality, the basic assumptions for a lumped TOPMODEL are here described.

The basic concepts expressed in the TOPMODEL are then:

- The precipitation is considered uniform over the catchment (or sub-catchment) area;
- All the precipitation falling over the soil infiltrates unless the soil has already reached saturation at that particular location;
- The total runoff is composed of surface runoff and sub-surface flow;
- The surface runoff mechanism is driven by the dynamical variation of the saturated contributing areas, which is determined on the basis of the topographic index curve;
- The soil moisture storage is depleted by the evapo-transpiration as

well as by the sub-surface flow towards the drainage network and the percolation to deeper layers;

- The permeability in a first soil layer is very large at the soil surface and decays exponentially with depth;
- Each point in the catchment has a uniquely defined contributing area according to topography;
- The downslope movement of water in the capillary fringe of the unsaturated zone is null;
- The downslope movement of water in the saturated zone is driven by gravity and is a function of the local surface topographic slope;
- The sub-surface flow is computed for each point, by integrating in space over the relevant contributing area, the basic equation expressing the motion in saturated porous media, under the assumption of steady state. This reduces the problem to a non linear reservoir model for which both a continuity of mass equation and a relation between storage and runoff are available.

The model has recently being used for many catchments where it gives reasonable to good results (Wood et al., 1988; Durand et al., 1992; Troch et al., 1993a,b; Franchini et al., 1993).

The major advantage of the TOPMODEL is the theoretical possibility of estimating the topographic index curve from the topography of a basin (Quinn et al., 1991); in addition all the other parameters are physically related to the soil and porous media characteristics, which is also a very useful property of the model when used in GCMs, where no direct calibration of parameters is really possible.

The major disadvantage of the TOPMODEL lays in the steady state assumption, which is advocated in order to derive the model integral equations. This assumption, which corresponds to a null travel time from one side to the opposite side of an elementary grid cell (pixel) of the digital terrain model used in the derivation of the Topographic Index Curve, becomes unrealistic for cells of the order of magnitude of hundreds of meters. As a consequence, in order to obtain a correct travel time, the horizontal saturated permeability parameter must be artificially increased by orders of magnitude, which in turn makes the infiltrated water immediately drop to the saturated zone leaving the unsaturated zone mostly depleted, with a general underestimation of the actual evapotranspiration (Franchini et al., 1993).

A SHOPLIST FOR A NEW LUMPED MODEL

From the schematic description of the two models, of their advantages and disadvantages, it seems reasonable to define the following properties to be included in a new model formulation to be used in GCMs for a simplified but sufficiently accurate representation of the latent heat exchanges between soil and atmosphere .

In the following list the properties reflected by the ARNO and the TOPMODEL are marked (A) and (T) respectively, the new ideal properties are marked (N):

i) When dealing with mesh sizes of the order of magnitude of hundreds of metres the precipitation always infiltrates into the soil, unless the soil has already reached saturation (A,T);

ii) The total water storage in the first soil layer is responsible for the control of the saturated surface dynamics, which must also depend on topography, soil properties and soil moisture content (A,T), and for the control of the actual evapo-transpiration losses (A);

iii) The total water stored in the soil is explicitly accounted for at each step in time (A,T) and the actual evapo-transpiration will depend upon it (A);

iv) The total catchment area must be described by means of elementary areas (pixels) which are small compared to the overall catchment area (A,T);

v) The position of a pixel must not be relevant: what matters is the total probability distribution which is used for obtaining the surface average value (A,T);

vi) Drainage and percolation must follow physical laws (T) and not empirical ones (A);

vii) The digital elevation model should provide information on the horizontal flow driving mechanism (T);

viii) The local horizontal flow as well as the local transmissivity must depend upon the overall soil moisture content obtained as the integral along the vertical direction of the soil moisture profile (N);

ix) Model parameters must be determined on physical grounds (T) and their values must be of physically meaningful orders of magnitude (A);

x) The discrete space model must be expressed in terms of a non-linear reservoir model invariant in its essential structure, with only the parameters varying from small to larger scales (N).

The first hypothesis, which is the same expressed by both ARNO and TOPMODEL, is a reasonable hypothesis mainly due to the extremely large values of permeability in the first soil layer (Dunnian infiltration); but even if the Hortonian infiltration holds, this property tends to be true with the increasing size of the pixel, due to the characteristic travel time needed to cross the pixel itself, which is no longer infinitesimal.

In addition, the major idea for the development of the new model is the concept of lumping. The original problem is obviously a three dimensional problem which can be reduced to a two dimensional problem if property (viii) applies and then reduced to a zero dimensional problem if property (v) applies.

Which brings up the following two major questions:

a) Is it possible to lump the horizontal flow properties in the vertical dimension? In other words can we express the total horizontal permeability (generally called transmissivity) as a function of the total soil moisture content?

b) Is the succession of hill-slopes important or can it be disregarded, while only the overall frequency distribution or its integral are retained?

These questions must be answered by comparing errors originating from other sources of uncertainty with the order of magnitude of the error one makes assuming that a and b are correct.

The answer to question a) is yes it is possible to lump the phenomenon in the vertical dimension without a significant loss of accuracy. This was demonstrated by showing that the integral of the horizontal permeability along the vertical dimension can be related to the vertically integrated soil moisture content, without great loss of accuracy.

The answer to question b) is also affirmative, and in fact both models (the ARNO and the TOPMODEL) are already implicitly assuming it.

In the next sections the description of the results from single experiments set up in order to demonstrate the validity of the assumptions, will be reported.

THE VERTICAL LUMPING

Description of horizontal flow in unsaturated conditions could be made based upon the knowledge of the vertical moisture content profile into the soil. Due to the high conductivity value, caused by macropores in the top of the soil (Beven and Germann, 1982), gravity will be the dominant mechanism driving water from the top of the soil to the bottom (impermeable or semi-impermeable lower boundary). The latter mentioned boundary will create a perched water table. In this zone a not negligible horizontal propagation (also involving unsaturated flow), will occur.

Nevertheless, the depth of this high conductive soil (one or two metres) will be negligible with respect to the horizontal grid dimensions (> 100×100 m^2). By assuming a number of vertical soil moisture profiles it was shown that the horizontal flow (or hydraulic conductivity) evaluated starting from a real vertical profile and that deriving from the integral moisture content were comparable. This allows the transient phase of vertical infiltration to be neglected; in other words, when interested in the horizontal movement of water, it is possible to avoid, within the range of reasonable errors, the integration of the unsaturated soil vertical infiltration equation, namely Richards equation, by assuming a transmissivity which is a function of the total soil moisture content.

In order to clarify the point, let us assume that the relative permeability can be expressed, with a more or less simple relationship, as a function of the reduced soil moisture content defined as:

$$\tilde{\vartheta} = \frac{\vartheta - \vartheta_r}{\vartheta_s - \vartheta_r} \qquad (1)$$

where: ϑ is the soil moisture content
ϑ_r is the residual soil moisture content
ϑ_s is the saturation soil moisture content

and where all the quantities are point values varying with the downwards directed vertical co-ordinate z.

The hypothesis is that the horizontal transmissivity T can be expressed in terms of the integral, along the vertical, of the reduced soil moisture and that the shape of this function, apart from the parameter values, is the same as the one relating the point relative permeability to the point reduced soil moisture content, namely:

$$T = \int_0^L k\left(\tilde{\vartheta}(z)\right) dz \cong T\left(\frac{\tilde{\Theta}}{L}\right) = T\left(\frac{1}{L}\int_0^L \tilde{\vartheta}(z)\, dz\right) \qquad (2)$$

where: $\tilde{\Theta} = \int_0^L \tilde{\vartheta}(z)\, dz$ is the integral, along the vertical dimension, of the relative moisture content
T is the horizontal transmissivity
k is the point relative permeability

To demonstrate the validity of this hypothesis, two different but very well known relationships describing hydraulic conductivity with moisture content were used; the first one due to Brooks and Corey (1964) and the second one given by van Genuchten (1980).

Using the Brooks and Corey expression, namely $k\left(\tilde{\vartheta}\right) = k_s\, \tilde{\vartheta}^a$, with k_s the permeability at saturation and a a parameter, the hydraulic permeability was calculated from a number of different approximated vertical profiles having a variation in moisture content with depth while keeping the same integral moisture content Θ. A number of different $\bar{\Theta} = \Theta / (\vartheta_s - \vartheta_r)$ values were taken into account, ranging from 0 to 1.

The results for a wide range of soil moisture profiles, both increasing and decreasing with depth, are shown in Figure 1 where also the flux index given by the expression $T(\bar{\Theta})/k_s = L\bar{\Theta}^{\alpha}$ is plotted clearly demonstrating that given the range of uncertainty in the soil properties, the assumption holds.

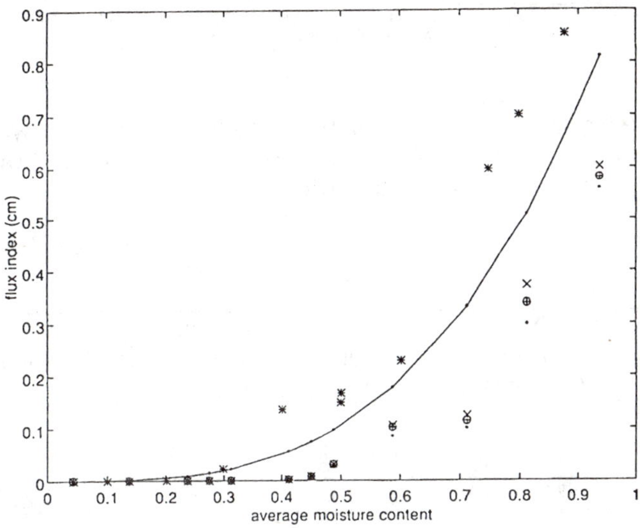

Figure 1 - The different values of the flux index, calculated as the integral of the point relative permeability for different soil moisture profiles or as a function of the total soil moisture content (solid line).

A similar analysis was done by using a suitable approximation of the van Genuchten expression (suitable in that it allowed for analytical integration); Figure 2 shows a number of different water profiles for a given terrain, and Table 1 shows a comparison of the two expressions this time matched in terms of transmissivity.

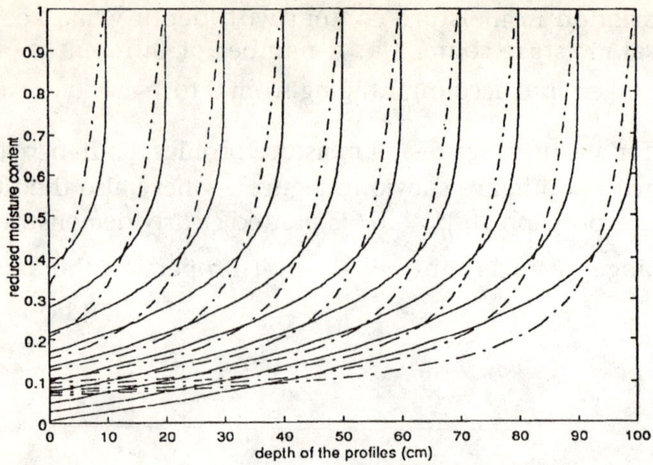

Figure 2 - Different soil moisture profiles, calculated with the van Genuchten expression and the approximated one allowing for analytical integration.

depth of profiles	moisture content van Genuchten	moisture content appr. expr.	transmiss. van Genuchten	transmiss approx. expression
	%	%		
100 cm.	16.8089	17.0488	0.0121	0.0269
90 cm.	21.0549	26.9006	0.2000	0.2267
80 cm.	28.4701	36.5355	0.4000	0.4262
70 cm.	36.6885	45.9320	0.6000	0.6254
60 cm.	45.2917	55.0630	0.8000	0.8242
50 cm.	54.1264	63.8934	1.0000	1.0225
40 cm.	63.1180	72.3755	1.2000	1.2204
30 cm.	72.2239	80.4400	1.4000	1.4175
20 cm.	81.4173	87.9757	1.6000	1.6138
10 cm.	90.6803	94.7661	1.8000	1.8087
0 cm.	100.0000	100.0000	2.0000	2.0000

Table 1 - Values of reduced moisture content and transmissivity as a function of depth, calculated with the van Genuchten expression and the approximated one.

In this case, from the following expression which was adopted for the point relative permeability:

$$k(\bar{\vartheta}) = k_s\left[1-(1-\bar{\vartheta})^c\right] \tag{3}$$

the following similar expression could be derived for the transmissivity, by analytical integration (Benning, 1994):

$$T(\bar{\Theta}) = k_s L\left[1-\lambda\left(1-\frac{\bar{\Theta}}{L}\right)^{\gamma}\right] \tag{4}$$

where γ and λ are two parameters that can be derived from the original ones, and again one can see that the shape of the function remains more or less unchanged.

Concluding, on the basis of the obtained results the author believes that it is possible to assume a simplified expression in terms of the total vertical soil moisture content for the description of the horizontal transmissivity, thus reducing the initially three dimensional problem to a two dimensional one, after lumping in the vertical dimension.

THE SUCCESSION OF SLOPES

If one assumes that the horizontal transmissivity is very high in a first soil layer and that can be derived from the Brooks and Corey expression, namely $T(\bar{\Theta}) = k_s L \bar{\Theta}^{\alpha}$, it is easy to show that the horizontal movement of water in the unsaturated zone, on the same assumptions of the TOPMODEL, can be approximated by means of a kinematic wave model, which gives:

$$(\vartheta_s - \vartheta_r)L\frac{\partial\bar{\Theta}}{\partial t} = p + \frac{\partial q}{\partial x} = p + \alpha k_s L \tan(\beta)\bar{\Theta}^{(\alpha-1)}\frac{\partial\bar{\Theta}}{\partial x} \tag{5}$$

where : $\tan(\beta)$ represents the terrain slope;
q is the interflow
p is the precipitation

The model is written in one dimension because it is thought that the movement of water will proceed along the hillslopes towards the closest drain, and afterwards it will proceed along the drainage network: this allows for the reduction of an additional dimension and reduces the original problem to the solution of a one dimensional kinematic wave problem.

The basic idea of a higher conductivity in the first layer of limited dimension (the rootzone) expressed in TOPMODEL, which was stated by Freeze (1972b) and Sloan and Moore (1984), is clearly preserved in this formulation. However, the exponential decay in which this phenomenon is presented in TOPMODEL is replaced by an equation that gives the relation of the transmissivity with the total moisture content while neglecting, after the vertical lumping, the dependence of saturated hydraulic conductivity with depth. This means that the kinematic model considers a vertical layer of limited dimensions characterised by a high constant saturated conductivity, but with a corresponding transmissivity varying upon the total moisture content.

Kinematic wave equations have been extensively used to model surface propagation. It was the American hydrologist, Horton who in 1933 carried out the earliest recorded scientific studies of surface runoff. Later Keulegan (1945) applied the continuity and momentum equations conjunctively for surface runoff analysis. He investigated the magnitude of the various terms in the dynamic equation of St. Venant and indicated that a simplified form of the equation, now termed the kinematic equation, would be adequate for surface runoff.

Starting with the formulation of the kinematic wave theory by Lighthill and Whitham (1955), kinematic surface runoff models have been utilised increasingly in hydrologic investigations. The first application of kinematic wave routing to surface runoff and groundwater flow was by Henderson and Wooding (1964). The conditions under which the kinematic flow approximation holds for surface runoff were first investigated by Woolhiser and Liggett (1967); they found it is an accurate approximation to the full equations for most surface runoff cases (cf. Stephenson et al., 1986).

In an earlier study, Henderson and Wooding (1964), provide solutions for horizontal kinematic subsurface flow through a porous medium of constant permeability both for steady state and a rising water-table. They compare the extended Dupuit-Forchheimer equation to a further simplification of the flow equation in which it is assumed that the hydraulic gradient at any point within the saturated zone is equal to the bed slope. Beven (1981, 1982), gave in two papers about kinematic subsurface storm flow a kinematic wave equation valid in a sloping soil mantle of constant saturated hydraulic conductivity overlying a relative impermeable bedrock sloping surface. In the first article, mainly devoted to the study of horizontal propagation, Beven stated that this equation was a good approximation to the more correct extended Dupuit-Forchheimer equation (see Beven (1981) for further details). In the second article of Beven (1982) the vertical propagation in the unsaturated zone was also taken into account, to evaluate the time at which the wetting front reaches the bottom of the profile. In addition a kinematic approximation in subsurface flows (both in vertical and horizontal directions) has been successfully tested by several other authors (see Borah et al. 1980; Charbeneau 1984; Sloan and Moore, 1984; Hurley and Pantelis, 1985; Stagnitti et al., 1986, Steenhuis et al., 1988).

In TOPMODEL a topographic index curve is used in order to synthesise the topographical information of the catchment. In doing so there is an implicit assumption that the overall behaviour does not depend upon the position of the single slope, but essentially on the overall distribution of the slopes. Also in the case of the kinematic wave approach a similar assumption must be made, and therefore it was felt essential to prove that the flow at the end of a cascade of slopes is not strongly modified by the relative position.

In order to prove this property of the kinematic scheme several slopes were generated at random in the range 0-10%, and the relevant resulting flows at the end of the different cascades are compared among them as in Figures 3a,b.

Given the results, as one can see in Figures 3a,b, the author believes that the alternative successions of slopes produce differences that can hardly be separated from other sources of errors, thus justifying the hypothesis that slopes can be treated in terms of their probability distribution, rather than in terms of their actual succession.

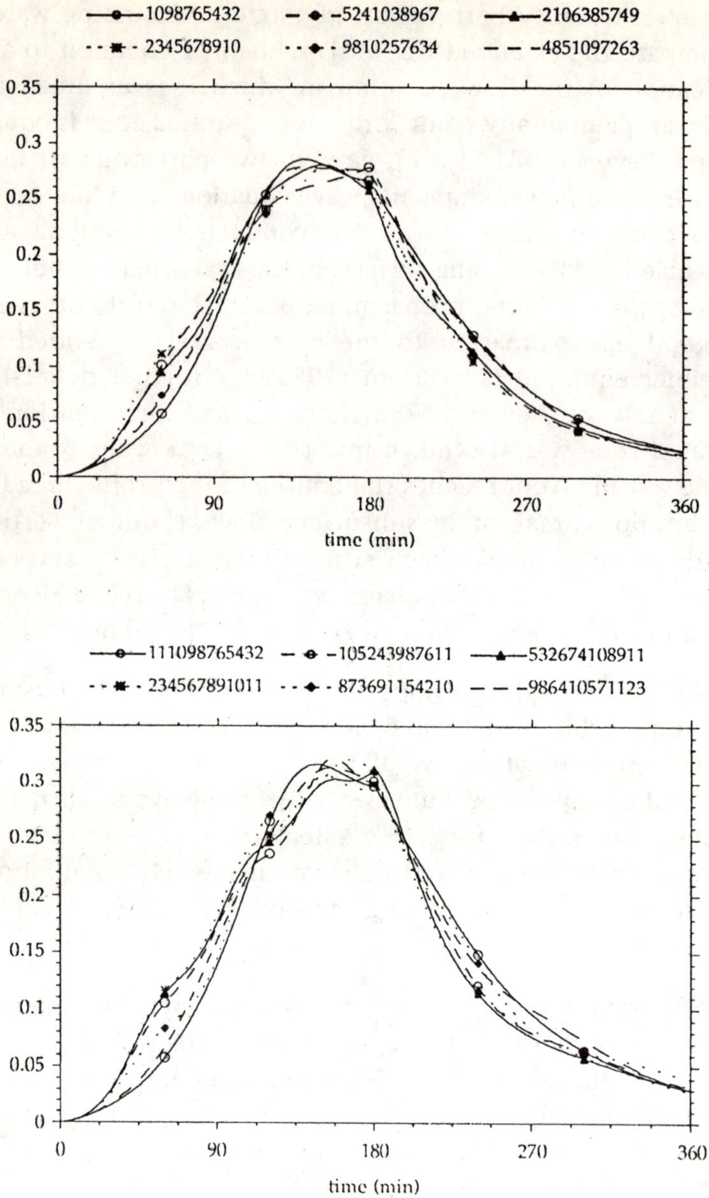

Figure 3 - Different kinematic waves resulting from permutation of 9 slopes (a) and 10 slopes (b).

THE PROPOSITION OF A NEW MODEL: THE TOPKAPI

The development of a new modelling approach at catchment scale is here sketched, in that it is not yet a real model but it constitutes a logical synthesis of what has previously been analysed.

The model name TOPKAPI derives from its main features, i.e. TOPographic Kinematic APproximation and Integration. Its development originates in fact from the integration of the kinematic approximation presented above over increasing size domains.

The continuity of mass can be expressed as:

$$(\vartheta_s - \vartheta_r) L \frac{\partial \bar{\Theta}}{\partial t} + \frac{\partial q}{\partial x} = p \qquad (6)$$

where p is the precipitation taken as constant in space and time, and in line with what expressed above, is considered to entirely infiltrate in the soil until it reaches saturation;

q is the flow, given as:

$$q = tan(\beta) T(\bar{\Theta}) = tan(\beta) k_s L \bar{\Theta}^{\alpha} \qquad (7)$$

where, similarly to the assumption made in TOPMODEL, the surface slope is taken instead of the head slope.

The main idea is then to combine the kinematic model with the topography, which is described by means of a number of elementary cells of fixed size (for instance 100×100 m or 400×400 m).

For this purpose, it is convenient to rewrite equations (6) and (7) in terms of the actual total vertical soil moisture content η, defined as $\eta = (\vartheta_s - \vartheta_r) L \bar{\Theta}$ as described in Beven et al.(1988), to give:

$$\begin{cases} \dfrac{\partial \eta}{\partial t} = p - \dfrac{\partial q}{\partial x} \\ q = T\, tan(\beta) = L k_s\, tan(\beta) \bar{\Theta}^{\alpha} = \dfrac{L k_s\, tan(\beta)}{(\vartheta_s - \vartheta_r)^{\alpha} L^{\alpha}} \eta^{\alpha} = C \eta^{\alpha} \end{cases} \qquad (8)$$

where the following substitution was applied for the sake of clarity:

$$C = \frac{L k_s \tan(\beta)}{(\vartheta_s - \vartheta_r)^\alpha L^\alpha} \tag{9}$$

The two equations (8) can then be combined together, to give the following kinematic wave model:

$$\frac{\partial \eta}{\partial t} = p - \frac{L k_s \tan(\beta)}{(\vartheta_s - \vartheta_r)^\alpha L^\alpha} \frac{\partial \eta^\alpha}{\partial x} = p - C \frac{\partial \eta^\alpha}{\partial x} \tag{10}$$

In the TOPMODEL, after integrating in space a similar expression, the time derivative is set to zero, thus implying that stationarity is immediately reached, which may be reasonable for small mesh sizes (in the range of one or few meters), but it is not when dealing with mesh sizes of hundreds of meters. Equation (10) can in fact be integrated by means of the method of characteristics (Eagleson, 1970) over a mesh of size x, and it is possible to derive the concentration time corresponding to a precipitation that equals the maximum possible discharge $p = L k_s \tan(\beta)$, which gives:

$$t_c = x^{1/\alpha} \frac{(\vartheta_s - \vartheta_r)}{k_s \tan(\beta)} \tag{11}$$

Unfortunately the solution provided by the method of characteristics is only valid along the characteristic lines and, given the non-linearity, does not provide an explicit solution in time and space, which is required to find a relationship between the point water content and the total volume stored in the mesh. It is therefore necessary to integrate equation (10) over the mesh on the basis of a reasonable simplifying assumption.

On these lines it is reasonable to assume that, although the water content η varies in space, its time partial derivative $\frac{\partial \eta}{\partial t}$ does not vary strongly in space at a given time t; therefore its space integral can also be approximately considered constant in space, which gives:

$$p - \frac{1}{x} \int_0^x \frac{\partial \eta}{\partial t} dx \cong q_s \tag{12}$$

with q_s a specific yield, which varies in time, but at any time is assumed practically constant in space. Under this assumption the flow at any time increases linearly in space, but not necessarily by an increment equal to the precipitation p as in TOPMODEL; this will only happen at saturation when $\frac{\partial \eta}{\partial t}$ is null. It should also be noted that this formulation is applicable when p is null and the specific yield q_s becomes negative.

At the end of a mesh of size x the discharge thus becomes:

$$q_x = q_0 + \int_0^x \left[p - \frac{\partial \eta}{\partial t} \right]_\xi d\xi = q_0 + q_s x \tag{13}$$

which, substituted for in the second of equation (8), allows for the derivation of the soil moisture horizontal profile:

$$\eta(\xi) = \left(\frac{q_\xi}{C} \right)^{1/\alpha} = \left(\frac{q_0 + q_s x}{C} \right)^{1/\alpha} \tag{14}$$

The total volume stored in a single mesh can thus be computed as:

$$V_x = \int_0^x \eta(\xi) d\xi = \int_0^x \left(\frac{q_0 + q_s \xi}{C} \right)^{1/\alpha} d\xi$$

$$y = \frac{q_0 + q_s \xi}{C} \qquad \xi = \frac{C y - q_0}{q_s} \qquad d\xi = \frac{C}{q_s} dy$$

$$\begin{cases} \xi = 0 & \Rightarrow \quad y = \dfrac{q_0}{C} \\ \xi = x & \Rightarrow \quad y = \dfrac{q_0 + q_s x}{C} \end{cases} \tag{15}$$

$$V_x = \int_{\frac{q_0}{C}}^{\frac{q_0 + q_s x}{C}} \frac{C}{q_s} y^{1/\alpha} dy = \frac{\alpha C}{(\alpha + 1) q_s} \left[y^{\frac{\alpha+1}{\alpha}} \right]_{\frac{q_0}{C}}^{\frac{q_0 + q_s x}{C}} = \frac{\alpha C}{(\alpha + 1) q_s} \left[\left(\frac{q_0 + q_s x}{C} \right)^{\frac{\alpha+1}{\alpha}} - \left(\frac{q_0}{C} \right)^{\frac{\alpha+1}{\alpha}} \right]$$

V_x can also be written as:

$$V_x = \frac{\alpha C^{-1/\alpha}}{(\alpha + 1) q_s} \left[(q_0 + q_s x)^{\frac{\alpha+1}{\alpha}} - (q_0)^{\frac{\alpha+1}{\alpha}} \right] \tag{16}$$

Up to this point the integration has been carried out on a single mesh. By extending the hypothesis of constancy of the specific yield to all the meshes contributing to a downstream one, it is possible to write an expression relating the outflow discharge from the ith mesh, to its contributing surface and to the specific yield.

For the sake of clarity the derivation will be done for meshes in series as in a hillslope, but a similar result holds for the general case of meshes in series and in parallel.

Given the constancy of the specific yield one easily gets for the ith mesh in series:

$$q_i = i\,x\,q_s \tag{17}$$

and bearing in mind the relationship between the discharge and the water content:

$$\eta_i = \left(\frac{i\,x\,q_s}{C_i}\right)^{1/\alpha} \tag{18}$$

which can be written as:

$$x\,q_s = \frac{\eta_i^{\alpha}\,C_i}{i} \tag{19}$$

The total volume of water stored in i^{th} mesh becomes:

$$\begin{aligned} V_i &= \frac{\alpha C_i^{-1/\alpha}}{(\alpha+1)q_s}\left[\left(q_{(i-1)}+q_s x\right)^{\frac{\alpha+1}{\alpha}} - \left(q_{(i-1)}\right)^{\frac{\alpha+1}{\alpha}}\right] = \frac{\alpha C_i^{-1/\alpha}}{(\alpha+1)q_s}\left[\left(i\,x\,q_s\right)^{\frac{\alpha+1}{\alpha}} - \left((i-1)\,x\,q_s\right)^{\frac{\alpha+1}{\alpha}}\right] \\ &= \frac{\alpha}{(\alpha+1)}\,x\left(\frac{x\,q_s}{C_i}\right)^{1/\alpha}\left[i^{\frac{\alpha+1}{\alpha}} - (i-1)^{\frac{\alpha+1}{\alpha}}\right] \end{aligned} \tag{20}$$

while the total volume stored in n meshes can finally be estimated as:

$$V_n = \frac{\alpha}{\alpha+1}\,x\left(x\,q_s\right)^{1/\alpha}\sum_{i=1}^{n}\frac{i^{\frac{\alpha+1}{\alpha}} - (i-1)^{\frac{\alpha+1}{\alpha}}}{C_i^{1/\alpha}} \tag{21}$$

In order to find a relationship between the outflow discharge and the total volume stored as well as with the average soil moisture content, one has to recall that equation (17) also applies to the nth mesh, which gives:

$$x\,q_s = \frac{\eta_n^{\alpha}\,C_n}{n} \tag{22}$$

Equation (22) can then be substituted for into equation (21) to give:

$$V_n = \frac{\alpha}{\alpha+1}\frac{x C_n^{1/\alpha}}{n^{1/\alpha}}\eta_n \sum_{i=1}^{n} \frac{i^{\frac{\alpha+1}{\alpha}} - (i-1)^{\frac{\alpha+1}{\alpha}}}{C_i^{1/\alpha}} = \frac{\alpha}{\alpha+1}\frac{n x C_n^{1/\alpha}}{n^{\frac{\alpha+1}{\alpha}}}\eta_n \sum_{i=1}^{n} \frac{i^{\frac{\alpha+1}{\alpha}} - (i-1)^{\frac{\alpha+1}{\alpha}}}{C_i^{1/\alpha}}$$
$$= \frac{\alpha}{\alpha+1} n x C_n^{1/\alpha}\eta_n \sum_{i=1}^{n} \frac{\left(\frac{i}{n}\right)^{\frac{\alpha+1}{\alpha}} - \left(\frac{i-1}{n}\right)^{\frac{\alpha+1}{\alpha}}}{C_i^{1/\alpha}} \tag{23}$$

Recalling the meaning of C_i, the summation term in equation (23) represents an average equivalent resistance weighted by a topological density function, given by the i/n terms (which will also contain summations for the parallel connections), and can be replaced with:

$$\frac{1}{\overline{C}_n} = \left[\sum_{i=1}^{n} \frac{\left(\frac{i}{n}\right)^{\frac{\alpha+1}{\alpha}} - \left(\frac{i-1}{n}\right)^{\frac{\alpha+1}{\alpha}}}{C_i^{1/\alpha}}\right]^{\alpha} \tag{24}$$

where $1/\overline{C}_n$ is the TOTOPS index, which can be computed on the basis of both the TOpography of the surfaces and the TOpology of the meshes connection graph, as well as on the Physical Soil properties.

Substituting for equation (24) into equation (23), the sought expression, relating the outlet water content to the total volume stored, can be finally given as:

$$V_n = \frac{\alpha}{\alpha+1} n x \eta_n \left(\frac{C_n}{\overline{C}_n}\right)^{1/\alpha} \tag{25}$$

or as:

$$\eta_n = \frac{\alpha+1}{\alpha n x}\left(\frac{\overline{C}_n}{C_n}\right)^{1/\alpha} V_n \tag{26}$$

On the other hand the spatial integration of equation (10) to all the contributing area gives:

$$\frac{\partial V_n}{\partial t} = n x p - C_n \eta_n^{\alpha} \tag{27}$$

Substitution of equation (26) into equation (27) allows for the derivation of a non-linear reservoir model at the scale of the hillslope whose parameters are entirely derived from the physical as well as from the topographical and topological characteristics:

$$\frac{\partial V_n}{\partial t} = nxp - \left(\frac{\alpha + 1}{\alpha n x}\right)^{\alpha} \overline{C}_n V_n^{\alpha} \tag{28}$$

Equation (28) shows that the kinematic model assumption results in the derivation of a zero dimensional non-linear reservoir model which extends its validity beyond the size of the elementary grid cells (pixels) spatially describing the geographical information, to larger sizes provided that the soil properties, the topographical and topological characteristic are appropriately synthesised. Integration of equation (28) allows for the determination of the volume stored in the soil at any time and can be used in the same form for all the contributing areas, only requiring the computation of the TOTOPS index for the different areas contributing to the drainage network, as is done by the TOPMODEL combined topography-soil index. Note that the TOTOPS index bears more information than the TOPMODEL index, in that it includes the topology of the drainage channel connections, thus allowing for a geomorphological interpretation.

Finally the explicit integration of equation (28) only exists in the form of an infinite series, but an approximated solution can be found in order to compute at any time the volume stored in the soil. Given the volume it is then possible, by means of equation (26), to compute the local soil moisture and to compare it with the maximum amount that can be locally stored in order to estimate the saturated areas, for which a similar kinematic model will be used in order to express the surface runoff.

CONCLUSIONS

Although much work is still needed in order to validate the hypotheses expressed in this paper, it is the belief of the author that a number of points can already be reasonably well established, allowing the problem of extending the high non-linearities inherent in the unsaturated soil horizontal flow phenomena to larger scales to be overcome.

The first hypothesis addresses the possibility of lumping in the vertical dimension the horizontal flow hydraulic conveyance characteristics and relating them to the total soil moisture content.

The second is the possibility of using a unique partial differential model (the kinematic model) from the micro scale to that of the hillslope and beyond it in order to derive a non linear reservoir type model also valid at macro-scales.

In addition, although it may well be that the present assumptions, made in order to extend the validity of the point model to larger scales, will have to be modified after verification with real world data, nevertheless it has been shown that:

1) the concentration time is an important characteristic to be preserved in the model in order to guarantee the physical meaning of the parameters;

2) a physically meaningful characteristic index must include not only the soil properties and the topography, but also the topological connection among the meshes (a well known principle in the derivation of the GIUH);

3) the kinematic wave model is one of the most appropriate schemes to study the horizontal flow on the surface slopes of the catchments but also in the unsaturated zone.

Finally it is hoped that experimental work will be carried out in order to verify and validate the assumptions made. In particular the assumption of the space invariance of the specific yield, which although certainly more realistic than the steady state assumption of the TOPMODEL, should still be verified and if necessarily improved.

ACKNOWLEDGEMENT

The numerical experiments described in the paper were carried out, in collaboration with Dr. P. Di Giammarco and Mr. R. Benning, within the frame the R&D contract EPOCH-CT90-0016, between the Commission of the European Communities and the University of Bologna.

REFERENCES

Abbott, M.B., Bathurst, J.C., Cunge, J.A., O'Connell, P.E. and Rasmussen, J., 1986a. An introduction to the European Hydrological System - Système Hydrologique Européen, "SHE", 1: History and philosophy of a physically-based, distributed modelling system. J. Hydrol., 87: 45-59.

Abbott, M.B., Bathurst, J.C., Cunge, J.A., O'Connell, P.E. and Rasmussen, J., 1986b. An introduction to the European Hydrological System - Système Hydrologique Européen, "SHE", 2: Structure of physically.based, distributed modelling system. J. Hydrol., 87: 61-77.

Amorocho, J. and G.T. Hurt, 1964. Nonlinear analysis of hydrologic systems, Water Resour. Res. Contrib. 40, University of California, Berkeley.

Bathurst, J.C. 1986. Physically based distributed modelling of an upland catchment using the Systeme Hydrologique Europeen, Journal of Hydrology 87 pp. 79-102.

Benning, R.G., 1994. Towards a new lumped parameterization at catchment scale. Master Thesis University of Wageningen in collaboration with the University of Bologna (in press).

Beven, K.J. and Kirkby M.J., 1979. A physically based, variable contributing area model of basin Hydrology. Hydrological Sciences-Bulletin des Sciences Hydrologiques, 24, 1-3.

Beven, K.J., 1981. Kinematic subsurface stormflow, Water Resour. Res., Vol. 17 no. 5, pp. 1419-1424.

Beven, K.J., 1982. On subsurface stormflow: Predictions with simple kinematic theory for saturated and unsaturated flows, Water Resour. Res., Vol. 18 no. 6, pp. 1627-1633.

Beven, K.J. and P. Germann, 1982. Macropores and water flow in soils, Water Resour. Res., Vol. 18, no. 5, pp. 1311-1325.

Beven, K.J., A. Calver and E.M. Morris, 1987. The Institute of Hydrology Distributed Model (IHDM). Inst. Hydrolo. Rep. no.98, Wallingford.

Beven, K.J., Wood E.F., and Sivapalan M., 1988. On hydrological heterogeneity-catchment morphology and catchment response. Journal of Hydrology, 100, pp. 353-375.

Beven, K.J., 1989. Changing ideas in hydrology - The case of physically-based models, Journal of Hydrology, 105, pp. 157-172.

Binley, A., J. Elgy and K.J. Beven, 1989. A physically based model of heterogeneous hillslopes 1. Runoff production, Water Resour. Res. Vol. 25, no. 6, pp. 1219-1226.

Borah, D.K., S.N. Prasad and C.V. Alonso, 1980. Kinematic wave routing incorporating shock fitting, Water Resour. Res. Vol. 16, no. 3, pp. 529-541.

Brooks, R.H. and A.T. Corey, 1964. Hydraulic properties of porous media,

Hydrol. Pap. 3, Colo. State Univ., Fort Collins.
Charbeneau, R.J., 1984. Kinematic models for soil moisture and solute transport, Water Resour. Res., Vol. 20, no. 6, pp. 699-706.
Cordova, J.R., I. Roderiguez-Iturbe, 1983. Geomorphologic estimation of extreme flow probabilities, Journal of Hydrology 65, pp. 159-173.
Dooge, J.C.I., 1957. The Rational Method for Estimating Floot Peak. Engineering (London), 184: 311-3374.
Dooge, J.C.I., 1973. The linear theory of hydrologic systems. Tech. Bull.U.S. Dep.Agric., No. 1468, U.S. Gov. Print. Off., Washington, D.C.
Dümenil, L and Todini, E., 1992. A rainfall-runoff scheme for use in the Hamburgh climate model. In: J.P. O'Kane (Editor), Advances in Theoretical Hydrology, a tribute to James Dooge. European Geophysical Society Series of Hydrological Sciences, 1. Elsevier, Amsterdam, pp. 129-157.
Dunne, T., T.R. Moore, and C.H. Taylor, 1975. Recognition and prediction of runoff-producing zones in humid regions, Hydrol. Sci. Bull., vol. 20, no. 3, pp. 305-327.
Dunne, T. 1978. Field studies of hillslope flow processes, in Hillslope Processes, edited by M.J. Kirkby, pp. 227-293, John Wiley, New York.
Durand, P., Robson A., and Colin N., 1992. Modelling the hydrology of submediterranean mountain catchments (Mont-Lozère, France) using TOPMODEL: initial results. Journal of Hydrology, 139 (1992) pp. 1-14.
Eagleson, P.S., 1970. Dynamic Hydrology, McGraw-Hill, New York.
ET&P, 1992. The Fuchun River Project - A Computer Based Real Time System, Final Report of EC Research Contract CI13-0004-I.
Franchini, M., J. Wendling, E. Todini and Ch. Obled 1993. Physical interpretation and sensitivity analysis of the TOP Model, submitted to Journal of Hydrology.
Freeze, R.A. and R.L. Harlan, 1969. Blueprint for a physically based digitally simulated hydrologic response model, Journal of Hydrology no. 9, pp. 237-258.
Freeze, R.A., 1972a. Role of subsurface flow in generating surface runoff, 1, Base flow contributions to channel flow, Water Resour. Res., 8 (3), 609-623.
Freeze, R.A., 1972b. Role of subsurface flow in generating surface runoff, 2, Upstream source areas, Water Resour. Res., 8 (6), 1272-1283.
Freeze, R.A., 1980. A stochastic conceptual analysis of rainfall-runoff processes on a hillslope, Water Resour. Res., Vol. 16 no. 8, pp. 1272-1283.
Henderson, F.M., R.A. Wooding, 1964. Overland flow and groundwater flow from a steady rainfall of finite duration, Journal of Geophysical Research, Vol. 69, no. 6 pp. 1531-1540.
Horton, R.E., 1933. The role of infiltration in the hydrologic cycle. Trans.

Am. Geophys. Union, Vol. 14, pp. 446-460.

Horton, R.E., 1939. Analysis of runoff plat experiments with varying infiltration capacity, Trans. Am. Geophys. Union, Vol. 20 pp. 693-711.

Hurley, D.G. and G. Pantelis, 1985. Unsaturated and saturated flow through a thin porous layer on a hillslope, Water Resour. Res., Vol. 21, no. 6, pp. 821-824.

Keulegan, G.H., 1945. Spatially varied discharge over a sloping plane. Amer. Geophys. Union Trans. Part 6, pp. 956-959.

Kirkby, M.J., 1985. Hillslope hydrology, in Hydrological Forecasting, edited by M.G. Anderson and T.P. Burt, John Wiley & Sons Ltd.

Lighthill, F.R.S. and G.B. Whitham, 1955. On kinematic waves, I, Flood measurements in long rivers. Proc. Royal Soc. of Londen, A, 229, pp. 281-316.

Moore R.J. and R.T. Clarke, 1981. A distribution function approach to rainfall-runoff modelling, Water Resour. Res., Vol. 17, no. 5, pp. 1367-1382.

Moore R.J., 1985. The probability-distributed principle and runoff production at point and basin scales, Hydrological Sciences Journal, Vol. 30, no. 2, pp. 273-297.

Mosley, M.P., 1979. Streamflow generation in a forested watershed, New Zealand, Water Resour. Res., Vol. 15, no. 4, pp. 795-806.

Naden, P.S., 1992. Spatial variability in flood estimation for large catchments: the exploitation of channel network structure. J. Hydrological Sciences, 37,1,2:53-71.

Quinn, P., K. Beven, D. Morris and R. Moore, 1991. The use of Digital Terrain Data in Modelling the Response of Hillslopes and Headwaters, Internal technical report, University of Lancaster.

Rowntree, P.R. and J. Lean, 1994. Validation of hydrological schemes for climate models against catchment data. J. Hydrology 155:301-323.

Sherman, L.K., 1932. Streamflow from rainfall by the unit graph method. Eng. News Rec., 108: 501-505.

Sivapalan, M.K., and Wood E.F., 1986. Spatial heterogeneity and scale in the infiltration response of catchments. Scale Problems in Hydrology, edited by V.K. Gupta, I. Rodriguez-Iturbe, and E.F. Wood, D. Reidel, Dordrecht, Holland, pp. 81-106.

Sivapalan, M., K. Beven and E.F. Wood, 1987. On Hydrologic Similarity 2. A Scaled Model of Storm Runoff Production. Water Resour. Research, Vol. 23 no. 12 pp. 2266-2278.

Sloan, G.P., I.D. Moore, 1984. Modelling subsurface stormflow on steeply sloping forested watersheds, Water Resour. Res., Vol. 20, no. 12, pp. 1815-1822.

Smith, R.E. and R.H.B. Hebbert, 1983. Mathematical simulation of interdependent surface and subsurface hydrologic processes, Water Resour. Res., Vol. 19, no. 4, pp. 987-1001.

Stagnitti, F., M.B. Parlange, T.S. Steenhuis and J.-Y. Parlange, 1986. Drainage from a uniform soil layer on a hillslope, Water Resour. Res., Vol. 22, no. 5, pp. 631-634.

Steenhuis, T.S., J.-Y. Parlange, M.B. Parlange and F. Stagnetti, 1988. A simple model for flow on hillslopes, Agric. Water Manage., 14, pp. 153-168.

Stephenson, D. and M.E. Meadows, 1986. Kinematic Hydrology and Modeling, Developments in Water Science 26, Elsevier, Amsterdam.

Todini, E. 1988. Rainfall-runoff modelling past, present and future. Journal of Hydrology, 100 (1988) 341-352. Elsevier Science Publishers b.v.

Todini, E., 1988. Un modello di previsione di piena per il fiume Arno, Regione Toscana, Firenze (in Italian).

Todini, E. 1989. Flood forecasting models. Excerpta, Vol 4, pages 117-162.

Troch, P.A., M. Mancini, C. Paniconi, and E.F. Wood, 1993a. Evaluation of a distributed catchment scale water balance model, Water Resour. Res., Vol. 29, no. 6, pp. 1805-1817.

Troch, P.A., J.A. Smith, E.F. Wood and F.P. de Troch, 1993b. Empirical and model based analysis of extreme flood events in a humid region, Extreme Hydrological Events: Precipitation, Floods and Droughts (Proceedings of the Yokohama Symposium, July 1993) IAHS Publ. no. 213, pp. 245-256.

van Genuchten, M.Th., 1980. A closed-form for predicting the hydraulic conductivity of unsaturated soils, Soil Sci. Soc. Am. J. no. 44, pp. 892-898.

Whipkey, R.Z, 1965. Subsurface stormflow from forested slopes, Int. Assoc. Sci. Hydrol. Bull., Vol. 10, no.2, pp. 74-85.

Wood, E.F., M.Sivapalan, K.J. Beven and L. Band, 1988. Effects of spatial variability and scale with implications to hydrologic modelling, Journal of Hydrology, 102, pp. 29-47.

Woolhiser, D.A. and J.A. Ligget, 1967. Unsteady one-dimensional flow over a plane - The rising hydrograph. Water Resour. Res., Vol., 3 no. 3, pp. 753-771.

Zhao, R.J., 1977, Flood forecasting method for humid regions of China, East China College of Hydraulic Engineering, Nanjing, China.

Zhao, R.J., 1984, Watershed Hydrological Modelling, Water Resources and Electric Power Press, Beijing.

River Runoff Data for the Validation of Climate Simulation Models

Nigel W. Arnell
Institute of Hydrology
Wallingford
Oxon OX10 8BB
UK

INTRODUCTION

Climate model validation involves the comparison of model simulations with observed data. River runoff is a major part of the water balance, and represents one of the most important diagnostic variables of a climate model. Compared to other elements of the water balance - precipitation, evaporation and storage - runoff is easy to measure, and furthermore provides data on water fluxes integrated over both time and space. Areal actual evaporation is very difficult to measure, but can be inferred from observed runoff and precipitation. The ability of a climate model to simulate the partitioning of energy at the land surface into latent and sensible heat fluxes can therefore be assessed using observed precipitation and runoff data.

Most model validation exercises have so far concentrated on surface temperature and precipitation (eg. IPCC, 1990), and there have been few attempts to use runoff data. This paper aims to summarise how runoff data might be used, with illustrative examples: the paper does not attempt to validate climate models with runoff data.

There are two levels of validation. The first level focusses on the land-surface parameterization in isolation, and is essentially concerned with ensuring that climate processes are being adequately simulated within a model. The second level examines the results obtained by running a complete climate model.

NATO ASI Series, Vol. I 31
The Role of Water and the Hydrological Cycle
in Global Change
Edited by Howard R.Oliver and Sylvia A.Oliver

Before reviewing these two levels of validation and the contribution of observed runoff data, it is necessary to summarise the sources of runoff data and outline some of the potential complications.

SOURCES OF RUNOFF DATA

There are four basic sources of runoff data for model validation.

The first is small research basins operated by universities and research institutes. Many such basins have been instrumented: 57 from Europe alone are summarised in Robinson (1993). These basins tend to be small (less than $10km^2$) and heavily instrumented.

The second source is national hydrometric archives. Data availability and quality vary considerably, and some national agencies are more willing to distribute data than others. Data are frequently available through yearbooks, occasionally available on magnetic tape or diskette, and in a few instances on CD-ROM. Data for the United States (Slack & Landwehr, 1992; Wallis *et al*, 1991), for example, are available on CD-ROM.

Data for climate model validation, however, might be required from a large region covering several countries. A third source of data is therefore multi-national regional archives, holding data from all or many of the countries in a region. The European Water Archive collected under the FRIEND (Flow Regimes from International Experimental and Network Data) project, for example, contains data from all countries in northern and western Europe (Roald *et al*, 1993), and has been used for some model validation tests in Europe (see below).

The fourth type of data archive is global. Maps and tables of average annual runoff by continent have been produced by Korzun *et al* (1977) and Baumgartner & Reichel (1975), but of most use for climate model validation are numerical data from selected river basins. Monthly runoff data from many river basins - mostly large - have been published in several volumes by Unesco (various dates). A subset of these data were tabulated by Milliman & Meade (1983), and subsequently used in climate model validation by Russell & Miller (1990) and Kuhl & Miller (1992), as reviewed below. Considerably more data are held by the Global Runoff Data Centre (GRDC) in Koblenz, Germany. The GRDC is run by the Bundesanstalt für Gewässerkunde for the World Meteorological Organisation (WMO), and is charged with collecting river flow data from "significant" basins world-wide. As of May 1994, the GRDC held data for 2941 rivers and 3347 gauging stations world-wide; daily flow data are held for 1327 of these gauging stations (Lüllwitz, pers. comm.). Dümenil *et al* (1993) identified a set of 50 large catchments from the GRDC suitable for model validation, and published time series of annual runoff and long-term monthly averages.

There are several potential complications with observed runoff data. First, river flows may be subject to considerable human intervention. The total volume of flow can be affected significantly by abstractions and returns (irrigation, which is a consumptive use, can have particularly large effects), and the timing of flow through the year may be altered by reservoirs. Data stored on multinational or global archives should (in principle) be free of major human influences, or have the effects of such influences removed.

Second, the accuracy of gauged flow data varies considerably. Although measurement accuracy is generally high, particularly when measurements are integrated over periods of a month or more (with a typical error of well under 10%), flows can be very difficult to measure in some environments. Flows in arid catchments, for example, are very difficult to gauge, as the range in flow is very variable and channels are frequently unstable. Flows in rivers which are seasonally ice covered may be difficult to measure.

Third, river gauging stations do not necessarily measure all the runoff that is generated upstream. In semi-arid and arid areas much of the runoff generated from hillslopes either infiltrates before it reaches a river channel - and may be either evaporated or stored in a slowly-responding aquifer - or infiltrates into the river bed and is either evaporated or passes the gauging station underground. Water may also be lost by evaporation from wetlands and lakes: approximately half of the inflows into the Sudd marshes from the Nile evaporate (Sutcliffe & Parks, 1987). The volume of runoff generated may therefore be considerably greater than the volume of runoff recorded at a gauging station.

A fourth complication lies in presenting river runoff data in a form compatible with gridded climate model output: this will be discussed in more detail below.

VALIDATION OF PROCESS PARAMETERIZATIONS

A great many parameterizations of land surface processes have been proposed, all aiming to simulate energy and water fluxes at the earth's surface. These parameterizations are based on different conceptualizations of the processes involved, different numerical schemes and different sets of input parameters. The ability of each parameterization to simulate realistically the processes involved in water and energy fluxes can only be determined by comparing simulation results with observed meteorological and hydrological data.

There are two different approaches to the inter-comparison and validation of land surface parameterizations: one runs the land surface parameterization code "off-line" using observed or modelled input data, whilst the other runs the parameterization embedded within a single-column atmospheric model, and therefore considers the interaction between land surface and atmosphere.

Off-line validation of land surface parameterizations

This approach takes a land surface parameterization and feeds it with the necessary input parameters. These input parameters will include precipitation, incident solar radiation, temperature, windspeed and vapour pressure. They may be derived from observational data or from the output of a climate model. Validation data - which should include evaporative fluxes and runoff - are best taken from observational data, because otherwise the results from one model will be compared with the output from another.

Abramopoulos *et al* (1988) ran a ground-hydrology model using output from the Goddard Institute for Space Studies (GISS) climate model, in four different parts of the world. They plotted the annual cycle of runoff (and other elements of the water and energy balance) for each site, but only for the Brazilian rainforest point did they have observed water balance data for validation. These observed data came from a small research basin, but unfortunately only one year's worth of data were available.

Land surface parameterizations are currently being assessed under PILPS, the Project for Intercomparison of Land-Surface Parameterization Schemes (Pitman *et al*, 1993). PILPS is a contribution to WMO's Working Group on Numerical Experimentation (WGNE) and the GEWEX (Global Water and Energy Cycle Experiment) Continental Scale International Project (GCIP), and essentially involves running different land surface parameterizations with the same sets of input data. Twenty schemes were included in Phase 1a of PILPS (Pitman *et al*, 1993), using input atmospheric data derived from a general circulation model for a tropical forest site and a grassland site. Different estimates of simulated runoff were compared, but could not be validated because of the lack of observed runoff data. Phase 2 of PILPS uses observational data sets to provide the climatic forcing, and therefore has the potential for incorporating runoff data.

Wood *et al* (1993) compared the performance of a simple bucket land surface parameterization with that of a "hydrological" rainfall-runoff model (VIC), using observed data from a small catchment in North Carolina. They found that the hydrological model, which considered spatial variability in soil properties and included drainage from the soil, simulated the short-term variability of water and latent heat fluxes in a much more realistic manner (Figure 1). In a similar way, Chiew *et al* (1993) compared one land surface parameterization scheme - BEST (Pitman *et al*, 1991) - and one hydrological rainfall-runoff model - MODHYDROLOG (Chiew & McMahon, 1994) - using two catchments in Australia. They concluded from their simulations that the land surface parameterization scheme worked well in the wet catchment, but simulated flows less adequately than the hydrological model in the dry catchment.

Validating land surface parameterizations with a single-column model

A single-column model simulates the behaviour of a column of atmosphere reaching upwards from the land surface. It is run using solar radiation inputs and time series of boundary conditions, which may be derived from observations or a larger three-dimensional climate model. A single-column model allows the assessment of the effects of different land surface parameterizations on the atmosphere above, especially precipitation and evaporation. A single-column model is much quicker to run than a three-dimensional climate model.

Rowntree & Lean (1994) used a single-column model, based on the UK Met. Office global climate model, to simulate runoff under varying land surface parameterizations. They applied the model using observed climatological inputs from southern England, and validated simulations using observed monthly runoff from the Thames in southern England and an average of several small catchments in eastern England. The original land surface parameterization was found to *underestimate* runoff, especially in summer. In fact, the model generated very little runoff from summer to mid-winter. A different land surface parameterization, accounting for sub-grid variability in soil moisture status (the ARNO model: Dümenil & Todini, 1992), was then considered, with

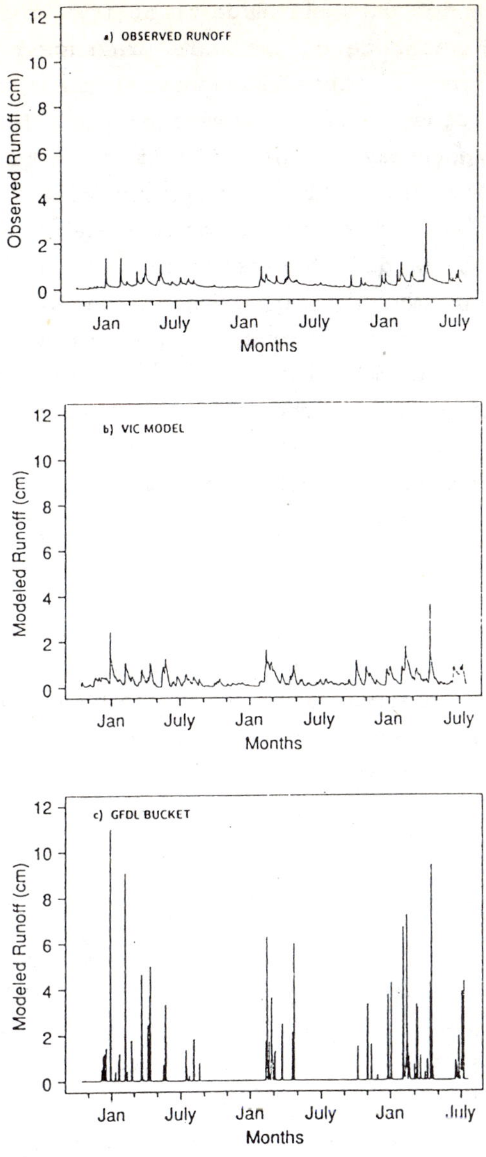

Figure 1 Comparison of observed streamflow with streamflow simulated by the VIC model and a bucket model, for French Broad River, North Carolina (Wood *et al*, 1993)

different values for parameters defining the degree of sub-grid variability and the rate of downward drainage out of soil. By careful selection of these parameters, it was possible to reproduce more closely the observed runoff. Rowntree & Lean (1994) then hypothesised that runoff was too low because evaporation was too high. Allowing for the effect of aerosols in reducing incoming solar radiation reduced evaporation and increased runoff, as did increasing plant stomatal resistance. Figure 2 shows the monthly variation in runoff, as a percentage of precipitation, for the two sets of observed data and three land surface parameterizations: C is the original parameterization, B0.5 is the ARNO model with a value of 0.5 describing sub-grid variability, and BA90 is the ARNO model with the same subgrid variability but higher stomatal resistance and allowing for aerosols.

An important lesson to draw from Rowntree & Lean's (1994) study is that validation against runoff data triggered investigations not just into the land surface parameterization, but also into other aspects of the atmospheric model. As Rowntree & Lean (1994) conclude, "it is evident that validation with hydrological data can be a valuable aid to improving atmospheric models" (p321).

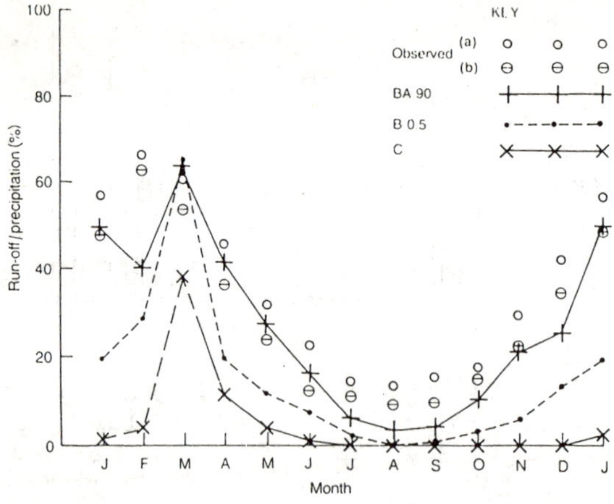

Figure 2 Variation in runoff in southern England as a percentage of precipitation for two observed data sets and three land surface parameterizations (Rowntree & Lean, 1994).

VALIDATION OF CLIMATE MODEL SIMULATIONS

Here, the aim is primarily to validate the ability of a global or regional climate model *as a whole* to simulate climate. All the processes involved in the climate system - atmospheric processes, land surface processes, interactions with the ocean - are being considered together.

The largest problem lies in presenting hydrological data and climate model results in a compatible form. Thomas & Henderson-Sellers (1992) compared observed and simulated *continental* water budgets (Figure 3), but other studies have worked at a finer spatial scale.

Climate models operate on a grid basis, so effort has gone into expressing hydrological data on a grid. Two approaches can be used: one is "top-down", using data from large basins covering several atmospheric grid cells; the other is "bottom-up", and attempts to produce data for each atmospheric model grid cell from smaller hydrological units.

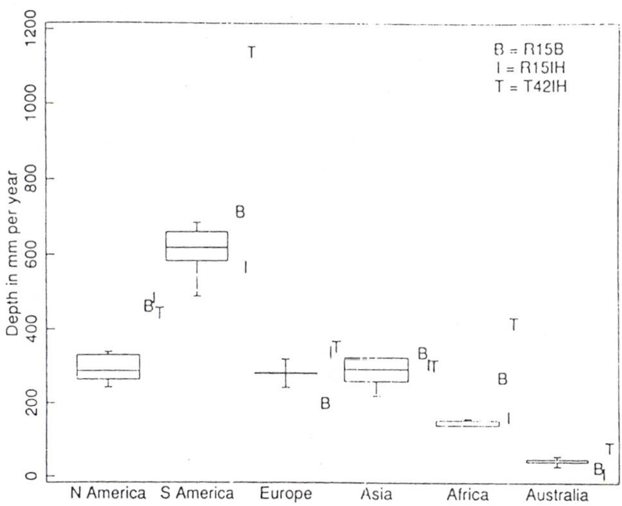

Figure 3 Observed and simulated continental runoff (Thomas & Henderson-Sellers, 1992). Three different versions of the NCAR GCM were used to simulate runoff.

Validation using runoff data from large river basins

This approach has been used by Russell & Miller (1990) and Kuhl & Miller (1992) to assess the performance of the GISS General Circulation Model, and by Dümenil & Todini (1992) and Sausen *et al* (1994) to evaluate the Max Planck General Circulation Model ECHAM. The first stage is to generalise large river basins onto a coarse grid. Russell & Miller (1990) generalised 33 basins onto a 2x2.5° grid and Sausen *et al* (1994) generalised 17 basins onto the T21 5.625x5.625° grid (Figure 4). The next stage is to compare the observed runoff from each large basin with the climate model runoff generated in the cells defining the basin.

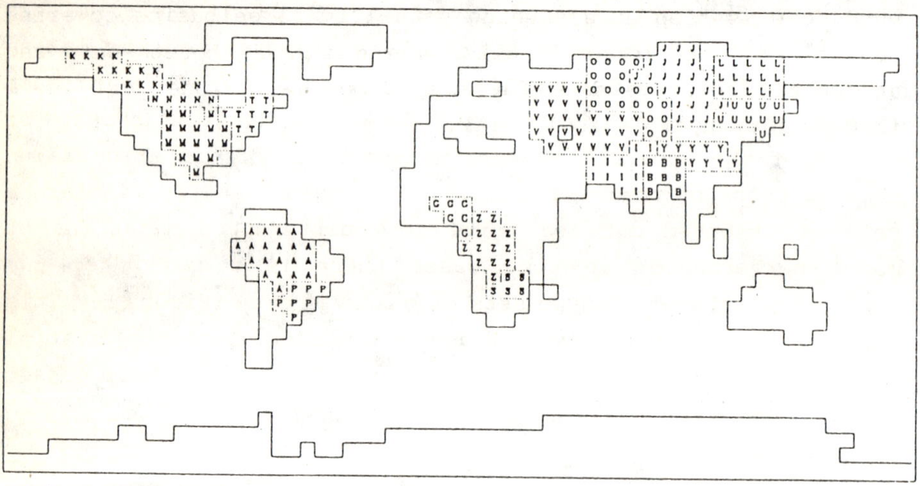

Figure 4 Major river basins generalised onto a coarse grid (Sausen *et al*, 1994)

Russell & Miller (1990) compared annual runoff as simulated with the GISS GCM with observed annual runoff, taken from Milliman & Meade (1983), for 33 basins. In a later paper, Kuhl & Miller (1992) assessed monthly runoff in a subset of these basins.

Figure 5 shows average annual runoff as estimated for major river basins by both the GISS and ECHAM-T21 GCMs, and as observed according to Milliman & Meade (1983) and the GRDC (Sausen *et al*, 1994). The aim here is not actually to evaluate the performance of the two GCMs, but rather to illustrate some general points about the use of observed runoff data for validation.

Whilst the models give close estimates in some basins, in others runoff may be underestimated by a factor of two or overestimated by a factor of five or ten. There are several reasons for this. Basin precipitation may be badly simulated, and Figure 5 also compares observed and simulated precipitation for the study basins. The processes controlling runoff generation may be poorly represented - although this can really only be assessed by evaluating the process parameterizations themselves. The generalisation of basins onto a coarse grid introduces errors. The streamflow data gauged close to the basin outlet misses runoff generated upstream but lost in evaporation from lakes, swamps and floodplains: the GCMs considered do not allow for such losses. These losses may be very substantial, and can account for a part of the apparent overestimation of runoff in arid basins such as those of the Nile, Murray and Orange. Finally, different sets of observed data can give very different indications of model performance: in some cases there is a factor of two difference between the two sets of observed data (Figure 5). This may occasionally be due to the use of a different gauging station (Sausen *et al* (1994) tabulate both sets of data for their subset of catchments, but do not give the locations of the gauging stations used), but may also arise because different periods of records are used. This is likely to be particularly significant in arid areas, where the year to year variability in runoff is high. Obviously, care needs to be exercised in the selection of period of record for analysis: when assessing simulated output from just a few model years, it is important to consider the short-term variability in observed runoff totals.

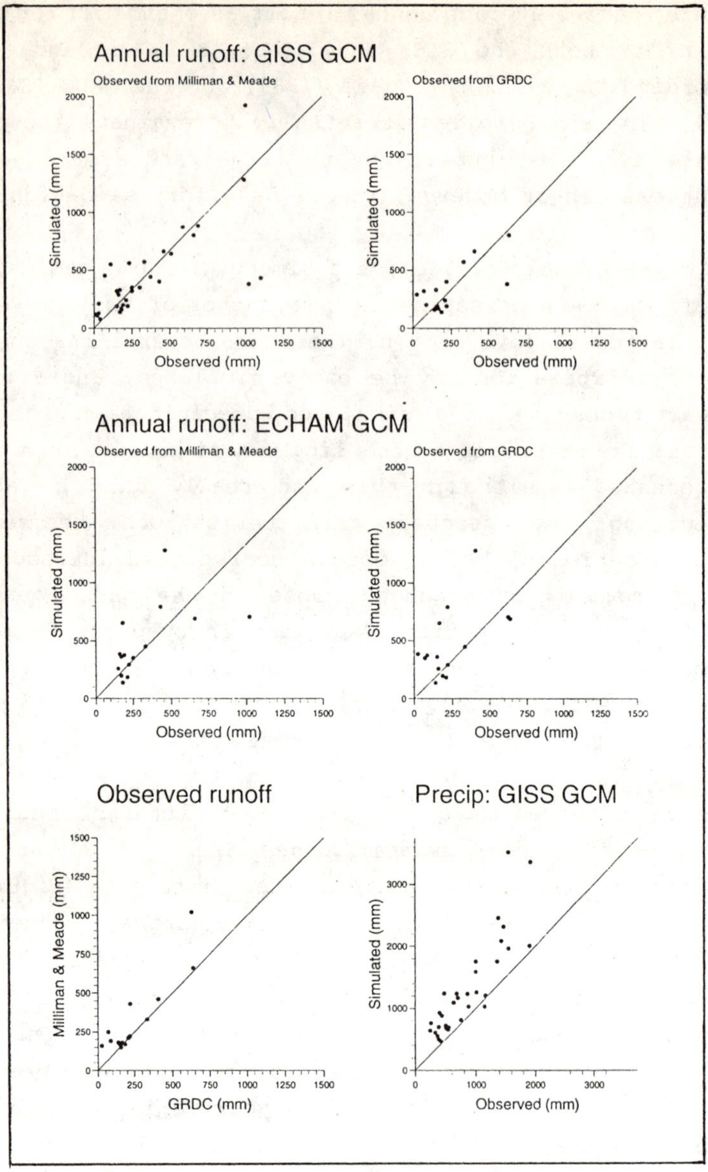

Figure 5 Observed and modelled average annual runoff and precipitation, with the GISS GCM (Russell & Miller, 1990) and the ECHAM GCM (Sausen *et al*, 1994), and two different sets of observations

Kuhl & Miller (1992) and Sausen *et al* (1994) both compared simulated monthly runoff with observed monthly runoff for a number of large basins, and a selection of their results are shown in Figures 6 and 7 respectively. Kuhl & Miller's simulated runoff does not allow for routing delays or lags due to slow movement of subsurface drainage to river channels, whilst Sausen *et al* do allow for these delays through the use of a linear advection scheme linking GCM grid cells. As with annual runoff, differences between observed and simulated runoff can occur for several reasons.

Validation using small and medium-sized basins

The previous approach groups those climate model grid cells which together make up a large basin. This approach uses observed hydrological data to define the "true" hydrological characteristics of each model grid cell; it builds up to the climate model grid resolution using information from small and medium-sized catchments.

There are five basic ways in which runoff data can be generalised onto a grid (Arnell & Gottschalk, 1993; Arnell, 1994). The first is simply to average data from all the small catchments which lie within a grid cell. This tends to overestimate grid cell average runoff because small catchments tend to have a higher than average runoff: they include headwaters, and have higher than average rainfall. In principle it is possible to weight by rainfall, but in practice there will rarely be enough available data, especially at a high grid resolution.

The second approach involves statistical interpolation around small catchments (eg Bishop & Church, 1992). This assumes a continuous and smooth variation in runoff across space, and is really only appropriate when the catchments are very small relative to the domain over which data are to be interpolated. A variant of this approach interpolates along the river network (Gottschalk and Krasovskaia, 1993).

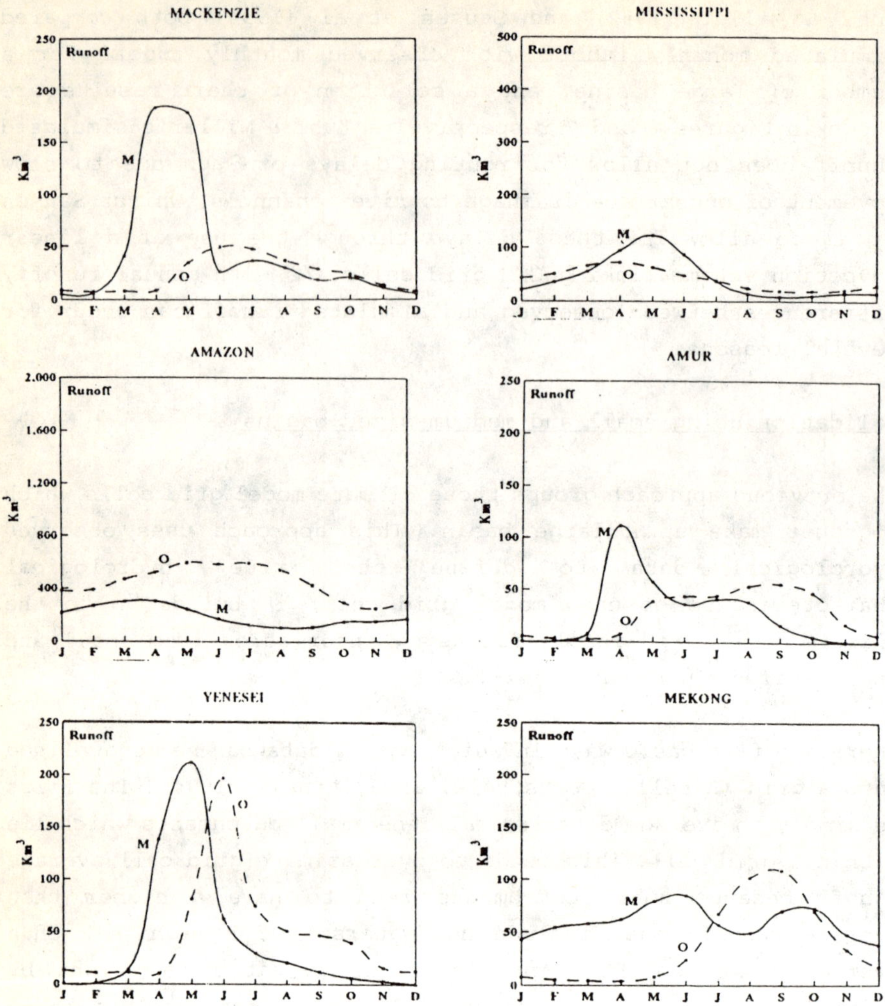

Figure 6 Observed (O) and modelled (M) monthly runoff and precipitation: GISS GCM (Kuhl & Miller, 1992).

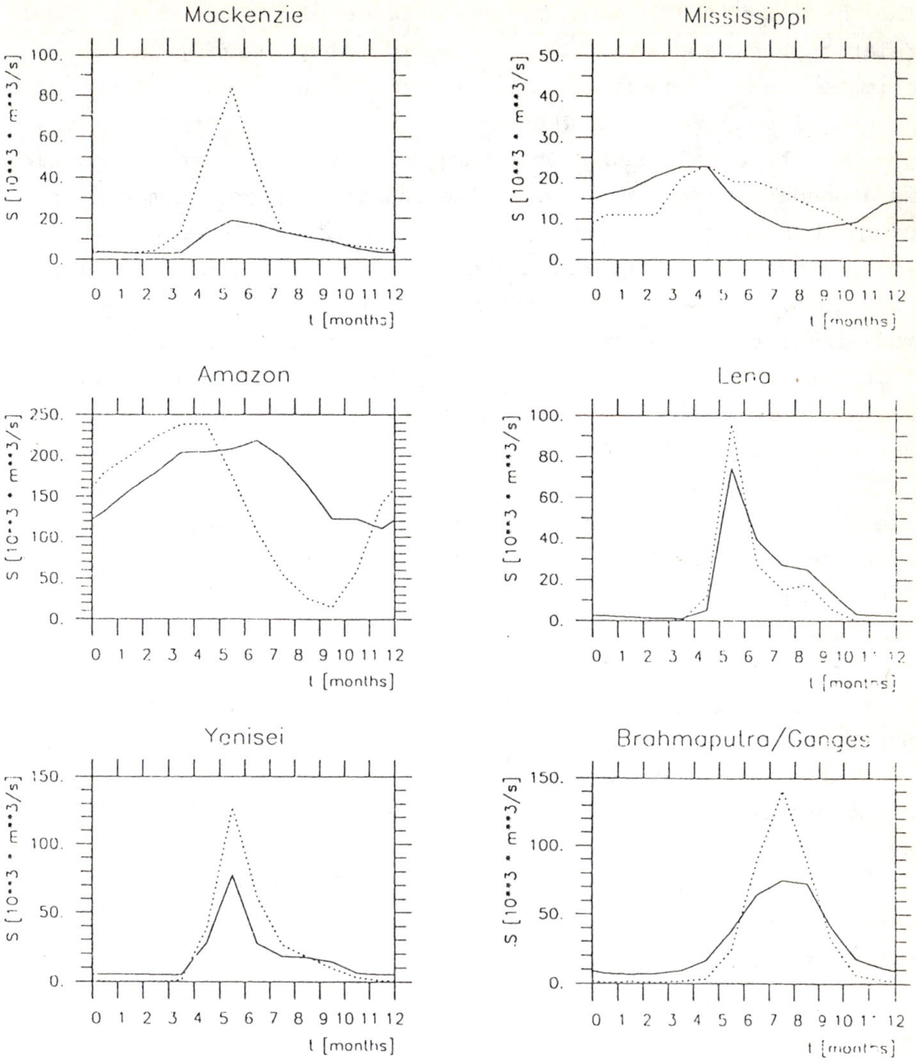

Figure 7 Observed (continuous lines) and modelled (dotted lines) monthly runoff: ECHAM-T21 GCM (Sausen *et al*, 1994)

The third and fourth methods involve the application of empirical relationships and water balance models respectively to gridded climate data (specifically precipitation, temperature and potential evaporation). The estimated gridded runoff is only as good as the model used - which may be poor at the margins - and is dependent on spatial scale. The runoff averaged from a number of grid cells, for example, may not be the same as the runoff from climate inputs averaged over the same cells, because of the non-linear form of relationships between climate and runoff. A more important problem for climate model validation, however, is that the estimated runoff is dependent on the input gridded climate data: the results do not constitute an independent validation data set.

The fifth method simply maps catchment runoff (eg Figure 8), superimposes a grid, and calculates weighted grid cell averages (where the weight is the proportion of the cell covered by each catchment). Figure 9 shows average annual runoff on a 0.5x0.5° grid resolution, estimated using this method and data from the FRIEND European Water Archive (Arnell, 1994). For comparative purposes, the grid network of the UK Hadley Centre High Resolution GCM is superimposed on Figures 8 and 9: note the considerable variability in sub-grid runoff. The approach can be used with different grid resolutions. The coarser the final grid required, the larger - and fewer - the catchments which can be used. Catchments contained within a catchment that is itself contained entirely within a grid cell are superfluous.

Using this last approach, the volume for a given grid cell represents the runoff generated within that cell, less the volume of runoff from upstream cells lost to evaporation within the cell. Of course, if these losses exceed in-cell runoff, then the cell will have *negative* runoff.

Data from small and medium-sized catchments can be used in other ways to describe hydrological ground truth within a climate model grid cell. Figure 10 shows monthly runoff, expressed as a percentage of the annual runoff, within each of the cells of the

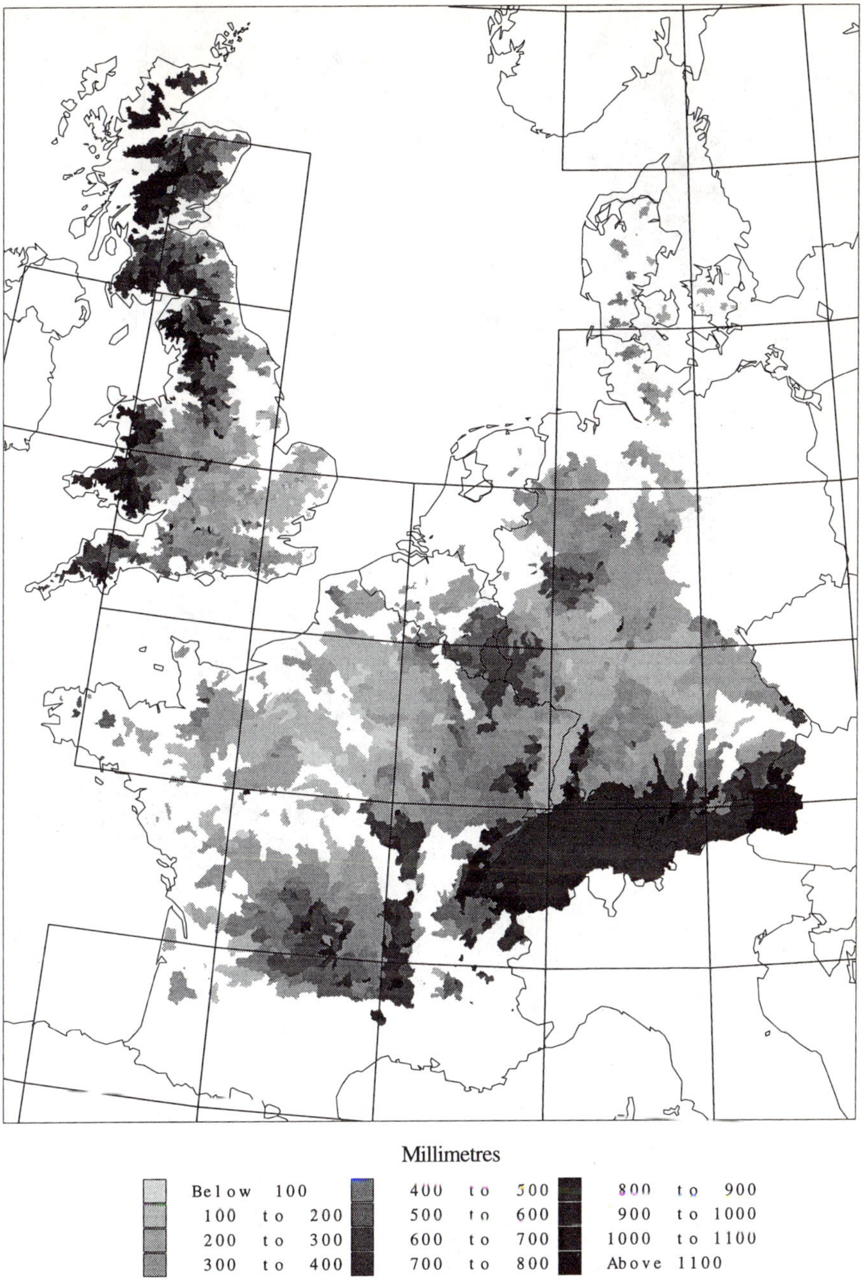

Figure 8 Average annual runoff by catchment (with 2.5x3.75° GCM grid superimposed)

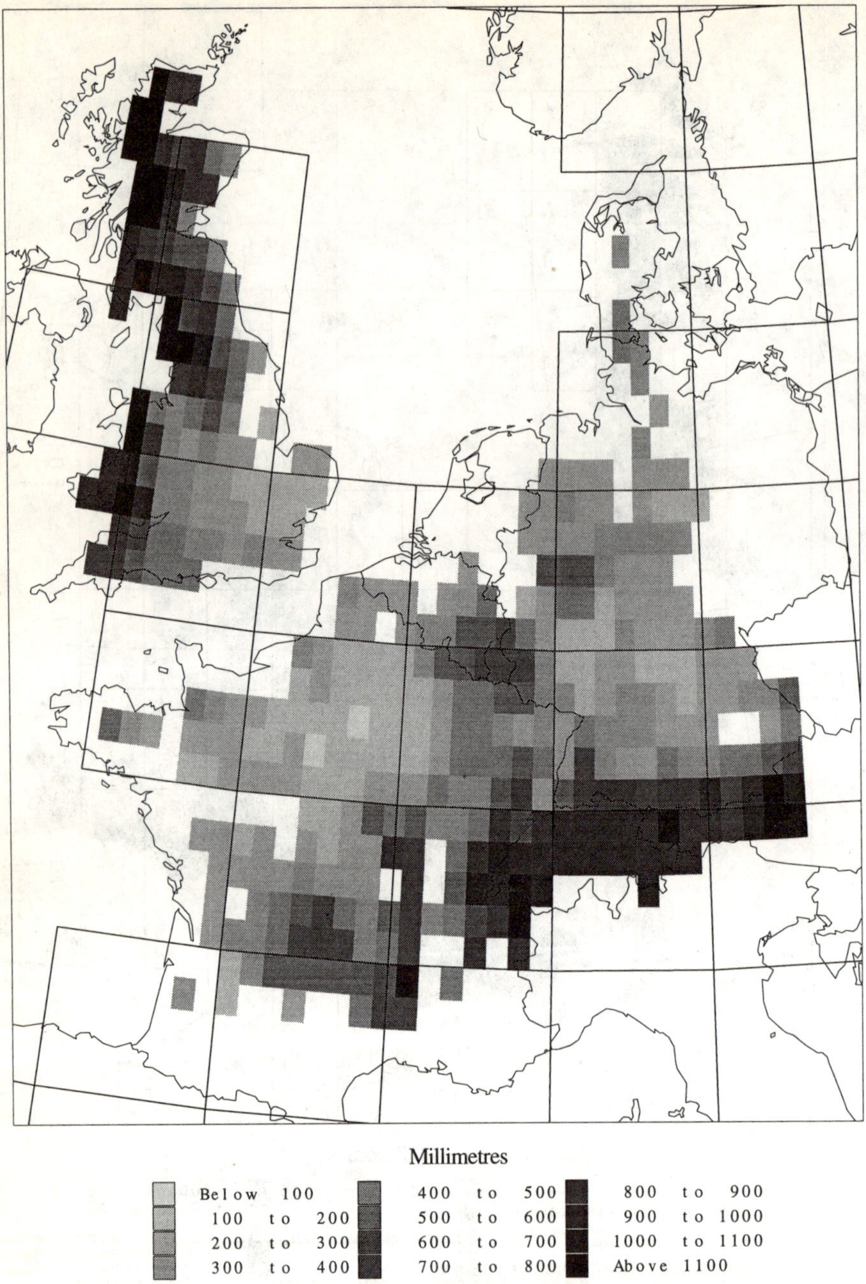

Figure 9 Average annual runoff by 0.5x0.5° grid, derived from catchment data (with 2.5x3.75° GCM grid superimposed)

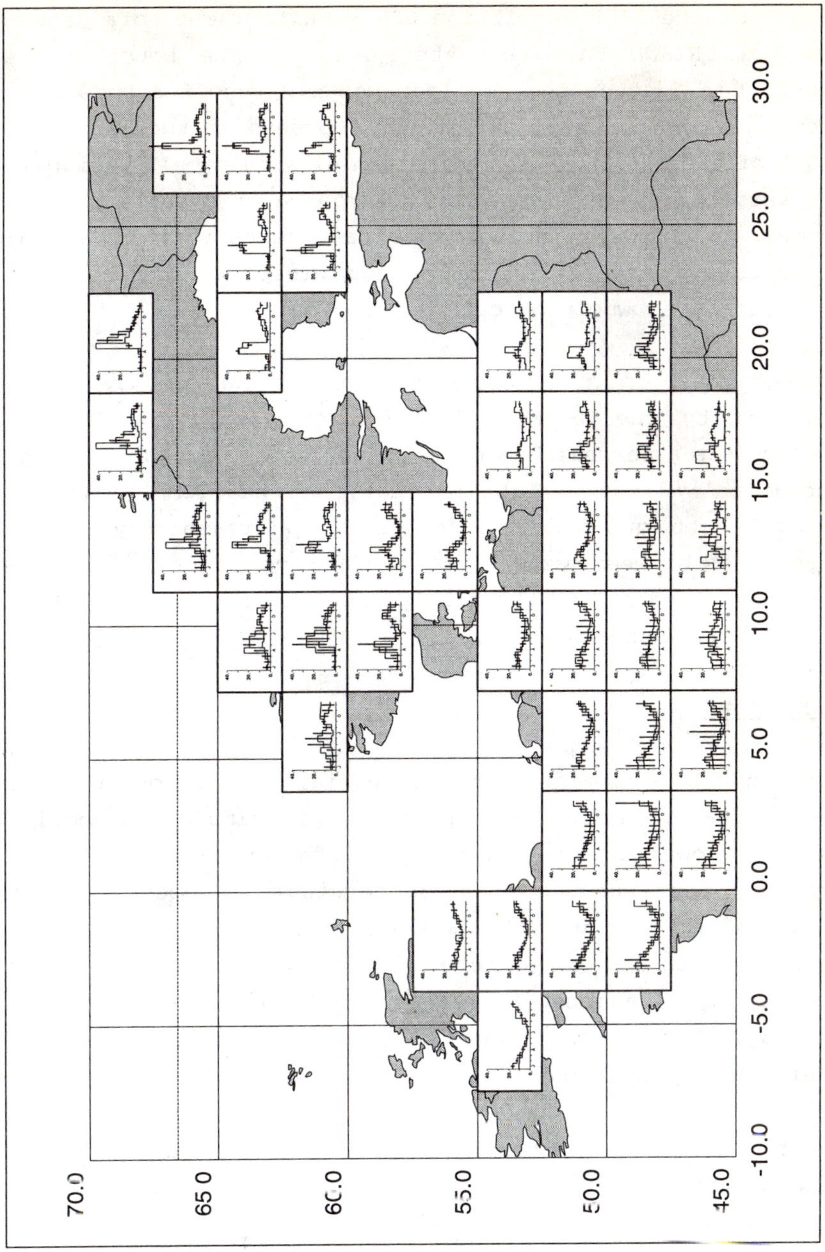

Figure 10 Monthly flow regimes by 2.5x3.75° grid box, as a percentage of annual total runoff. The top of the thick bar denotes the maximum value across all the catchments in the box, and the bottom denotes the minimum value. Runoff as modelled by the UKMO High Resolution GCM is shown by the stepped line.

UK Met. Office High Resolution GCM (UKHI). The runoff data come from the FRIEND European Water Archive. The bottom of each monthly bar denotes the smallest monthly runoff across all the catchments in the cell, whilst the top denotes the largest. The length of the bar gives an indication of within-cell variability. This variability is particularly large in grid cells containing mountains, as these include some catchments with flow regimes dominated by rainfall - and hence a winter maximum - and some dominated by snowmelt or even glacier melt.

For comparative purposes, Figure 10 also shows monthly runoff as simulated by the UK Met. Office High Resolution GCM, again expressed as a percentage of the annual total. Without attempting a detailed validation, it is clear that the GCM simulates the seasonal pattern of runoff across Europe reasonably well; the differences between maritime western Europe, the mixed regimes in Norway and Sweden, and the highly continental regimes in Finland are well reproduced.

CONCLUSIONS

This paper has summarised the potential uses of observed river runoff data in the validation of climate simulation models. It has been shown that observed runoff data can be very valuable in assessing the representation of particular processes within a model, and in evaluating the ability of an entire global model to simulate water and energy budgets. Validation using hydrological data can provide insights not just into the representation of hydrological processes, but also into other aspects of a climate model (as shown by Rowntree & Lean, 1994). Validation over a large geographic domain - such as a continent - can also provide valuable information, complementing that derived from rainfall and temperature observations, on where a model simulates climate well, and where it does not.

Validation of the performance of complete climate models requires observed hydrological data over a large region, and these data must be expressed in a form compatible with gridded climate model output. The easiest way appears to be to plot maps of catchment runoff (for example Figure 8), and then to superimpose a grid within a GIS (for example Figure 9): the size of catchment to use depends on the resolution of the final grid.

There are, however, a number of complications with using observed runoff data for model validation. Different periods of record may give very different indications of "ground truth", especially in semi-arid and arid regions where there is considerable year-to-year variability. The major problem is that climate model runoff does not account for loss of water along the river network due to human activities, channel transmission losses or evaporation from lakes and wetlands. In principle, the effects of human interventions on the observed data can be corrected for, at least at the monthly scale. Evaporation from major lakes and swamps can also be determined in some cases from observed data and corrected for (Sutcliffe & Parks, 1989, for example), but transmission losses along the channel are very difficult to estimate. These losses make it very difficult to compare observed and simulated runoff in semi-arid and arid areas.

The validation of process parameterizations requires runoff data together with surface and atmospheric meteorological data. Mesoscale field experiments such as FIFE and HAPEX provide an major source for such meteorological data. It is important that river runoff - as well as soil and groundwater storage changes - are measured as part of such experiments, and that measurements are made over several seasons.

Acknowledgements

Simulated runoff from the UK Met. Office High Resolution GCM were provided by Dr Peter Rowntree and Dr Jenny Lean. The author would like to thank Dr Andy Pitman (Macquarie University) for his comments.

References

Abramopolous, F., Rosenzweig, C. & Choudhury, B. (1988) Improved ground hydrology calculations for global climate models (GCMs): soil water movement and evapotranspiration. J. Climate 1, 921-941.

Arnell, N.W. (1994) Grid mapping of river discharge. J. Hydrol. in press

Arnell, N.W. & Gottschalk, L. (1993) Mapping average annual runoff. in A. Gustard (ed.) Flow Regimes from International Experimental and Network Data (FRIEND). Volume 1 Hydrological Studies. Institute of Hydrology, Wallingford. pp122-131.

Baumgartner, A. & Reichel, E. (1975) The World Water Balance. Elsevier: Amsterdam

Bishop, G.D. & Church, M.R. (1992) Automated approaches for regional runoff mapping in the northeastern United States. J. Hydrol. 138, 361-383.

Chiew, F.H.S. & McMahon, T.A. (1994) Application of the daily rainfall-runoff model MODHYDROLOG to twenty-eight Australian catchments. J. Hydrol. 153, 383-416.

Chiew, F.H.S., Pitman, A.J. & McMahon, T.A. (1993) Land-surface parameterization schemes and conceptual rainfall-runoff models. Bureau of Meteorology Conference Proceedings. Australia.

Dümenil, L. & Todini, E. (1992) A rainfall-runoff scheme for use in the Hamburg climate model. in J.P. O'Kane (ed.) Advances in Theoretical Hydrology: A Tribute to James Dooge. European Geophysical Society Series on Hydrological Sciences, 1. Elsevier: Amsterdam.

Dümenil, L., Isele, K., Liebscher, H.-J. *et al* (1993) Discharge data from 50 selected rivers for GCM validation. Max-Planck-Institut für Meteorologie. Report 100. Hamburg.

Gottschalk, L. & Krasovskaia, I. (1993) Interpolation of annual runoff to grid networks. in Macroscale Modelling of the Hydrosphere. IASH Publ. 214, 81-89.

IPCC (1990) Climate Change: The IPCC Scientific Assessment. Intergovernmental Panel on Climate Change. Cambridge University Press: Cambridge

Korzun, V.I., Sokolov, A.A., Budyko, M.I. *et al* (1977) Atlas of World Water Balance. USSR National Committee for the International Hydrological Programme. Unesco, Paris.

Kuhl, S.C. & Miller, J.R. (1992) Seasonal river runoff calculated from a global atmosphere model. Water Resources Research 28, 2029-2039.

Milliman, J.D. & Meade, R.H. (1983) World-wide delivery of river sediment to the oceans. J. Geology 91, 1-21.

Pitman, A., Yang, Z.-L., Cogley, J.G. & Henderson-Sellers, A. (1991) Description of Bare Essentials of Surface Transfer for the Bureau of Meteorology Research Centre AGCM. Australian Bureau of Meteorology Research Report No. 32.

Pitman, A., Henderson-Sellers, A., Abramopoulous, F. *et al* (1993) Project for Intercomparison of Land-Surface Parameterization Schemes (PILPS). Results from off-line control simulations (Phase 1a). International GEWEX Project Office (IGPO) Publication Series 7. Washington D.C. 47pp

Roald, L., Wesselink, A.J., Arnell, N.W. *et al* (1993) European water archive. in A. Gustard (ed.) Flow Regimes from International Experimental and Network Data (FRIEND). Volume 1 Hydrological Studies. Institute of Hydrology, Wallingford. pp7-20.

Robinson, M. (ed.) (1993) Flow Regimes from International Experimental and Network Data (FRIEND). Volume 3 Inventory of Streamflow Generation Studies. Institute of Hydrology, Wallingford.

Rowntree, P.R. & Lean, J. (1994) Validation of hydrological schemes for climate models against catchment data. J. Hydrol. 155, 301-323.

Russell, G.L. & Miller, J.R. (1990) Global river runoff calculated from a global atmospheric general circulation model. J. Hydrol. 117, 241-254.

Sausen, R., Schubert, S. & Dümenil, L. (1994) A model of river runoff for use in coupled atmosphere-ocean models. J. Hydrol. 155, 337-352.

Slack, J.R. & Landwehr, J.M. (1992) Hydro-climatic data network (HCDN): a U.S. Geological Survey streamflow data set for the United States for the study of climate variations, 1874-1988. U.S. Geological Survey Water Supply Paper 2406. Reston, Virginia.

Sutcliffe, J.V. & Parks, Y.P. (1987) Hydrological modelling of the Sudd and Jonglei Canal. Hydrol. Sci. J. 32, 143-159.

Sutcliffe, J.V. & Parks, Y.P. (1989) Comparative water balances of selected African wetlands. Hydrol. Sci. J. 34, 49-62.

Thomas, G. & Henderson-Sellers, A. (1992) Global and continental water balance in a GCM. Climatic Change 20, 251-276.

Unesco (various dates) Discharge of selected rivers of the world. Studies and Reports in Hydrology 5. Unesco: Paris

Wallis, J.R., Lettenmaier, D.P. & Wood, E.F (1991) A daily hydro-climatological data set for the continental U.S. Water Resources Research 27, 1657-1663.

Wood, E.F, Lettenmaier, D.P. & Zartarian, V.G. (1993) A land-surface parameterization with subgrid variability for general circulation models. J. Geophys. Res. 97, D3, 2717-2728.

Introduction to Climate Impact Assessment

Stewart J. Cohen
Atmospheric Environment Service
Environment Canada
4905 Dufferin Street
Downsview, Ontario
Canada M3H 5T4

PURPOSE

The possibility that global climatic change could arise because of human activities has been the subject of considerable research for the entire 20th century. The translation of this possibility into actual "pictures" of a future altered climate became feasible with the development of General Circulation Models (GCM) in the 1960s and 1970s. As the results of these GCM simulations of an atmosphere with increased concentrations of CO_2 were published during the 1970s and 1980s, questions about ecological and societal implications were asked, and the scientific question of global warming quickly became a political issue.

The Intergovernmental Panel on Climate Change (IPCC) was created by the United Nations to provide state of the art reviews on global warming science, impacts, and responses. The impacts component has been a broad review of many climate-sensitive issues, including agriculture, forestry, natural terrestrial and freshwater ecosystems, coastal and marine areas, the cryosphere, human settlements, and water resources (Tegart et al., 1990; Tegart and Sheldon, 1993). The IPCC has also undertaken a compilation of impact assessment methods (Carter et al., 1992; revision in progress).

What is climate impact assessment, and in the context of global warming "futures" or scenarios, what is its purpose? There are two main objectives:

a) identification and evaluation of effects of climatic variability and scenarios of climatic change, and

NATO ASI Series, Vol. I 31
The Role of Water and the Hydrological Cycle
in Global Change
Edited by Howard R.Oliver and Sylvia A.Oliver

b) description of policy implications, including the assessment of potential adaptation strategies.

Since climate is a complex agent of change, and GCMs and other sources of scenarios include many uncertainties, it is clear that such assessments are not designed to provide forecasts, nor to prescribe specific changes in policy. They do represent, however, an important opportunity to learn about climate-related sensitivities and vulnerabilities. These learning exercises have three main attributes:

- linkage of regional concerns with global scale atmospheric science;
- development of interdisciplinary approaches, and
- development of new partnerships between disciplines and jurisdictions (Cohen, 1993a).

FROM EXCLUSION TO INCLUSION -- A HIERARCHY OF ASSESSMENTS

A wide range of impact assessment methodologies have been tried. The simplest of these has focussed on a single sector (e.g. agriculture, forestry, water resources), with the assumption that only climate would change and that population, technology, institutional arrangements, etc., would remain unchanged. In order to overcome this shortcoming, other more complex approaches have been developed in order to utilize parallel scenarios of socio-economic changes (for the same time period as the climatic change scenarios). These include:

- national and global sectoral assessments (e.g. agriculture--Rosenzweig and Parry, 1994; river basin management--Wescoat and Leichenko, 1992);
- spatial and temporal analogues (e.g. Glantz, 1988, 1992; Ninh et al., 1991);
- parallel sectoral assessments (e.g. UKCCIRG, 1991; Hulme et al., 1992; Magalhaes and Glantz, 1992; Nishioka et al., 1993; Henderson-Sellers and Colls, 1994);
- integrated assessment models, which can be applied at various scales;
- integrated regional/national assessments (e.g. Parry et al., 1992; Rosenberg et al., 1993; Cohen, 1993b), and

- integrated continental/global assessments (Rotmans, 1990; Alcamo et al., in press).

Analogue studies don't require GCM-based scenarios. However, spatial analogues may have to account for the differences in geography and history between the analogue site and the study site, while technological and institutional changes between the analogue time period and study period could affect temporal analogue studies. Parallel sectoral assessments have common scenarios, but are not necessarily linked within a model framework. Integrated assessments include methodologies that explicitly account for simultaneous changes in climate sensitive sectors, so that the net effects of climatic change can be determined. At continental and global scales, however, the resolution is necessarily coarse, and regional details may be missing. At the regional and national scales, intersectoral linkages can be quite complex, and there may be differing views on how they should be represented.

If one considers the range of climate impact assessments that have been attempted, it is useful to place them on a scale that indicates the degree of integration over space and time (Figure 1). Single sector studies are the simplest, particularly if they assume that Everything Else Remains Equal (EERE). These studies may not incorporate external forces that can influence the sector, such as international trade. Global studies that focus on a single sector can provide good spatial integration for a future period, but unless it utilizes a transient scenario, temporal integration would still be weak. Analogues and parallel sectoral assessments provide higher levels of integration, and if they are linked using integrated assessment models (e.g. Rosenberg et al., 1993), some level of integration can be achieved.

The highest level of integration would require the development of global scale modelling frameworks, such as IMAGE (Rotmans, 1990; Alcamo et al., in press). These include linked sub-models of physical, biological and socio-economic processes. One example of model output is a series of new continental scale land cover patterns resulting from various strategies to limit greenhouse gas emissions.

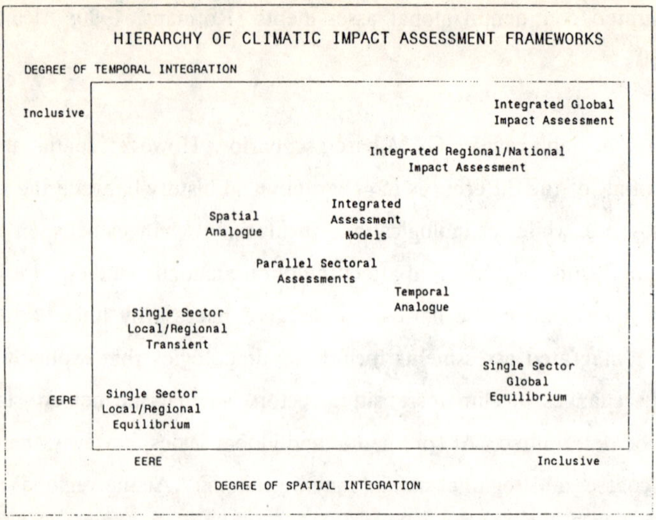

Figure 1. Hierarchy of climatic impact assessment frameworks. Source: Cohen (in press), after Burton and Cohen (1992). A variation on this theme is provided in Henderson-Sellers and Colls (1993).

The regional assessment methodology is gradually being developed from the experiences of sectoral and analogue studies. The global scale modelling effort is being approached from the global level, but these currently lack regional detail. Integrated regional/national assessment therefore constitutes an important methodological challenge that needs to be addressed so that the mismatch of scales between local and global scale approaches can be overcome. This is also the level where stakeholders are most likely to see the relevance of global climatic change.

ROLE OF WATERSHEDS

The focus of this series of lectures is on the role of water and the hydrological cycle in Global Change. Climate impact assessments can be performed in various ecological and administrative settings (Carter et al., 1992), but there are distinct advantages to choosing watersheds. They are usually easy to define, and act as natural integrators of physical, biological and human systems, including the hydrologic cycle, landscape changes, channel and flow modifications, industrial water users and other stakeholders.

Environmental concerns (e.g. point source pollutants) which can affect water quality can also be included. Many different kinds of stakeholders can be affected by impacts on watersheds, including hydroelectric utilities, shoreline properties and infrastructure, navigation, cooling requirements (industry, electricity generation), agriculture (irrigation), municipalities, health services, ecosystem maintenance (fisheries, wetlands), tourism and recreation.

STICKING POINTS

There are four main problems associated with climate impact assessments, particularly if the study is on global climatic change:

1. Scenarios of climatic change

Uncertainties associated with GCMs at regional scales lead to uncertainties in regional climate scenarios, especially regarding precipitation. Analogues are not always available or appropriate, however, and GCMs are the only real source of climatic "futures."

2. Scenarios of socio-economic changes

Population, technology, prices, and institutional arrangements are difficult to predict, but these are needed in order to provide appropriate context for the assessment of climate-related impacts.

3. Assessment of impacts

Models that might be used for impact assessments (e.g. hydrology) may not be calibrated for GCM-based scenarios, especially if there is no analogue for the study area that is similar to such scenarios. It is therefore difficult to determine how well the model will perform under these circumstances. Efforts to carry out integrated assessments have not yet demonstrated which methodology would be the most appropriate for describing the net effects of climatic change, so there is no consensus particularly at the regional scale.

4. Linkage with policy

Stakeholders and decision makers have considerable experience with scenarios of various kinds, but they have little or no experience in adapting to scenarios of climatic change. A clear signal of global warming has not yet been detected, so many are taking a "wait and see" approach. The "real" costs of current climatic variability are still not known, and this may be affecting economic analyses of climatic change scenarios, which have so far resulted in cost estimates of about 1% of Gross Domestic Product. Given the ratification of the Framework Convention on Climate Change in 1994, however, there is considerable interest in establishing various policy instruments (Amano et al., 1994) despite incomplete and uncertain information on impacts.

APPROACHES

The IPCC (Carter et al., 1992) have identified three main approaches to conducting climate impact assessments: impact, interaction, and integrated.

Impact approach

This term describes a simple 'cause and effect' approach, or 'If-Then-What', in which the exposure unit is affected by climate alone, assuming that everything else remains equal. It usually starts with studies of first-order effects on individual activities or organisms in the study area (e.g. plant growth, runoff, permafrost). These direct effects lead to indirect impacts (e.g. change in plant growth leading to change in wildlife productivity), eventually extending to higher-order economic and social effects (e.g. agricultural production, tourism). This approach requires uncertain assumptions at each level, including the exclusion of possible feedback effects. If the assessment is meant to examine higher-order issues, other approaches will be needed.

Interaction approach

Climate and other factors are considered together in a 'What-Then-If' approach. What points of a system are sensitive to changes in climate, and what might the impacts of a scenario be? These sensitivities are likely to be sectorally and regionally specific, so the same climatic change scenario could have different effects depending on the specific biophysical and socioeconomic characteristics of the study area. Both climate and socio-economic scenarios are utilized, and feedbacks are included. In a study on agriculture, for example, the entire agricultural production system for a specific region could be examined, from crop yield to processing to marketing, with external forces (e.g. international market conditions) being considered. This would require the development of socio-economic scenarios (for the sector of interest, e.g. agriculture) that would parallel the climatic scenarios.

Integrated approach

Interactions within and between sectors, including adjustments, are included. Results of impact and interaction studies are utilized as inputs to integration exercises. These can be in the form of a model, but not necessarily. Examples include:

- single model or framework, as in the MINK study (Rosenberg et al., 1993)
- model with linked sectoral modules, such as IMAGE (e.g. Alcamo et al., in press)
- policy analysis, such as effects on water management authorities (e.g. Arnell et al., 1994).

A study could include more than one method that can provide complementary analyses, or "family" of integrators (Cohen, in press).

SEVEN STEPS FOR CONDUCTING AN ASSESSMENT

The IPCC (Carter et al., 1992) provide a 7-step approach for conducting climate impact assessments.

1. Define problem

This can be specified using several criteria:

- goals (e.g. power generation, regional economy)
- sector (single, several in parallel, several interactively)
- study area
    - administrative unit (e.g. district, province, country)

- geographical unit (e.g. river basin, plain)
- ecological zones (e.g. savannah, wetland)
- climatic zones (e.g. monsoon zone)
- sensitive regions (e.g. tree line, coastal zones, semi-arid margins)
- representative unit (e.g. farm, river catchment)

- time frame (depends on climate scenario)
- data needs (existing sources, new data)
- wider context (must account for other changes, studies. etc.).

2. Select method

There are a wide range of options, and a study can include more than one from among the following:

- experimentation
  - performance of plants, building materials, etc. under controlled conditions
- impact projections from models
  - biophysical (e.g. hydrology)
    - empirical-statistical
    - process-based simulation
  - economic
    - microsimulation (e.g. farm level)
    - market (e.g. farm commodity)
    - economy-wide (e.g. input-output, macrosimulation)
  - integrated systems models (e.g. IMAGE)
- analogue studies
  - temporal (historical)
  - spatial (regional)
- expert judgement
  - can include probability judgements (depends on representativeness of sample of experts)

3. Test method

These assessments are based on scenarios, which represent possible future conditions that cannot be verified at the time of the research. Testing and validation of methods become sensitivity analyses, in which climatic input variables are altered to represent the range of climatic conditions likely to occur. Information is obtained on:

- model sensitivity to changes in inputs
- ability of model to behave realistically (robustness)
- transferability of model to other climates.

Validation requires the use of historic data from outside the calibration period.

4. Select scenarios

There are several types of scenarios required for a climate impact assessment that aspires to go beyond the EERE assumption. Baseline conditions need to be specified in order to provide reference points for current conditions. Scenarios are then constructed as changes from these assumed baselines.

An assessment of future climate utilizes one or more scenarios of climate change, which can be derived from various sources:

- historical anomalies or analogues (instrumental record)
- paleoclimatic analogues (proxy record)
- arbitrary adjustments (hypothetical)
- general circulation models (GCM, equilibrium or transient).

Scenarios of environmental change may be constructed from biophysical models using the above scenarios of climate change. Socio-economic change scenarios may be similarly obtained using techniques from economics and other social sciences. In both cases, however, a distinction should be made between projections of trends in the absence of climate change and those which would be affected by the climate change scenarios used in the assessment. Official projections may already exist for land use changes, unemployment, agricultural production, etc., and these could form the baseline.

Other changes (e.g. technology, institutional arrangements) are more difficult to estimate, so collaboration with experts and stakeholders in these areas should be sought.

More information on scenarios is provided by Arnell elsewhere in this volume.

5. Assess impacts

There are several measures of impacts that may be appropriate, depending on the purpose of the assessment. Studies which focus on ecological changes may be interested in climate-sensitive indicators, such as changes in vegetation that are sensitive to water level fluctuations. Infrastructure, which is designed and operated according to a set of approved standards, can be assessed for compliance to standards under scenario conditions.

Changes in spatial and temporal patterns, and economic impacts such as changes in costs and benefits, can be important measures as well, but these are more complicated challenges. Accurate description of current patterns are dependent on the representativeness of current monitoring networks. Economic assessments must include measures that may be difficult to quantify in monetary terms (e.g. cultural aspects, barter). It will be difficult to provide an impact assessment if there is uncertainty regarding baseline descriptions.

6. Evaluate response options -- technical adjustments

Impact models can be used to test climate change scenarios with technical adjustments, compared with scenarios without adjustments (e.g. change in irrigation technology or practice). This exercise provides the foundation for assessments of possible reponse options.

7. Consider policy response options

The result of a climate impact assessment can be used in a policy simulation or other exercise. A wide range of options can be considered, including limitation of greenhouse gases (e.g. economic instruments, energy efficiency standards), and adaptation policies (e.g. increase flexibility or reduce vulnerability through changes in zoning, price supports, harvesting quotas, operating rules, etc.).

FIVE STEPS FOR ORGANIZING AN INTEGRATED REGIONAL ASSESSMENT

Integrated assessment can achieve a more comprehensive analysis by including all within-sector and between-sector interactions. Since this is more ambitious than traditional sectoral studies, it is therefore a more difficult exercise to undertake, primarily because of insufficient knowledge of the direct and indirect linkages between climate and society. There are several good reasons for attempting to conduct such an assessment, particularly at the regional scale:

- global-scale climatic anomalies (e.g. ENSO) have resulted in impacts which were region-specific (Glantz et al., 1987)
- information on regional net effects of climate change scenarios would be identified; these may be hidden by continental/global assessments
- this scale is closer to the interests of many stakeholders (including decision makers)
- suggested approach combines science with consultation of stakeholders -- *vertical integration*
- promotes interdisciplinary research -- *horizontal integration.*

A 5-step process has been suggested by Cohen (in press). What follows is a brief description of each step, as illustrated in Figure 2:

1. Define study area and problems

Although this appears to be self evident, it is a step that may require two or more increments before these are firmly established. The study area and policy questions may be redefined after consultation with regional experts and various stakeholders.

2. Establish integration framework

The choice depends on policy questions (or 'targets' for research), study area and choice of scenarios. The 'integrator' could be an economist, water planner, decision support modeller, systems analyst or policy analyst. Each integrator becomes a client of one or more sectoral experts (e.g. hydrologist, plant ecologist, forest economist) participating in the study group.

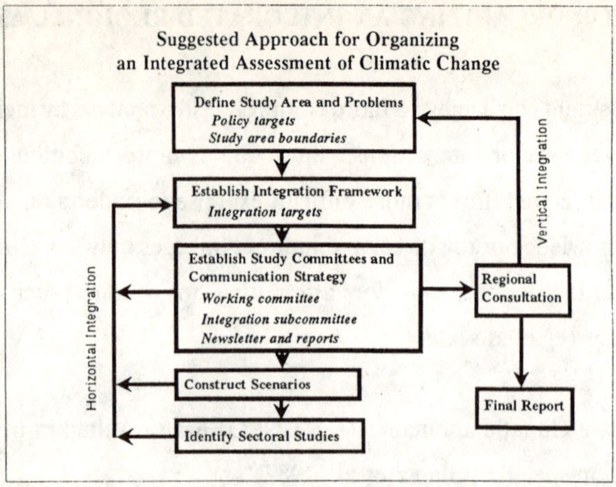

Figure 2. Suggested approach for organizing an integrated assessment of long term climatic change. Source: Cohen (in press).

3. Establish study committees and communication strategy

A key element in the organization of a multidisciplinary research group is the creation of an interagency working committee composed of government and non-government stakeholders. This committee facilitates both vertical integration (science - policy) and horizontal integration (science - science) through workshops, review of proposals and assistance in communication. Subcommittees could be created along disciplinary (e.g. hydrology) and sectoral (e.g. forestry, including biological and economic components) lines. There may also be value in creating an integration subcommittee, which would include the 'integrators' and a subset of investigators from various sectoral studies.

4. Construct scenarios

This step is similar to the Carter et al. (1992) model, and it also serves the integration process.

5. Identify sectoral studies

In an interdisciplinary exercise, particularly an assessment that aims to combine science with stakeholder inputs, good science is not the only condition that a research proposal must satisfy. Other criteria could include one or more of the following:

- direct applicability to study area
- linkage with climate
- linkage with critical subregions and issues
- linkage with other study components
- includes consultation within the study area.

SUMMARY

Climate impact assessment is a complex exercise, and integrated regional assessment is even more ambitious. The purpose is to determine the net effect of climatic change and variability, the "what if" and "so what" questions. Attributes of integrated assessment include flexible study design and scientific enquiry combined with stakeholder partnership. The former is important because interdisciplinary research can lead to shifts in paradigms among the participants. The latter represents the linkage between science and its non-technical clients.

The main challenges that need to be addressed by anyone participating in climate impact assessments are:

- accounting for complexities and uncertainties
- the long-term issue of climate change competing with short-term agendas
- the need to identify ongoing activities that can fit in so that the assessment can build on experience (e.g. monitoring, modelling, research networks)
- maintaining internal consistency in a large multidisciplinary group (e.g. scales, units, assumptions)
- appropriateness of integration methods.

Modellers of atmospheric and hydrologic processes can play important roles in these exercises. They provide baseline climatic and hydrologic data, climatic change scenarios, assessments of hydrologic impacts, and inputs to higher order assessments of implications for resource management. These are important learning opportunities which can lead to greater appreciation of the many dimensions of global environmental change.

REFERENCES

Alcamo J, Kreileman GJJ, Krol M, Zuidema G (in press) Modelling the global society-biosphere-climate system: Part 1: Model description and testing. Water, Air and Soil Pollution, Special Issue - IMAGE 2.0.

Amano, A, Fisher B, Kuroda M, Morita T, Nishioka S (eds) (1994) Climate Change: Policy Instruments and their Implications. Proceedings of the Tsukuba Workshop of IPCC Working Group III, Tsukuba Japan, 17-20 January, 1994. Center for Global Environmental Research, Tsukuba.

Arnell, NW, Jenkins A, George DG (1994) *The Implications of Climate Change for the National Rivers Authority.* Institute of Hydrology R&D Report 12, National Rivers Authority, Bristol, United Kingdom.

Burton, I, Cohen SJ (1992) Adapting to global warming: Regional options. Presented at International Conference on Impacts of Climatic Variations and Sustainable Development in Semi-Arid Regions (ICID), January - February 1992, Fortaleza.

Carter TR, Parry ML, Nishioka S, Harasawa H (1992) *Preliminary Guidelines for Assessing Impacts of Climate Change*. Environmental Change Unit, Oxford, and Center for Global Environmental Research, Tsukuba.

Cohen SJ (1993a) Climate change and climate impacts: Please don't confuse the two! Global Environmental Change 3: 2-6.

Cohen SJ (ed) *(1993b) Mackenzie Basin Impact*: Study Interim Report No.1. Environment Canada, Downsview, Canada.

Cohen SJ (in press) Integrated assessment of long term climatic change. In Munn, RE (ed), *Keeping Ahead: The Inclusion of Long-Term "Global" Futures in Cumulative Environmental Assessments.* Environmental Monograph, Institute for Environmental Studies, University of Toronto, Toronto.

Glantz MH (ed) (1988) *Societal Responses to Regional Climatic Change: Forecasting by Analogy.* Westview Press, Boulder.

Glantz MH (ed) *(1992) Climate Variability, Climate Change and Fi*sheries. Cambridge University Press, Cambridge.

Glantz MH, Katz R, Krenz M (eds) (1987) *The Societal Impacts Associated with the 1982-83 Worldwide Climate Anomalies.* National Center for Atmospheric Research, Boulder.

Henderson-Sellers A, Colls K (eds) (1993) Climatic impacts in Australia. Climatic Change 25: 3-4 (special issue), 201-438.

Hulme M, Wigley T, Jiang T, Zhao Z-c, Wang F, Ding Y, Leemans R, Markham A (1992) *Climate Change due to the Greenhouse Effect and its Implications for China.* World Wide Fund for Nature, Gland, Switzerland.

Magalhães AR, Glantz MH (eds) (1992) *Socioeconomic Impacts of Climate Variations and Policy Responses in Brazil.* Fundação Grupo Esquel Brasil, Brasilia.

Nishioka S, Harasawa H, Hashimoto H, Ookita T, Masuda K, Morita T (eds) (1993) *The Potential Effects of Climate Change in Japan.* Center for Global Environmental Research, Tsukuba CGER-I009-'93.

Ninh, NH, Glantz MH, Hien HM (1991) *Case Studies of Climate-Related Impact Assessment in Vietnam.* UNEP Project Document No. FP/4102-88-4102. United Nations Environment Programme, Nairobi

Parry ML, Blantan de Rozari M, Chong AL, Panich S (1992) *The Potential Socio-Economic Effects of Climate Change in Southeast Asia.* United Nations Environment Programme, Nairobi.

Rosenberg NJ, Crosson PR, Frederick KD, Easterling WE III, McKenney MS, Bowles MD, Sedjo RA, Darmstadter J, Katz LA, Lemon KM (1993) The MINK methodology: background and baseline. Climatic Change, 24, 1-2, 7-22.

Rosenzweig C, Parry ML (1994) Potential impacts of climate change on world food supply. Nature, 367: 133-138.

Rotmans J (1990) *IMAGE: An Integrated Model to Assess the Greenhouse Effect.* Kluwer, Dordrecht.

Schmandt J, Clarkson J (eds) (1992) *The Regions and Global Warming: Impacts and Response Strategies.* Oxford University Press, New York.

Strzepek KM, Smith JB (eds) (forthcoming) *As Climate Changes: The Potential International Impacts of Climate Change.* Cambridge University Press, New York.

Tegart WJ McG, Sheldon GW, Griffiths DC (eds) (1990) *Climate Change: The IPCC Impacts Assessment* Report to the IPCC by Working Group II, Australian Government Publishing Service, Canberra.

Tegart WJ McG, Sheldon GW (eds) (1993) *Climate Change 1992: The Supplementary Report to the IPCC Impacts Assessment.* Australian Government Publishing Service, Canberra.

United Kingdon Climate Change Impacts Review Group (UKCCIRG) (1991) *The Potential Effects of Climate Change in the United Kingdom.* DoE, London.

Westcoat JL Jr, Leichenko RM (1992) *Complex River Basin Management in a Changing Global Climate: The Indus River Basin in Pakistan, a National Assessment.* Centre for Advanced Decision Support for Water and Evironmental Systems, University of Colorado, Boulder.

Scenarios for Hydrological Climate Change Impact Studies

Nigel W. Arnell
Institute of Hydrology
Wallingford
Oxon OX10 8BB
UK

INTRODUCTION

Climate change scenarios are the link between process studies of climate change and impact assessments. It is currently difficult to use climate models directly to estimate the impacts of increasing greenhouse gas concentrations on environmental, social and economic systems, largely due to the coarse spatial resolution of current climate models. Impact assessments must therefore be based on the use of climate change scenarios. A scenario can be defined as "a coherent, internally-consistent and plausible description of a possible future state of the world" (Carter *et al*, 1994). Scenarios are not forecasts: scenario studies are intended to give an indication of what the future might feasibly be like.

This paper outlines methods that have been used to define scenarios for hydrological and water resources studies, illustrating the application of the methods with examples from the literature.

SCENARIOS FOR HYDROLOGICAL AND WATER RESOURCES IMPACT STUDIES

In principle, a climate change impact assessment should consider four levels of scenario (Figure 1). The first is a scenario for change in greenhouse gas concentrations, as expressed in an *emissions scenario*. Emissions scenarios make assumptions about the rate of change in human activity and the rate at which natural land, ocean and atmosphere systems can absorb greenhouse gases. In 1990 the Intergovernmental Panel on Climate Change (IPCC, 1990) defined four emissions scenarios,

NATO ASI Series, Vol. I 31
The Role of Water and the Hydrological Cycle
in Global Change
Edited by Howard R. Oliver and Sylvia A. Oliver

one representing a "business-as-usual" case assuming the continuation of present economic and population trends, and three representing different policies for restricting the growth of greenhouse gas concentrations. In 1992, the IPCC introduced six "business-as-usual" scenarios, each making different assumptions about population growth and economic development, but all assuming no explicit response to global warming (IPCC, 1992; Wigley & Raper, 1992). All published hydrological and water resources impact studies have been based, usually implicitly, on some form of business-as-usual scenario, although Kwadijk & Middelkoop (1994) did compare the hydrological implications of an emissions scenario assuming maximum effort in greenhouse gas mitigation with those of a business-as-usual scenario.

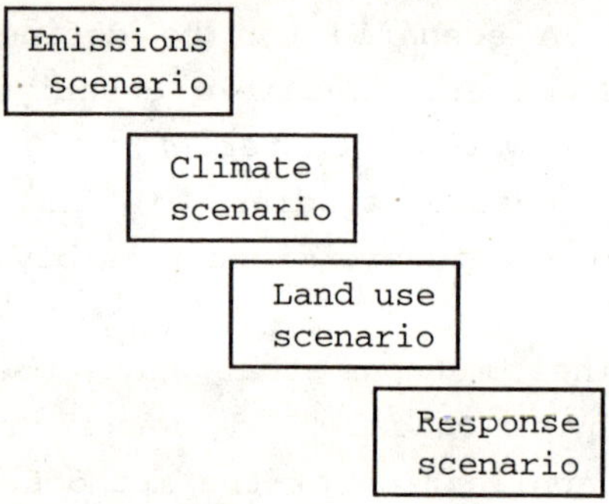

Figure 1 Scenarios for hydrological and water resources impact studies

Climate scenarios define the changes in climate (such as temperature and rainfall) associated with a given change in greenhouse gas concentrations. Catchment hydrological regimes are a function of climate and catchment land use, and land use may also change as climate changes; vegetation patterns and dynamics might alter, and agricultural practices might change. Other factors, such as urbanisation and change in agricultural policies, might also change independently of global warming. Very few studies have considered *land use change* scenarios alongside climate change scenarios when assessing possible future hydrological conditions (see for example Kwadijk, 1991 and Kite, 1993), largely because it is difficult to define credible land use change scenarios.

The fourth level of scenario concerns assumptions about water management *responses* to change, and should define changes in objectives, pressures and demands placed on a system. Such scenarios are not of course needed in hydrological impact studies, but are an essential component of assessments of possible future water resources.

This paper focuses on the construction of climate scenarios.

METHODS FOR DEFINING CLIMATE CHANGE SCENARIOS FOR HYDROLOGICAL STUDIES

Table 1 gives a classification of methods for defining climate change scenarios. The rows represent the source of the assumed changes, whilst the two columns differentiate between the direct estimation of hydrological changes and the application of changes in climatic inputs to a hydrological model.

Table 1 Methods for defining hydrological climate change scenarios

	Direct estimation of hydrological impacts	Application of changes in climate inputs with hydrological model
Arbitrary changes		
Temporal analogue		
Spatial analogue		
Climate model simulations		

Studies based on arbitrary change scenarios consider the implications of specified changes in either climatic inputs to the hydrological system or hydrological properties directly. The temporal analogue approach assumes that information from the past can provide an analogue for future conditions, whilst the spatial analogue method assumes that the future climate of a region can be described by the current climate of another region. A range of methods have been developed to use climate model simulations to define change scenarios.

Before examining these four basic types of scenarios in turn, it is important to outline briefly the different types of hydrological model which may be used to translate changes in climate into hydrological effects. There are three main types: empirical/statistical, conceptual and physics-based.

Empirical or statistical models are based, as their name implies, on an empirical relationship between some hydrological characteristic and a number of controlling variables; many are constructed through regression analysis. They are simple, but can give misleading indications of the effect of climatic change. The implied sensitivity to change is dependent on the form of the empirical model and the data used. Empirical models internalise relationships between independent variables, and the effects of other, possibly important, variables are hidden in calibrated coefficients. These internalised relationships (such as that between potential and actual evaporation) may not hold under changed climatic conditions. Also, the effect of a change in one of the controlling variables may depend on how that change is distributed: a given percentage change in annual rainfall, for example, has a greater effect on runoff in the UK if it is increasingly concentrated in winter (Arnell, 1992).

Conceptual models are based on an idealised representation of hydrological processes within a catchment, and generally operate by accounting for water as it moves through a variety of stores. Movement from store to store is determined by parameters which may have some notional physical interpretation, but which cannot necessarily be measured in the field. These parameters have to be calibrated using observed river flow data, so introducing an element of empiricism, but the conceptual framework behind such models means that they are more suitable for simulating the effect of changes in inputs than empirical/statistical models. Conceptual models have been very widely used in hydrological impact analysis (eg Bultot *et al*, 1988).

Physics-based models are based on equations which simulate the movement of water into and through the soil, and along the hillslope into the river (eg Abbott *et al*, 1986). They are spatially-distributed, and operate at a spatial resolution of the order of a few tens of metres. In principle, parameters can be measured in the field, but in practice this is rarely possible because of extreme spatial heterogeneity. Physics-based models are very demanding of data, both to define boundary conditions and for operation: they need climatic inputs at very high temporal and spatial resolutions. Very few catchment-scale climate change studies have therefore used physics-based hydrological models.

ARBITRARY CHANGES IN CLIMATE INPUTS

Many hydrological and water resources impact studies have considered the effects of arbitrary changes in climatic and hydrological characteristics. Examples include Nemec & Schaake (1982), Mimikou *et al* (1991) and Wolock *et al* (1993). Such studies cannot be seen as indications of possible future conditions, but they do provide very valuable information on the sensitivity of a hydrological or water resources system to change in climatic inputs.

TEMPORAL ANALOGUES

The temporal analogue approach assumes that information from the past can be used to define scenarios for possible future conditions. There are two types of analogue, one based on instrumental data and one derived from palaeoclimatic information.

Instrumental analogues

These analogues use data from the instrumental record, which rarely extends back more than 100 years. Examples include Palutikof (1987), who looked at simulated river flows in different decades, and Krasovskaia & Gottschalk (1992) who compared observed river flow regimes. The approach involves either comparing climate or hydrological regimes in a "warm" period (perhaps a decade) with those in a "cool" period or, more rarely, comparing behaviour in all the "warm" years, regardless of when they occurred, with characteristics of all the "cool" years. There are several problems with this approach.

It is assumed that the differences between the warm and cool periods do represent a reasonable analogue for differences between current "cool" conditions and future "warm" conditions, even though the factors causing the differences are not the same. Differences between warm and cool periods in the instrumental record, for example, are largely due to random fluctuations, not some systematic forcing: the spatial patterns and even direction of change in particular locations may be very different.

The analogue periods selected will depend on the extent of data considered. Taking the Northern Hemisphere as a whole, for example, the 20-year warm period 1934-1953 was 0.5°C warmer than the 1901-1920 cool period, but in Europe 1934-1953 was *cooler* than 1901-1920 (Palutikof, 1987).

The difference in temperature between the warm and cool period may be smaller than that anticipated under a global warming greenhouse gas emissions scenario and, finally, short records may make it very difficult to find analogue periods.

Instrumental analogues, however, can provide information on the behaviour of an environmental or water resources system under extreme conditions. Observations on response to drought, for example, could be used to develop hypotheses about impacts of future water shortages, and the robustness of a system to future change could be modelled by exposing it to a past extreme event. The MINK project, for example, investigated the effect that a repetition of the 1930s drought would have on the present and future economy and water use of Missouri, Iowa, Nebraska and Kansas (Frederick, 1993).

Palaeoclimatic analogues

The palaeoclimatic analogue approach assumes that conditions during some warmer period in geological history can be used as an analogue for climatic conditions under global warming. The approach is based on the reconstruction of past environments, using geological and botanical evidence. Geological evidence includes land forms, sediments and the age of groundwater, whilst botanical evidence may be based on pollen analysis and dendrochronology. The Holocene climatic optimum (6.2 to 5.3 ka Before Present), the last interglacial (125 ka BP) and the Pliocene (3 to 4 ma BP) have been proposed as analogues for global warming of 1, 2 and 4° respectively (IPCC, 1990). There are two major assumptions behind the use of palaeoclimatic analogues.

It is assumed that the relationships between form and process operating today are the same as those operating in the past (the uniformitarian assumption), and that it is possible to infer process - such as rainfall and river flow characteristics - from past form. More importantly, it is assumed that the effects of a change in climate are independent of the causes of the change in climate. Variations in global climate over geological time scales are a result of changes in the earth's orbital patterns, and hence changes in the receipt of solar radiation. The resulting climate changes will not necessarily be the same as those arising from an increased greenhouse effect.

A major practical problem with palaeoclimatic analogues lies in the derivation of quantitative information. Climatic reconstructions rely on the quality of the assumed links between form and process. Even when it is possible to reconstruct point climate, it can be very difficult to infer regional climate. Some studies, however, have used over-bank sedimentological evidence to reconstruct flood chronologies (eg Knox, 1993), showing large changes in flood risk for small changes in climate; these can best be seen as sensitivity studies.

SPATIAL ANALOGUES

The spatial analogue approach assumes that the future climate of a region can be represented by the present climate of another. Parry *et al* (1988), for example, used a number of spatial analogues to estimate possible changes in agricultural production: the future climate of Iceland, for example, was assumed to be similar to the current climate of northern Scotland. However, there are some major problems with spatial analogues.

The climate of an area is a function of large-scale climatic characteristics and local conditions, such as topography and the proximity to the sea and other water bodies. It is often not therefore possible simply to transfer the climate from one area to another; local conditions may be very different. Transfer of hydrological data is even more difficult, because hydrological regimes are determined by both climatic inputs and catchment physical characteristics.

Empirical relationships predicting hydrological characteristics from catchment physical and climatic properties can provide the basis for a spatial analogue; variability over space is taken as a proxy for variability over time. Example relationships include regression relationships used to predict flood characteristics from catchment area and rainfall, and the Langbein (1949) relationships predicting annual runoff in American catchments from annual rainfall and annual temperature. The Langbein relationships have been used to estimate the effects of climate change (Revelle & Waggoner, 1983). However, the implied sensitivity to change is dependent on the form of the model and the values of the coefficients, which themselves internalise relationships between controlling variables and variables not explicitly considered. Estimates of the sensitivity to change based on such spatial equations are to be treated with extreme caution, and Schaake (1990) noted that Revelle & Waggoner's (1983) estimated sensitivities were quantitatively different to those based on studies using the same changes in inputs applied to hydrological models.

CLIMATE MODEL OUTPUT

There are essentially two problems with using climate model simulations to define scenarios for hydrological climate change impact studies. The first is the inherent credibility of climate model simulations, and the other is the spatial resolution at which current models operate. The spatial resolution of current climate models is too coarse for catchment-scale impact studies (although continental-scale hydrological impact assessments have used global model output: Miller & Russell, 1992), and before scenarios can be derived from model output this output needs to be expressed at the correct spatial and temporal scale. The "correct" scale depends on the size of the catchment to be studied, but conceptual hydrological models rarely operate in units greater than $1x10^3$ km^2 or at time scales greater than one week; physics-based models operate at much finer space and time scales.

There are four ways of deriving finer spatial resolution data out from global climate models (a process known as "downscaling"), but before going through these methods it is necessary to review the information that is actually downscaled. There are three choices. The first is to use the downscaled model simulations of weather directly as inputs to the hydrological model, but this assumes that the climate model simulates current climatic regimes reasonably well. The remaining two approaches use downscaled *changes* in climate, and perturb historical data series accordingly. This essentially assumes that even if the climate model does not simulate climate very well, it gives useful information on the possible direction and magnitude of change. The first variant simply perturbs an observed record, by adding for example a given percentage change to observed rainfall. This approach changes the mean of the initial input climate series, but does not change relative variability, either from day to day or from year to year. In principle it is possible to add a random component to introduce noise and alter variability, but this has not been done in practice. The second variant perturbs the parameters of a stochastic weather generation model (Wilks, 1992). It is possible to alter temporal structures and variability with this approach, but it requires a realistic stochastic weather generation model which has parameters which can be altered with the information from climate model output.

<u>Interpolation</u>

The simplest approach to downscaling involves some form of interpolation, ranging from use of the nearest grid point (Cohen, 1991) to objective interpolation using geostatistical methods. The approach is generally applied to climate model estimates of change in monthly or seasonal rainfall and temperature; there is too much spatial variability for the approach to be feasible with daily model output.

Expert judgement

This approach combines the interpolation of climate model output with expert meteorological and climatological interpretation. This expert judgement may take the form of using local knowledge in the interpolation of large-scale changes (as used in the UK Climate Change Impacts Review Group (1991)). A more complicated methodology is presented by Robock *et al* (1993). In one of their examples, they developed scenarios for changes in rainfall in the Sahel region from the observation that seasonal rainfall totals in the past tended to vary as a result of changes in the length of the wet season, rather than changes in rain intensities; instead of using climate model output to alter rainfall intensities, they therefore used it to alter the duration of the wet season.

Downscaling from large-scale climate and weather features

This approach develops relationships between point weather and large-scale regional climate. It assumes that global climate models simulate regional weather well, and also that relationships between local and regional weather and climate not only can be developed, but also will remain constant during climate change. There are several variations.

One variation on the approach uses regression relationships between point monthly weather and regional average monthly weather (eg Wigley *et al*, 1990). Different regression relationships need to be derived for each location of interest and each month. Wigley *et al* (1990) found in Oregon that the strength of relationship between point and regional climate varied over space, and that in some parts of the state it was not possible to estimate local climate from regional spatial averages.

A second variation uses empirical relationships between free atmospheric variables and point weather to simulate local daily weather patterns (Karl *et al*, 1990). This is analogous to methods used for point weather forecasting. However, although Karl *et al* (1990) were able to simulate local daily rainfall regimes from climate model simulations of free atmospheric characteristics, the method is very time consuming to apply and has not yet been used to create scenarios for impact studies.

A third variation is based on the use of discrete weather or circulation types (Hay *et al*, 1990; Bardossy & Plate, 1992). Daily weather is classified into types, and for each type weather properties, such as probability of rainfall and mean rain intensity, are determined for each point of interest. Weather types can be defined climatologically (Hay et al, 1990; Lamb, 1972), or statistically (Matyasovszky *et al*, 1993). Climate change scenarios can be developed either by determining a sequence of daily weather types directly from climate model simulations (Hay *et al*, 1992), or by creating a stochastic model to simulate weather type occurrence and perturbing the parameters of this model using changes in the frequency of occurrence of different weather types as simulated by a climate model (Bardossy & Plate, 1992). Given a sequence of weather types, daily point weather can then be simulated stochastically using the parameters based on observed data. The approach assumes that stable empirical relationships can be developed between point weather and weather type. Hay *et al* (1992) found that the intensity of point rainfall for a given weather type differed in $1xCO_2$ and $2xCO_2$ simulation experiments; Matyasovszky *et al* (1993) accounted for this by introducing the height of the 500mb pressure isobar into the stochastic relationship between weather type and point weather.

The third of these three variants appears to offer the greatest prospects for the creation of climate change scenarios appropriate for hydrological impact assessments, as long as it is possible to develop stable relationships between weather type and point weather which will remain valid under altered climatic conditions.

Nested regional models

The final approach to downscaling uses a high-resolution regional climate model nested within a coarse-resolution global climate model, and running with boundary conditions provided by the global model (Giorgi & Mearns, 1990). One-way nesting means that the regional model does not send information back to the global model, whilst in two-way nesting there is some feedback. Two-way nesting is very complicated and resource-intensive; one-way nesting has been implemented in several global climate models. Careful attention needs to be paid to the domain over which the regional model is applied. This domain must be large enough to allow meso-scale features, such as depressions, to develop, but not so large that regional-scale features would be significantly affected. The appropriate regional model domain for studies within Europe, for example, extends from the west of Iceland to the Ural Mountains.

The use of nested regional models with a ground resolution of the order of 50x50km enables better simulation of within-region and local climate. Spatial patterns are much better simulated, particularly in areas with complicated topography or landscapes. A regional model nested within the NCAR global climate model, for example, was able to simulate accurately precipitation regimes around the Great Lakes (Giorgi *et al*, 1994). Climate change scenarios based on nested model simulations should therefore be more credible, although it does not necessarily follow, of course, that a model which simulates current climate well will simulate future climate well. Nested models can also provide information, at hydrologically-relevant spatial scales, on possible changes in daily weather characteristics (and it might even be feasible to use nested model output directly as input to catchment hydrological models). Nested model output has not yet been used in hydrological impact studies, although Mearns & Rosenzweig (1994) used scenarios derived from the NCAR nested regional model with a crop growth model.

A variation on this approach uses a global model with a variable resolution grid; such models are currently under development both at Metéo-France and the Canadian Meteorological Centre.

DEFINING SCENARIOS FOR SPECIFIC TIME PERIODS

The vast majority of climate change impact assessments have compared current ($1xCO_2$) average conditions with future ($2xCO_2$) average conditions, and have not explicitly defined *when* these impacts might be felt. Increasingly, however, studies have begun to attach dates to scenarios, as interest in rates of change, adaptation and economic costs develops.

Two types of global climate simulation experiments have been run. *Equilibrium* simulations represent stable climatic conditions with current and doubled CO_2 concentrations. This is rather artificial, for two main reasons. First, CO_2 concentrations will not stabilise at double current levels, so climate will not be able to reach a corresponding equilibrium. Second, the pattern of change might be very dependent on the gradual evolution of change. This problem is addressed through *transient* experiments, which simulate the response to a gradual increase in CO_2 concentrations. Equilibrium scenarios, however, are very valuable because they allow a comparison of current "stable" conditions with some notional period of stability in the future. Most hydrological assessments evaluate changes in statistics which ideally should be calculated over a long period, of at least 30 years (average annual runoff, or reliability of supply, for example), but it is difficult to define these statistics against a moving background. It is therefore convenient to define a set of data which represent average conditions over a number of years, even if it is accepted that in practice there will not be a stable period of that length.

It is, however, not possible to assign dates directly to equilibrium climate model experiments, because the date at which CO_2 concentrations reach double current values is not the same as the date by which climate would be in equilibrium with doubled CO_2 concentrations. A technique for defining scenarios for particular time horizons from equilibrium model output has nevertheless been developed by Viner & Hulme (1993). The first stage is to standardise the changes in climate as simulated by an equilibrium model by the model climate sensitivity (the change in global mean temperature resulting from a doubling of CO_2 concentrations). Changes for a given year are then determined by multiplying the standardised changes by the global temperature change as estimated for that year from a single-column energy balance model (such as that developed by Wigley & Raper, 1992) and a defined emissions scenario. The approach makes the assumptions that the rate of change is linear over time, and also that the spatial pattern of change does not vary over time. This last assumption in particular is unlikely to be true, and transient simulations have shown that some parts of the globe take a long time to react to increasing global mean temperatures (IPCC, 1992). The method does, however, enable scenarios to be defined for particular time periods.

In principle, it is possible to define scenarios for particular time periods by selecting corresponding periods from transient model experiments. In practice, however, this is not feasible for three main reasons. First, transient experiments suffer from the "cold-start" problem (Hasselmann *et al*, 1992), whereby it takes a few years for the model to settle down and start to show the effects of an increase in CO_2 concentrations; the model simulations are therefore out of phase with real time. Second, there is considerable year-to-year and decade-to-decade variability in transient climate simulations. The differences between control and transient simulations by a given decade will represent the combined effects of greenhouse signal and climatic noise. Third, climate model simulations suffer from model drift, and even control simulations with no change in boundary conditions show some trend over time. This model drift needs to be removed in some way before changes can be determined from transient simulations. There is a clear need for techniques to create scenarios from transient simulation output.

CONCLUSIONS

This paper has outlined methods for creating climate change scenarios for hydrological impact studies, summarising arbitrary change scenarios, temporal and spatial analogues, and reviewing methods for downscaling climate model output to scales appropriate for hydrological studies. There are several important points to make in conclusion:

1. A scenario is not a forecast: it is important to use several different scenarios, generated perhaps from different climate models, in an impact assessment. Sensitivity to different scenarios is as important a system characteristic as sensitivity to one scenario.

2. The assumptions behind the scenario need to be stated explicitly. What emissions scenarios are assumed? Are there assumed changes in land use and, for water resources impact studies, management policies? Does the scenario refer to a particular date?

3. It is important to define clearly the baseline period to which the climate change scenarios are applied. Different studies have used different baselines, and this may make it difficult to compare impacts in different studies.

References

Abbott, M.B., Bathurst, J.B., Cunge, J.A., O'Connell, P.E. & Rasmussen, J. (1986) An introduction to the European Hydrological System - Systéme Hydrologique Européen "SHE". 2. Structure of the physically-based distributed modelling system. J. Hydrol. 87, 61-77.

Arnell, N.W. (1992) Factors controlling the effects of climate change on river flow regimes in a humid temperate environment. J. Hydrol. 132, 321-342.

Bardossy, A. & Plate, E.J. (1992) Space-time model for daily rainfall using atmospheric circulation patterns. Wat. Resour. Res. 28, 1247-1259.

Bultot, F., Coppens, A., Dupriez, G.L., Gellens, D. & Meulenberghs, F. (1988) Repercussions of a CO_2 doubling on the water cycle and on the water balance: a case study from Belgium. J. Hydrol. 99, 319-347.

Carter, T.R., Parry, M.L., Nishioka, S. & Harasawa, H. (1994) Technical Guidelines for Assessing Climate Change Impacts and Adaptations. Intergovernmental Panel on Climate Change.

Climate Change Impacts Review Group (1991) The Potential Effects of Climate Change in the United Kingdom. HMSO: London.

Cohen, S.J. (1991) Possible impacts of climatic warming scenarios on water resources in the Saskatchewan River sub-basin, Canada. Climatic Change 19, 291-317.

Frederick, K.D. (1993) Climate change impacts on water resources and possible responses in the MINK region. Climatic Change 24, 83-115.

Giorgi, F. & Mearns, L.O. (1990) Approaches to the simulation of regional climate change: a review. Rev. Geophys. 29, 191-216.

Giorgi, F., Brodeur, C.S. & Bates, G.T. (1994) Regional climate change scenarios over the United States produced with a nested regional climate model. J. Climate 7, 375-399.

Hasselman, K., Sausen, R., Maier-Reimer, E. & Voss, R. (1992) On the cold start problem in transient simulations with coupled ocean-atmosphere models. Max Planck Institüt für Meteorologie Report 83, Hamburg.

Hay, L.E., McCabe, G.J., Wolock, D.M. & Ayers, M.A. (1990) Simulation of precipitation by weather type analysis. Wat. Resour. Res. 27, 493-501.

Hay, L.E., McCabe, G.J., Wolock, D.M. & Ayers, M.A. (1992) Use of weather types to disaggregate general circulation model predictions. J. Geophys. Res. 97 (D3) 2781-2790.

Intergovernmental Panel on Climate Change (1990) Climate Change. The IPCC Scientific Assessment. Houghton, J.T., Jenkins, G.J. & Ephraums, J.J. (eds.) Cambridge University Press: Cambridge.

Intergovernmental Panel on Climate Change (1992) Climate Change 1992. The Supplementary Report to the IPCC Scientific Assessment. Houghton, J.T., Callendar, B.A. & Varney, S.A. (eds). Cambridge University Press: Cambridge.

Karl, T.R., Wang, W., Schlesinger, M.E., Knight, R.W. & Portman, D. (1990) A method of relating general circulation model simulated climate to the observed local climate. Part 1: Seasonal statistics. J. Climate 3, 1053-1079.

Kite, G.W. (1993) Application of a land class hydrological model to climatic change. Wat. Resour. Res. 29, 2377-2384.

Knox, J.C. (1993) large increases in flood magnitude in response to modest changes in climate. Nature 361, 430-432.

Krasovskaia, I. & Gottschalk, L. (1992) Stability of river flow regimes. Nordic Hydrology 23, 137-154.

Kwadijk, J. (1991) Sensitivity of the River Rhine discharge to environmental change: a first tentative assessment. Earth Surface Processes and Landforms 16, 627-637.

Kwadijk, J. & Middelkoop, H. (1994) Estimation of impact of climate change on the peak discharge probability of the River Rhine. Climatic Change 27, 199-224.

Lamb, H.H. (1972) British Isles weather types and a register of the daily sequence of circulation patterns, 1861-1971. Geophysical Memoir 116. HMSO: London.

Langbein, W. (1949) Annual runoff in the United States. U.S. Geological Survey Circular 5. U.S. Dept. of the Interior, Washington D.C.

Matyasovszky, I., Bogardi, I., Bardossy, A. & Duckstein, L. (1993) Space-time precipitation reflecting climate change. Hydrol. Sci. J. 38, 539-558.

Mearns, L.O. & Rosenzweig, C. (1994) Use of nested regional climate model output with changed daily climate variability to test related sensitivity of dynamic crop models. American Meteorological Society Preprints. Applied Climatology Meeting, Nashville, Tennessee.

Miller, J.R. & Russell, G.L. (1992) The impact of global warming on river runoff. J. Geophys. Res. 97 (D3), 2757-2764.

Mimikou, M., Kouvopoulos, Y., Cavadias, G. & Vayiannos, N. (1991) Regional hydrological effects of climate change. J. Hydrol. 123, 119-146.

Nemec, J. & Schaake, J.C. (1982) Sensitivity of water resource systems to climate variation. Hydrol. Sci. J. 27, 327-343.

Palutikof, J. (1987) Some possible impacts of greenhouse gas-induced climatic change on water resources in England and Wales. The Influence of Climatic Change and Climatic Variability on the Hydrologic Regime and Water Resources. IAHS Publ. 168, 585-596.

Parry, M.L., Carter, T.R. & Konijn, M. (eds.) (1988) The Impact of Climatic Variations on Agriculture. Vol. 1 Cool and Temperate Regions. Kluwer: Dordrecht.

Revelle, R.R. & Waggoner, P.E. (1983) Effects of a carbon dioxide-induced climatic change on water supplies in the western United States. in Changing Climate. National Academy of Sciences, National Academy Press, Washington D.C.

Robock, A., Turco, R.P., Harwell, M.A., Ackerman, T.P., Andressen, R., Chang, H. & Sivakumar, M.V.K. (1993) Use of general circulation model output in the creation of climate change scenarios for impact analysis. Climatic Change 23, 293-335.

Schaake, J.C. (1990) From climate to flow. in Waggoner, P.E. (ed.) Climate Change and U.S. Water Resources. Wiley: New York. 177-206.

Viner, D. & Hulme, M. (1993) Construction of climate change scenarios by linking GCM and STUGE output. Technical Note 2. Climatic Research Unit, University of East Anglia, Norwich, England.

Wigley, T.M.L. & Raper, S.C.B. (1992) Implications for climate and sea level of revised IPCC emissions scenarios. Nature 357, 293-300.

Wigley, T.M.L., Jones, P.D., Briffa, K.R. & Smith, G. (1990) Obtaining sub-grid-scale information from coarse-resolution general circulation model output. J. Geophys. Res. 95 (D2) 1943-1953.

Wilks, D.S. (1992) Adapting stochastic weather generation algorithms for climate change studies. Climatic Change 22, 67-84.

Wolock, D.M., McCabe, G.J., Tasker, G.D. & Moss, M.E. (1993) Effects of climate change on water resources in the Delaware River basin. Wat. Res. Bull. 29, 475-486.

Potential Changes to Hydrological Systems

Stewart J. Cohen
Atmospheric Environment Service
Environment Canada
4905 Dufferin Street
Downsview, Ontario
Canada M3H 5T4

INTRODUCTION TO ASSESSMENT OF HYDROLOGIC IMPACTS

Climatic variability and change, including scenarios of global warming, may affect hydrological systems and actions taken to manage them. This section covers the first-order hydrologic effects, while water resource management issues are considered by Arnell elsewhere in this volume.

The earlier section by Cohen on climate impact assessment referred to the Intergovernmental Panel on Climate Change (IPCC) impacts assessment reports. These include updates on hydrologic and water resources impacts of global warming scenarios (Tegart and Sheldon, 1993). Recent reviews have also been provided by Gleick (1989) and Dooge (1992). It is not the intention of this section to repeat these reviews. The purpose here is to provide an overview of some of the methods that have been used to estimate hydrologic impacts from GCM-based global warming scenarios, and to compare results of various case studies. There have also been cases which used scenarios derived from arbitrary adjustments and analogues (e.g. Frederick, 1993), but these are not included here.

Assessments of hydrologic impacts can accomplish several objectives:

- identification and evaluation of effects of climatic variability and scenarios of climatic change;
- inclusion of secondary effects, such as CO_2 enrichment, changes in land cover, and

NATO ASI Series, Vol. I 31
The Role of Water and the Hydrological Cycle
in Global Change
Edited by Howard R. Oliver and Sylvia A. Oliver

changes in consumptive water use and transfers;

- provision of input to multisectoral and integrated assessments, including alternative strategies for facility operation, changes in water management (e.g. hydroelectric production, drought policy), effects on water users (e.g. agriculture, transportation, tourism, shoreline properties, energy production), and effects on ecosystems.

Specific outputs from hydrologic studies generally consist of quantitative information on specific characteristics of water cycle components (runoff, evaporation, soil moisture, etc.), such as mean annual totals, seasonal aspects, return periods, identification of new thresholds, and sensitivities to climate. The IPCC (Tegart and Sheldon, 1993) has noted, however, that case studies have not adopted a uniform methodology for impact assessments, which makes it difficult to judge their relative merit. Even with the substantial increase in the number of case studies in recent years, few new insights have been offered because of the continuing uncertainties associated with General Circulation Model (GCM) projections of future climate, especially the simulations of precipitation. However, the effort has raised awareness of the potential implications of scenarios of climatic change, and this could lead to institutional responses that are anticipatory and adaptive in nature (e.g. establishing drought management plans).

The challenge of translating climatic scenarios into hydrologic scenarios is extremely important since we are all dependent on water resources. We have to continually express various caveats (e.g. uncertainties in GCMs and hydrologic models, representativeness of field observations) so that case study results are not misinterpreted as definitive forecasts. These case studies represent important opportunities to learn about sensitivities of watersheds, and watershed models, to scenarios of climate change which could be outside the range of model calibration and historical observation.

Although GCMs produce simulations of evaporation, runoff and other hydrologic parameters, case studies have generally applied only the GCM simulations of temperature, precipitation, wind speed and other climatic parameters, as inputs to hydrologic models of various kinds, such as water balance methods, deterministic models, or models "customized" for individual basins. It is important to remember that climatic change scenario data may not be available for the specific input requirements of hydrologic and related models. Intermediate steps and assumptions may be required in order to use these models to estimate impacts of scenarios.

APPLICATION OF CLIMATE CHANGE SCENARIOS

Case studies have been carried out in various settings, with the majority done in the mid-latitudes. The main focus of these investigations is seasonal and annual runoff (R), but information on open water evaporation (E), actual evapotranspiration (AEt), potential evapotranspiration (PEt), soil moisture and snow regime is also provided. It is useful to examine these various constituents because of their role in determining impacts on R, streamflow and average water availability ("Net Basin Supply"). What follows is a review of various cases, with results provided for those which used GCM-based scenarios of climate change.

Evaporation and Evapotranspiration

Several studies have reported results for E, AEt and PEt for GCM-based scenarios (Table 1). In every case, a warmer climate leads to increased evaporation and evapotranspiration.

Cohen (1986) and Croley (1990) provide scenario estimates for E for the Great Lakes using a mass transfer approach, and both indicate substantial increases. The approach is described in Quinn and den Hartog (1981). E is considered a function of stability, wind speed and vertical differences in vapour pressure:

$$E = M(e_s \quad e_a)\, U_a$$

M = mass transfer coefficient at 3 metres

e_s = sat. vapour pressure at surface water temperature

e_a = vapour pressure at 3 metres

Table 1. Results of Case Studies: Evaporation, Actual and Potential Evapotranspiration (expressed as % change)

SITE/Ref	MODEL	GFDL80	GFDL86	GISS	OSU	UKHI
GREAT LAKES/ Cohen, 1986	E: mass transfer	+14; w=+10		+33		
Croley, 1990	E:LBRM		+44	+27	+26	
Cohen, 1986	AEt: Thornth.	+7		+18		
Croley, 1990	AEt:LBRM		+19	+21	+19	
CONNECTICUT/ Fennessey & Kirshen, 1993	AEt:P-M		+57	+20; r_c=+5	+23	+10 UKMO
Fennessey & Kirshen, 1993	PEt:P-M		+41	+17	+13	+32 UKMO
U.K./Arnell & Reynard, 1992	PEt:P-M (winter)					+51; r_c=+29
Arnell & Reynard, 1992	PEt:P-M (summer)					+24; r_c=+7
BELGIUM/ Bultot et al., 1988	AEt:IRMB	+10				
Bultot et al., 1988	PEt:IRMB	+12				
WARTA/ Kaczmarek & Krasuski (K&K), 1991	AEt:K&K PEt:B-Z		+13			

* Results provided for the following GCMs:

GFDL - Geophysical Fluid Dynamics Lab (1980, 1986, Qflux)
GISS - Goddard Institute for Space Studies
OSU - Oregon State University
UKMO - United Kingdom Meteorological Office (UKHI = high resolution model)

M increases with wind speed (U_a) and bulk evaporation coefficient (C_E). The latter is a function of stability (based on air - water temperature difference), and is highest for unstable conditions. Since current GCM grids are generally classified as "land", overlake wind speed and water temperature cannot be obtained directly from GCM outputs, so additional steps are required. For example, the Large Basin Runoff Model (LBRM), described by Croley elsewhere in this volume, includes a water temperature module in which air temperature is used as input.

Other models which have been used include the Priestley-Taylor model (1972). This requires net radiation, surface heat flux, and temperature. It also includes a constant non-dimensional surface evaporability factor which may be affected by warm air advection. The use of this model would require an assumed value for the water surface's albedo and net radiation, since GCM grids are generally classified as "land".

E can also be obtained directly from GCMs, but only for grids classified as water.

Actual Evapotranspiration (AEt), Potential Evapotranspiration (PEt):

Cohen (1986) provides scenario estimates for the Great Lakes using the Thornthwaite climatic water balance (Mather, 1978). PEt is computed as a function of temperature and daylength:

$$PEt = ADJ * 0.553 (10T/I)^a$$

PEt = potential evapotranspiration

ADJ = daylength adjustment (units of 12 hours per month)

T = mean air temperature

I = Thornthwaite annual heat index

-sum of 12 monthly heat indices, where $i = (t/5)^{1.514}$

a $= (6.75 * 10^{-7} I^3) - (7.71 * 10^{-5} I^2) + (1.792 * 10^{-2} I) + 0.49239$

When rain ≥ PEt, and soil moisture storage (SM) is at capacity,

$$AEt=PEt$$

while for rain < PEt, at time t_1

$$AEt_{t1}=rain_{t1}+snowmelt_{t1}+\Delta SM_{t1-t0}$$

Some case studies have employed other water balance methods derived from Thornthwaite. Examples include Gleick's (1987) study of the Sacramento River watershed and the Delaware River Basin study by Wolock et al. (1993).

Water balance models have modest data requirements, so it is relatively easy to apply a climate change scenario. There are, however, several important constraints. One is the uncertainty in determining a representative soil moisture storage capacity for a grid. Estimates of AEt and R are sensitive to this. A second problem is that there is no direct way for assessing the potential effects of CO_2 enrichment on plant transpiration. Finally, the model has generally been used in mid-latitude settings, and model performance in high and low latitudes is uncertain.

The relative effects of changes in climate and CO_2 enrichment on evapotranspiration can be determined with the Penman-Monteith (P-M) approach. Rosenberg et al. (1990) provide the following description:

$$\lambda E_t=\frac{\Delta(R_n-G)+\frac{\rho_a C_p[e_s(z)-e(z)]}{r_a}}{\Delta+\gamma\left[\frac{(r_a+r_c)}{r_a}\right]}$$

where λ = latent heat of vapourization, Δ = gradient of sat. vap. pressure curve, R_n = net radiation, G = soil heat flux, ρ_a = density of air, C_p = specific heat of air, r_a = aerodynamic resistance, r_c = canopy resistance,
$e_s(z)$ -e(z) = sat. vapour pressure deficit, and γ = psychrometric constant.

In a scenario of elevated CO_2, stomatal resistance increases but leaf area also increases. Overall, however, r_c increases, thereby reducing AEt (Rosenberg et al., 1990). In case studies from the United Kingdom (Arnell and Reynard, 1993) and the Connecticut River in the United States (Fennessey and Kirshen, 1993), r_c was assumed to increase. Both studies reported reductions in projected increases in AEt and PEt for all GCM-based scenarios with inclusion of assumed r_c (Table 1). This illustrates both the advantages and difficulties in applying P-M to a climate change scenario, since identifying representative values for energy budget terms, leaf area index and resistances will be difficult for a GCM grid.

An energy budget approach was also used in a case study of several watersheds in Belgium and Switzerland (Bultot et al., 1988a, 1992). This was part of a conceptual daily step hydrological model (IRMB) described in Bultot et al. (1988a).

The Budyko-Zubenok (B-Z) approach has been used by Kaczmarek and Krasuski (1991) in a study of the Warta River, Poland (Table 1). They provide the following description:

$$PEt=86400\left(\frac{0.622\rho}{Ap\rho_w}\right)D[e_s(Tw)-e]$$

where Ap = air pressure (month), ρ = air density, ρ_w = density of water, Tw = land surface temperature (month) derived from energy balance, and D = coefficient of diffusion (3 to 10 mm s^{-1}).

Tw requires shortwave, longwave and soil heat fluxes, and albedo. D depends on catchment aridity and temperature, and is derived from annual temperature and P, with higher values indicating increased aridity.

Soil Moisture

GCMs provide information on soil moisture changes for each grid identified as land. Kellogg and Zhao (1988) and Zhao and Kellogg (1988) compare the results of four GCMs for North America and Asia, respectively (Table 2). There is a tendency for increases during the winter and decreases during the summer, but the GCMs do not agree on the regional details in either season.

Table 2 also lists results, for GCM-based scenarios, of several case studies from North America and Europe which employed the Thornthwaite water balance or a variation (e.g. Cohen, 1986, 1991; Gleick, 1987; McCabe and Ayers, 1989), or an integrated basin model which includes a soil moisture component and where soil moisture was reported (e.g. Bultot et al., 1988b; Croley, 1990). Soil moisture declines in almost every case.

When the North American case studies are compared with direct outputs from GCMs, the cases exhibit much larger decreases than the direct outputs. It would be interesting to do a similar comparison with the newer GCMs which include more sophisticated land surface parameterizations.

Snow Regime

Table 3 lists two cases (Arnell and Reynard, 1993; Croley, 1990) which provide GCM-based scenario results for snow water equivalent. Large decreases are found since a greater proportion of winter precipitation would fall as rain instead of snow.

For high latitudes, the situation might be quite different. No cases are available for high latitudes, and although GCMs tend to project considerable increases in temperature, it is likely that winter precipitation will continue to be dominated by snow. An increase in snow water equivalent is a likely outcome, but the regional and temporal details are still not clear.

Table 2. Results of Case Studies*: Soil Moisture (expressed as change in mm, except where indicated)

BASIN/Ref	MODEL	GFDL80	GFDL86	GISS	OSU	UKHI
GREAT LAKES/ Cohen, 1986	Thornth. deficit	-66%		-16%		
Croley, 1990	LBRM		-43% to -19%	-57% to -3%	-33% to -8%	
DELAWARE/ McCabe & Ayers, 1989	Thornth.		-70 to -55	-27 to -22	-25 to +5	
SACRAMENTO/ Gleick, 1987	Thornth. derived		-36%	-14%		
SASKATCHEWAN/ Cohen, 1991	Thornth. deficit	-94 to +71	-234 to -78	-71 to -9	-53 to -6	
NORTH AMERICA/ Kellogg & Zhao, 1988	GCM, Dec. to Feb.		0 to +20	-40 to +20	0	-40 to +20 UKMO
NORTH AMERICA/ Kellogg & Zhao, 1988	GCM, June to Aug.		-20	-40 to +20	0	-40 to 0 UKMO
BELGIUM/ Bultot et al., 1988a	IRMB	-3%				
ASIA/ Zhao & Kellogg, 1988	GCM, Dec. to Feb		-10 to +40	-30 to +30	0 to +10	-40 to +40 UKMO
ASIA/ Zhao & Kellogg, 1988	GCM, June to Aug.		-20 to +10	-40 to 0	0	-50 to +30 UKMO

* see Table 1 for list of GCMs

Table 3. Results of Case Studies*: Snow Regime
(expressed as % change in snow water equivalent)

BASIN/Ref	MODEL	GFDL8 0	GFDL8 6	GISS	OSU	UKHI
U.K./Arnell & Reynard, 1993	Harding & Moore					-61 to -45
GREAT LAKES/ Croley, 1990	LBRM		-94 to -75	-92 to -60	-83 to -37	

* see Table 1 for list of GCMs

Basin or System Runoff (R):

GCMs provide estimates for R for each grid classified as land. Miller and Russell (1992) examined outputs from the GISS GCM for several major watersheds and found that there was a tendency for the control run to overestimate mean annual R. Scenario results generally indicated increases.

A case study from Australia (Morassutti, 1992) illustrates an application of the Wigley-Jones approach (Wigley and Jones, 1985):

$$\frac{R_1}{R_0} = \frac{\alpha - (1-\gamma_o)\,\beta}{\gamma_o}$$

where R_1 = runoff (scenario), R_0 = runoff (base case),
α = % change in precipitation P (scenario), and $\gamma_0 = R_0/P_0$.
β = % change in evaporation = $\beta_1\beta_2\beta_3$, where
β_1 = % change due to scenario, β_2 = % change due to change in % vegetation cover,
and β_3 = % change due to CO_2 enrichment.

Other approaches for estimating R include the Thornthwaite water balance model (Cohen, 1986, 1991), other water balance models derived from Thornthwaite (Gleick, 1987, Wolock et al., 1993), integrated basin models (e.g. Bultot et al., 1988a, 1992; Croley, 1990; Wescoat and Leichenko, 1992; Nash and Gleick, 1993), a probability distributed model (PDM) such as ARNO (e.g. Arnell and Reynard, 1993), and stochastic models (e.g. Revelle and Waggoner, 1983). Results are shown in Table 4 for studies that used GCM-based scenarios. Note that there is one example from the U.K. where assumed changes in stomatal resistance were included (Arnell and Reynard, 1993).

When case studies are compared with outputs from the GISS GCM (Miller and Russell, 1992) for several watersheds (e.g. Great Lakes - St. Lawrence, Colorado, Zambezi, Indus, Murray-Darling), results are mixed. For the Great Lakes - St. Lawrence and Colorado watersheds, the GISS output shows increased R while the case studies produce mostly decreases. The reverse occurs in the Indus case study, because the authors would not use precipitation outputs from GCMs due to uncertainties in GCM simulations of the Asian moonsoon (Wescoat and Leichenko, 1992). They chose a hypothetical precipitation increase of 30%, which resulted in 11% higher R, while Miller and Russell (1992) indicate a 43% decrease in R.

It should also be noted that the Indus case study considered water resource management issues by using hydrologic information as inputs to the World Bank Indus Basin Model (Revised), or IBMR. This provided an opportunity to assess future water development plans (Wescoat and Leichenko, 1992). Arnell provides other examples elsewhere in this volume.

Streamflow or Net Basin Supply:

Net Basin Supply (NBS) can be defined as follows:

$$NBS = P_{wat} + R - E \pm \text{groundwater} - \text{consumptive use} \pm \text{transfers}$$

where P_{wat} represents P falling directly on a water surface. Each parameter can be

Table 4. Results of Case Studies*: Runoff (expressed as %change)

BASIN/Ref	MODEL	GFDL80	GFDL86	GISS	OSU	UKHI
GREAT LAKES/ Croley, 1990	LBRM		-23	-24	-11	
Cohen, 1986	Thornth.	-8		-11		
ST. LAWRENCE/ M&R[y]	GCM			+6		
DELAWARE/ McCabe & Ayers, 1989	Thornth. derived		-7	-39 to -25	+2 to +9	
SACRAMENTO/ Gleick, 1987	Thornth. derived		+34 djf[#] -48 jja[#]	+81 djf[#] -53 jja[#]		
COLORADO/M&R[y]	GCM			+13		
Nash & Gleick, 1993	NWSRF[z]		-16 to -13	-14 to +10		
SASKATCHEWAN/ Cohen, 1991	Thornth.		-19	+38	+4	
MACKENZIE/ M&R[y]	GCM			+24		
U.K./Arnell & Reynard,1993	PDM[zz]					-25to5; r_c-15to5
BELGIUM/ Bultot et al., 1988a	IRMB	+12				
DANUBE/M&R[y]	GCM			+19		
NILE/M&R[y]	GCM			-13		
ZAMBEZI/M&R[y]	GCM			+4		
Kaczmarek & Salewicz,1991	IIASA[zzz]		-6	+6		
INDUS/M&R[y]	GCM			-43		

Wescoat & Leichenko, 1992	UBC[zzzz]		+16,hyp +20P[##]	+11,hyp +30P[##]		
MURRAY-DARLING/M&R[y]	GCM			+25		
Morassutti, 1992	Wigley-Jones			+16to+89		

* see Table 1 for list of GCMs

y - Miller and Russell (1992)

\# - djf = December to February; jja = June to August

\## - hyp = hypothetical adjustment (e.g. +20P = +20% from current P)

z -National Weather Service River Forecasting model (U.S.)

zz -Probability Distributed Model

zzz -International Institute for Applied Systems Analysis (Austria)

zzzz -University of British Columbia watershed model

expressed as an equivalent discharge ($m^3 sec^{-1}$). P_{wat} becomes important in watersheds with large lakes, such as the Great Lakes - St. Lawrence Basin. Almost 1/3 of the Great Lakes portion is lake surface, and it is estimated that P_{wat} is of similar magnitude to R on an annual basis.

In the absence of an integrated basin model, some studies estimated NBS by combining outputs of unlinked component models (e.g. Cohen, 1986, 1991; Marchand et al., 1988). In the Saskatchewan River case (Cohen, 1991), scenarios from GCMs and arbitrary adjustments to the 1951-80 mean were used to estimate a) R from Thornthwaite water balance, and b) irrigation derived from an assumed change in irrigated area, and PEt-weighted soil moisture deficit (from Thornthwaite).

The analysis showed that basin R is sensitive to changes in the source region's P, and since GCM-based scenarios do not agree on projected changes in P in this region, there are large differences in NBS estimated for each scenario (Table 5). Peak R, which is driven by snowmelt, is projected to occur earlier in the year. This is consistent with results obtained in other watersheds where snowmelt is a factor in the seasonal cycle (e.g. Gleick, 1987; Croley, 1990; Arnell and Reynard, 1993). Soil moisture deficit

Table 5. Results of Case Studies*: Streamflow -- Net Basin Supply (expressed as % change)

BASIN/Ref	MODEL #	GFDL80	GFDL86	GISS	OSU	UKHI
GREAT LAKES/ Croley, 1990	LBRM		-51	-37	-23	
Cohen, 1986	separate	-17		-24		
Marchand et al., 1988	separate			-15		
CONNECTICUT/ Kirshen & Fennessey, 1993	NWSRF & SacSM[x]		-33 (Qflux)	-16	+6	+30 UKMO
SASKATCHEWAN/ Cohen, 1991	separate	-70 to -65	-36 to -27	+29 to +40	-4 to +2	
BELGIUM/ Bultot et al., 1988a	IRMB	+3 to +10				
ZAMBEZI/ Kaczmarek & Salewicz	IIASA		-6	+7		
INDUS/ Wescoat & Leichenko, 1992	IBMR[xx]		+31 hyp+20P	+22 hyp+30P		

* see Table 1 for list of GCMs

\# see Table 4

x -Sacramento Soil Moisture Model

xx -World Bank Indus Basin Model (Revised); see Wescoat and Leichenko (1992; Table 27)

becomes larger in summer (Table 2) so irrigation is projected to increase substantially unless more water-efficient technology or change in operating policy occurs.

In addition to the above, Table 5 lists results from several case studies which used integrated basin models. Note that for those cases which reported R separately in Table 4 (Great Lakes, Saskatchewan, Belgium, Zambezi, Indus), it appears that the magnitude of impact for annual streamflow or NBS is greater than for R alone (except in the Zambezi case), though the direction of impact does not change. This is a reminder that when determining impacts on streamflow, it is important to look beyond R.

Several points regarding impacts on flow regime should also be made:

- annual R appears to be more sensitive to changes in P, but in watersheds with seasonal snowmelt, the seasonal distribution of R and soil moisture is more sensitive to T;
- earlier higher winter or spring peaks, with lower summer flows have been reported for the Connecticut, Great Lakes, Colorado, Sacramento and UK cases; and,
- higher winter or spring peaks, with no change in timing but an earlier recession occur in the Belgium and Indus cases.

The geographical extent of this sample of case studies does not include subtropical arid, high latitude and tropical humid regions. Several case studies for these regions have been reported in Tegart and Sheldon (1993).

SUMMARY

From the above discussion, one could take either an optimistic or pessimistic view of current capabilities to describe hydrologic impacts of climatic change scenarios. Uncertainties are evident in all aspects of the exercise, from scenario development to model application. It should be remembered, however, that the purpose of the exercise is to provide information that could not only improve the science, but ultimately assist governments in planning and managing water resources for a future climate that could be different from the past. Indeed, as a learning experience, there are several points that have emerged.

The various case studies have shown results that vary by region, and sometimes by scenario within a watershed. For comparable scenarios, different hydrologic methods produce similar results, but these may differ substantially from hydrologic outputs obtained directly from GCMs.

Most watersheds are largely controlled by changes in P. However, in regions where snowmelt is important, or where P and PEt are of similar magnitudes, changes in T may result in a shift in the water balance and/or the timing of peak and low flow periods. High latitude watersheds may experience very different impacts from snowmelt changes than those described in the mid-latitude case studies, particularly if winter snow cover increases.

Some methods or study areas require data not currently available (or available but with large errors) from GCMs. Results may be sensitive to assumptions regarding water temperatures, stomatal resistance, precipitation in mountainous regions and monsoonal climate zones, etc. Scale differences between GCMs and watershed processes must be accounted for. This represents a major component of the downscaling challenge.

Use of basin models with linked subcomponent modules should be encouraged. Case studies provide an opportunity to test system sensitivities to scenarios of climatic change (e.g. Colorado, Indus). However, many watersheds are not yet modelled in this fashion, so there will continue to be a requirement for simpler methods (e.g. Wigley-Jones, Thornthwaite).

REFERENCES

Arnell, N.W. 1992. Factors controlling the effects of climate change on river flow regimes in a humid temperate environment. Journal of Hydrology, 132, 321-342.

Arnell, N., and N. Reynard. 1993. *Impact of climate change river flow regimes in the United Kingdom.* Report to the Dept. of Environment, Water Directorate. Institute of Hydrology, Wallingford.

Bultot, F., A. Coppens, G.L. Dupriez, D. Gellens and F. Meulenberghs. 1988a. Repercussions of a CO_2 doubling on the water cycle and on the water balance--a case study for Belgium. Journal of Hydrology, 99, 319-347.

Bultot, F., G.L. Dupriez and D. Gellens. 1988b. Estimated annual regime of energy-balance components, evapotranspiration and soil moisture for a drainage basin in the case of CO_2 doubling. Climatic Change, 12, 39-56.

Bultot, F., Gellens D., Spreafico M., Schädler B. 1992. Repercussiosns of a CO_2 doubling on the water balance - a case study in Switzerland. Journal of Hydrology, 137, 199-208.

Cohen, S.J. 1986. Impacts of CO_2-induced climatic change on water resources in the Great Lakes Basin. Climatic Change, 8, 135-153.

Cohen, S.J. 1991. Possible impacts of climatic warming scenarios on water resources in the Saskatchewan River Sub-basin, Canada. Climatic Change, 19, 291-317.

Croley, T.E., II. 1990. Laurentian Great Lakes double-CO_2 climate change hydrological impacts. Climatic Change, 17, 27-47.

Dooge, J.C.I. 1992. Hydrologic models and climate change. Journal of Geophysical Research, 97, D3, 2677-2686.

Fennessey, N.M. and P.M. Kirshen. 1993. Evaporation and evapotranspiration under climate change in New England. ASCE Journal of Water Research and Management, in press.

Frederick, K.D. 1993. Climate change impacts on water resources and possible responses in the MINK region. Climatic Change, 24, 83-116.

Gleick, P.H. 1987. Regional hydrologic consequences of increases in atmospheric CO_2 and other trace gases. Climatic Change, 10, 137-161.

Gleick, P.H. 1989. Climate change, hydrology and water resources. Reviews of Geophysics, 27, 3, 329-344.

Kaczmarek, Z. and D. Krasuski. 1991. *Sensitivity of water balance to climate change and variability.* International Institute for Applied Systems Analysis, Working Paper WP-91-047. International Institute for Applied Systems Analysis, Laxenburg.

Kaczmarek, Z. and K.A. Salewicz. *Water resources management in the face of climatic/ hydrological uncertainties: outline of the Zambezi case study.* International Institute for Applied Systems Analysis, Laxenburg, in preparation.

Kellogg, W.W. and Z.-c. Zhao. 1988. Sensitivity of soil moisture to doubling of carbon dioxide in climate model experiments: Part I: North America. Journal of Climate, 1, 348-366.

Kirshen, P.H. and N.M. Fennessey. 1993. Potential climate change impacts on water supply in the Boston metropolitan area. ASCE Journal of Water Research and Management, in press.

Marchand, D., M. Sanderson, D. Howe and C. Alpaugh. 1988. Climatic change and Great Lakes levels: the impact on shipping. Climatic Change, 12, 107-133.

Mather, J. 1978. *The Climatic Water Balance in Environmental Analysis.* Lexington Books, Lexington, Massachusetts.

McCabe, G.J. and M. Ayers. 1989. Hydrologic effects of climate change in the Delaware River Basin. Water Resources Bulletin, 25, 1231-1242.

Miller, J.R. and G.L. Russell. 1992. The impact of global warming on river runoff. Journal of Geophysical Research, 97, D3, 2757-2764.

Morassutti, M.P. 1992. Australian runoff scenarios from a runoff-climate model. International Journal of Climatology, 12, 797-813.

Nash, L.L. and P.H. Gleick. 1993. *The Colorado River Basin and Climatic Change: The Sensitivity of Streamflow and Water Supply to Variations in Temperature and Precipitation.* U.S. Environmental Protection Agency, EPA230-R-93-009, Washington.

Priestly, C.H.B. and R.J. Taylor. 1972. On the assessment of surface heat flux and evaporation using large-scale parameters. Monthly Weather Review, 100, 81-92.

Quinn, F.H. and den Hartog, G. 1981. *Evaporation Synthesis.* In E.J. Aubert and T.L. Richards (eds.), IFYGL - The International Field Year for the Great Lakes, 221-245, U.S. Dept. of Commerce, Ann Arbor.

Revelle, R.R. and P.E. Waggoner. 1983. *Effects of a Carbon Dioxide-Induced Climatic Change on Water Supplies in the Western United States.* In U.S. National Research Council, Changing Climate: Report of the Carbon Dioxide Assessment Committee, 419-432. National Academy Press, Washington, DC.

Rosenberg, N.J., B.A. Kimball, P. Martin and C.F. Cooper. 1990. From Climate and CO_2 Enrichment to Evapotranspiration. In Waggoner, P.E. (ed.) Climate Change and U.S. Water Resources, 151-176, John Wiley & Sons, New York.

Tegart W.J.McG., Sheldon G.W. (eds). 1993. *Climate Change 1992: The Supplementary Report to the IPCC Impacts Assessment.* Australian Government Publishing Serive, Canberra.

Wescoat, J.L., Jr., and R.M. Leichenko. 1992. *Complex River Basin Management in a Changing Global Climate: The Indus River Basin in Pakistan, a National Assessment.* Centre for Advanced Decision Support for Water and Environmental Systems, and the Institute of Behavioral Science, U. of Colorado, Boulder.

Wigley, T.M.L. and P.D. Jones. 1985. Influences of precipitation changes and direct CO_2 effects on streamflow. Nature, 314, 149-152.

Wolock, D.M., G.J. McCabe, Jr., G.D. Tasker and M.E. Moss. 1993. Effects of climate change on water resources in the Delaware River Basin. Water Resources Bulletin, 29, 475-486.

Zhao, Z.-c. and W.W. Kellogg. 1988. Sensitivity of soil moisture to doubling of carbon dioxide in climate model experiments: Part II: The Asian monsoon region. Journal of Climate, 1, 367-378.

Socio-economic impacts of changes in water resources due to global warming

Nigel W. Arnell
Institute of Hydrology
Wallingford
Oxon OX10 8BB
UK

INTRODUCTION

Perhaps the largest impact of global warming on mankind will be through its effects on water resources and water availability. Many studies have investigated potential changes in water balance and hydrological regime, but rather fewer have examined in a systematic or quantitative way the effects of these changes on the users of water. The aim of this paper is to outline the institutional, economic and social effects of changes in water resources, summarising potential areas of impact and illustrating them with examples from the literature.

Before going into specific detail, however, it is important to make four general points. First, the impact of global warming on an institution in the water sector may be very complicated. Figure 1 (generalised from Arnell *et al*, 1994) shows how the effect of global warming will be affected by other changes, by the response of the institution to climate and other changes, and by the response of other institutions and sectors of the economy to change. An institution will make progressive adaptations to changes in the conditions under which it operates, and these could mitigate or amplify impacts of global warming. Second, the progression from global warming to impact on water user is not necessarily linear (Figure 2): the steps are determined by the sensitivity of the hydrological and water management system to change. A given change in climate would probably have a greater impact on a water management system currently under stress - demands close to supply, for example - than on a system with greater slack. Third, there may be critical levels or limits

NATO ASI Series, Vol. I 31
The Role of Water and the Hydrological Cycle
in Global Change
Edited by Howard R.Oliver and Sylvia A.Oliver

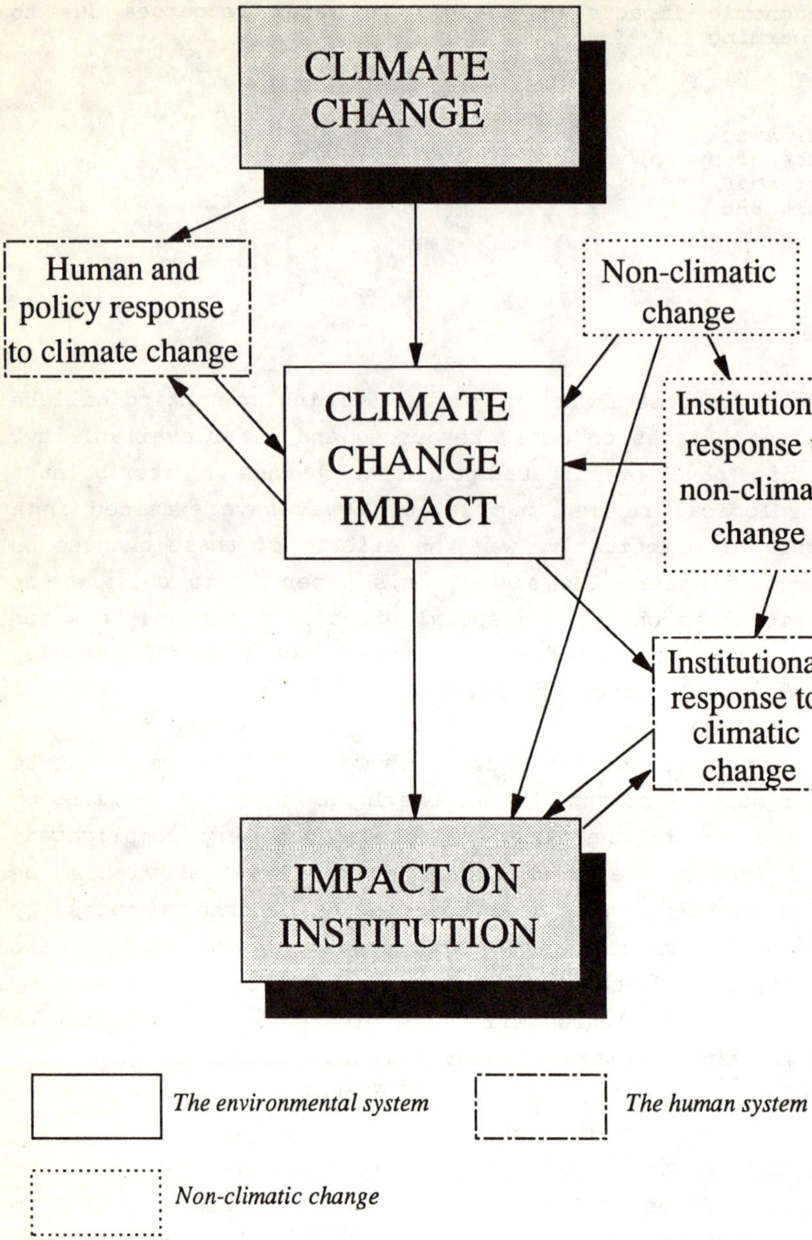

Figure 1 The impact of global warming on an institution (adapted from Arnell *et al*, 1994)

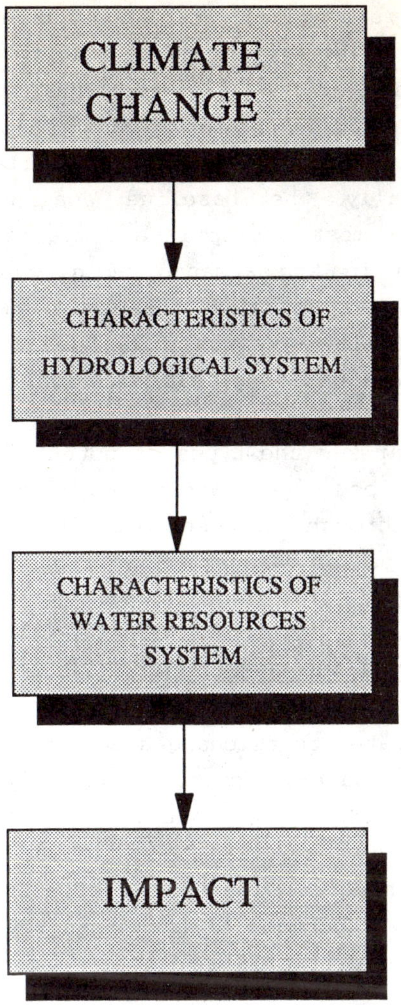

Figure 2 The transition from global warming to impact

within a water use system, such as statutory standards of service and protection. The system may be able to cope relatively easily with some change - by changing operating rules, for example - but a slightly greater change might cause major problems leading for example to significant investment. Fourth, it is important to define carefully the baseline against which impacts are evaluated. The most appropriate comparison is with what would have happened in the absence of global warming: this would include responses to other changes - such as improvements in river water quality - which might have the effect of mitigating (or possibly exaggerating) global warming effects.

Figure 3 summarises the types of potential impacts that changes in water resources might have in a variety of water-use sectors. Before going through these sectors, it is helpful to review very briefly the types of changes in water resources that might occur.

CHANGES IN WATER RESOURCES

Figure 3 indicates three changes in water resources: a change in the *quantity* of water, a change in water *quality*, and a change in the offstream *demands* for water.

Many studies have considered the effects of global warming on water quantity, and many of these are reviewed by Cohen (1994). Although it is difficult to compare studies because different scenarios were used, some generalisations can be drawn. River flows and reservoir contents may change by a greater percentage than precipitation and evaporation, but in other cases increased evaporation may be almost entirely compensated for by increased precipitation. Changes in seasonal or monthly runoff have been found in most studies to be greater than changes in annual totals. In regimes dominated by snowfall and snowmelt, an increase in temperature has a major effect on the timing of river flows. There have been very few studies of the sensitivity of groundwater recharge to changes in precipitation and evaporation.

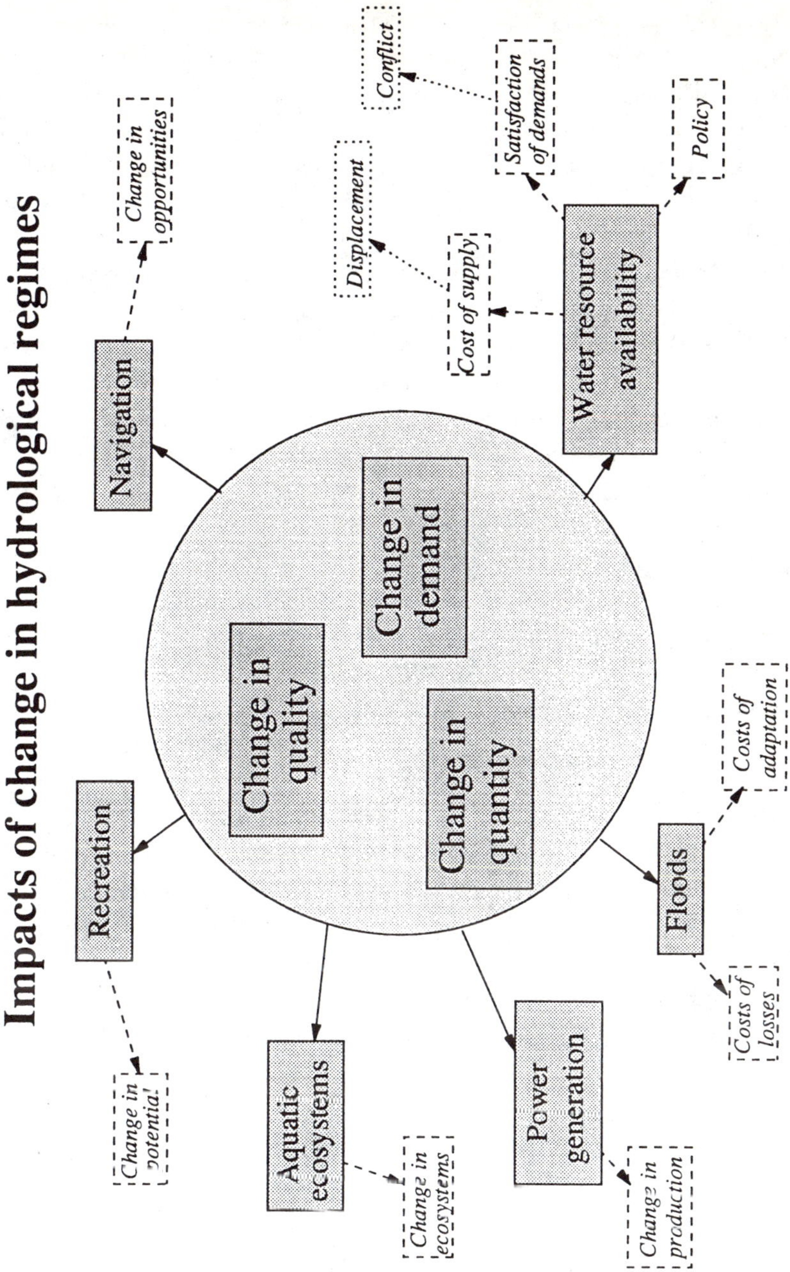

Figure 3 Potential impacts of global warming on different water-use sectors

Water quality depends on a complex system of inputs and feedbacks, including natural factors such as catchment physical characteristics, rainfall and runoff regime and water temperature, as well as anthropogenic factors such as effluent inputs, abstractions and returns, and land use. All these factors may change following global warming, with consequent implications for water quality. Changes will, of course, be very site specific, depending on the amount of change and current water quality characteristics. Studies in the UK (Jenkins *et al*, 1993) suggest that changes in water temperature will have less effect on water quality than either changes in flow regimes or changes in land use. Longer residence times due to lower flows, for example, would lead to increased growth of algal blooms in heavily-utilised rivers, resulting in a reduction on dissolved oxygen contents. Nitrate concentrations would be affected by flow volume and hence dilution, and also by agricultural inputs; higher peaks might be expected following autumn rains after a longer, drier summer.

A change in climate might also result in a change in the offstream demands for water. The most obvious potential increase is in agricultural demands. Both rain-fed and irrigated crops would require more water if temperature and evaporative demands were to increase, although this might be offset by the effects of increased rainfall. Allen *et al* (1991) investigated potential changes in irrigation demand in the Great Plains region of the US, using the Penman evaporation formula. They simulated increased demand for irrigation by alfalfa, due largely to increases in the length of the crop-growing season and crop-water requirements during summer. Smaller increases, or even decreases, were found for winter wheat and corn due to reductions in the length of crop life cycles and possible increases in plant stomatal conductance. Cohen (1991) simulated an increase in irrigation demands in part of the Saskatchewan river basin in Canada, even under scenarios with increased precipitation, using a regression relationship between irrigation demands and soil moisture deficit derived from historical data: some of his results are summarised in Table 1, under two scenarios of growth in irrigated area.

Table 1 Percentage change in irrigation consumptive use, Saskatchewan basin, under two irrigation growth scenarios and five climate change scenarios (Cohen, 1991)

	Change in precipitation: temperature +2°C.				
	-20%	-10%	-	+10%	+20%
Low growth	39	30	24	20	13
High growth	133	117	107	98	87

Demand for domestic water may also increase. Herrington & Hoschatt (1993) estimated that higher temperatures would result in an increase in per capita demand in southern England of around 4%, above that due to non-climatic factors, as a result largely of increases in water use in the garden. Other sectors which might demand more water include public parks, golf courses and air conditioning. Cohen (1987) estimated an increase in summer municipal water withdrawals in communities in the Great Lakes basin of approximately 5% due to increased temperatures and hence higher potential evaporation.

WATER SUPPLY

This is the most obvious area of potential impact, and the one which has been given the most attention so far. Several studies have considered possible changes in water supply in specific study areas. These studies fall into three groups.

The first set of studies infer changes in supply from modelled changes in the water balance, without looking specifically at water supply systems. In fact, most water balance studies infer potential changes in water supply from their results. Arnell & Reynard (1993), for example, inferred problems in maintaining summer supplies from direct river abstractions from their results showing reductions in summer flows in the UK. Cohen (1986) and more recently Croley (1993) modelled changes in average annual water availability ("Net Basin Supply") in the Great Lakes Basin. There are many other such studies.

The second set of studies consider the sensitivity of hypothetical supply systems, usually comprising just one reservoir, to change in inputs. Nemec & Schaake (1982) in one of the earliest climate change impact studies, for example, showed the large change in yield reliability following relatively small changes in reservoir inputs. Cole *et al* (1991) examined the sensitivity of yield, reliability and required storage capacity in hypothetical UK reservoirs to changes in inputs. They found, for example, that for one size of reservoir with inputs typical of catchments in southern England, it would be necessary either to reduce the yield by 10% or increase storage by 18% in order to maintain a reliability of supply of 99%; they also showed that the sensitivity of a reservoir to change depended significantly on the capacity of the reservoir relative to total annual runoff and the yield relative to natural runoff.

The third group of studies examine the potential changes in real, specific water supply systems. Four examples illustrate the types of study and conclusions. The first two consider single reservoirs; the last two look at entire integrated water supply systems, and include explicitly real operating rules for the allocation of water amongst users.

i. McMahon *et al* (1989) investigated future changes in the reliability of the Myponga reservoir, located in South Australia. Under the worst case, reliability - for the current yield - would fall from 99.4% to 81.5%, with no change in operating rules.

ii. Weir Wood reservoir is a small direct supply reservoir in Sussex, south east England (Hewett *et al*, 1993). The current storage of the reservoir is 5500 megalitres (Ml), which is regarded as being too small for current demands; the reservoir fills too frequently spilling excess water. Under current climatic conditions, the reservoir can supply a yield of 13.9

Ml/d with a risk of failure of 2%. Figure 4 shows the yield which could be supplied with the same reliability for different storage volumes under current climatic conditions and two change scenarios. Both scenarios assume an increase in winter rainfall, a reduction in summer rainfall and higher evaporation, but both result in *higher* yields for a given storage capacity and level of reliability; with a capacity of 5500 Ml, yield could be increased by 1.6 Ml/d (11.5%) under one scenario and 0.6 Ml/d (4%) under the other. This occurs because winter and spring runoff are higher under both scenarios, allowing the reservoir to remain full for longer. Other scenarios could result in lower yields, but the example indicates that water supplies might in some cases be increased by the climate changes associated with global warming.

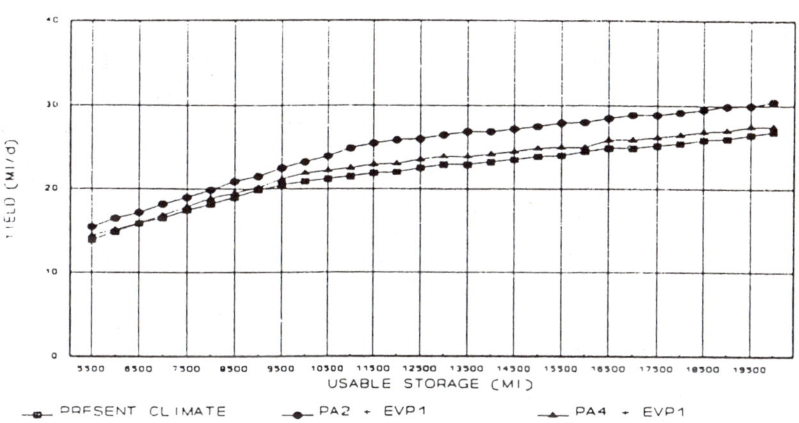

Figure 4 Yield of Weir Wood reservoir, southern England, with a failure risk of 2%, for different storage capacities and climate change scenarios (Hewett *et al*, 1993)

iii. Mimikou *et al* (1991) examined potential effects of climate change on a proposed system of four reservoirs in the Acheloos River basin in central Greece. The system is designed both to supply water and to provide hydroelectric power. Calculations showed that climate change could have very significant impacts on the risk associated with design values of water supply and energy production, and that the downstream reservoirs in the system were more sensitive than the upstream ones. A 20% reduction in precipitation through the year would increase the probability of failure of one upstream reservoir to 33% (from less than 1% at present), and would increase the risk of failure of the most downstream reservoir to 54%. This variation in impacts reflects the difference in hydrological impacts of global warming along the river network, and also the particular characteristics of the four component reservoirs.

iv. The Delaware River basin in the eastern United States supplies water to communities in Pennsylvania, but is also the major supply source for New York City. The water supply system consists of a number of linked reservoirs, and Table 2 shows the percentage of time in which the New York City supply system is in what is defined as a "crisis" state, under different climatic conditions and assuming no change in operating rules (Wolock *et al*, 1993). A substantial proportion of the water supplies to Philadelphia are abstracted directly from the lower reaches of the Delaware, and the study also examined changes in the frequency with which the abstraction point is threatened by salt water due both to sea level rise and reduced flows in the Delaware: the Philadelphia supplies were found to be much less affected by climate change than New York supplies.

Table 2 Percentage of time in which the New York City supply system in the upper Delaware River Basin is in "crisis", with no change in operating rules (Wolock *et al*, 1993)

Change in temperature	Change in precipitation		
	-20%	0	20%
0	27	4	3
+2°C	36	7	3
+4°C	42	11	3

These four examples have illustrated potential changes in the yield and reliability of water supply systems. The detailed results are of course a function of the characteristics of the study sites, but three general points can be drawn. First, relative changes in reliability of supply may be considerably larger than relative changes in inputs, and vary considerably between water supply systems. Second, global warming does not necessarily lead to a reduction in water supply. Third, studies have tended to assume no management response to a changing climate, and can be expected therefore to overestimate the magnitude of impacts (although of course incremental adaptation will involve costs).

There are three possible effects of changes in the reliability and yield of water supply systems.

Change in the cost of supply

The costs of supplying a given amount of water will be different if global warming changes both supplies and demands. Increased availability might lead to a reduction in supply costs, but reduced availability will result in higher costs of service provision.

It may be possible to cope with climate change simply by altering the operation of a water supply system. Costs may be increased because more water is pumped along a distribution network, for example, or because groundwater pumps are worked harder and for longer. The total cost of supplying water in an integrated system may increase as more expensive sources of water - such as river water which needs treatment or groundwater which is expensive to pump - need to be tapped. This is in practice what happens during droughts.

Alternatively, demand management measures might be introduced to keep demand within system capabilities. Here, the economic costs will be the additional costs of operating such a system, but individual users might incur significant financial costs.

In some cases it will be impossible to cope with altered supplies and demands by a change in operating rules or demand management, and additional capital investment may be necessary. This may take the form of additional reservoirs, additional wells, water transfer schemes or improvements in the efficiency of existing systems (by, for example, leakage control or greater interconnections).

In practice, of course, it is not necessarily the case that water managers will continue to match existing demand patterns, and global warming might trigger fundamental changes in methods used to allocate scarce water resources: this is discussed further in the next section.

Change in the degree to which demands are satisfied

It may not always be possible, for economic or technical reasons, for water management systems to be adjusted fully to changed climatic conditions. This would mean that the system would not be able to meet demands as effectively or fully as before.

At the most trivial, this would simply be inconvenient. Supplies might be rationed during certain times of the year or at times during the day, as occasionally happens at present in many areas. At the other extreme, an inability to satisfy demands for water may lead to the creation of environmental refugees and international conflict over the allocation of scarce resources.

Agricultural productivity in some regions may be highly sensitive to fluctuations in water availability, and a reduction in water can therefore have very significant impacts on local economies and famine risk. This may result in the creation of famine refugees (Tickell, 1993), who may move away from a famine area putting additional stress on other regions: this has been seen repeatedly in Africa during the 1980s and 1990s.

Much has been written in the 1980s and 1990s about conflict over scarce environmental resources (eg Thomas & Howlett, 1993). A change in the availability of water in upstream countries has the potential for triggering conflict (Gleick, 1988; 1993). Thirty-one nations currently receive more than a third of their freshwater from upstream countries (Gleick, 1992).

Disputes over the allocation of water resources can also, of course, occur within a country. There are 21 agreements (or "compacts") in the western United States defining the allocation of water between upstream and downstream states (Stein, 1993). In nearly all cases, the allocations are based on historic assessments of dependable supply. Droughts have, unsurprisingly, led to several legal disputes over the interpretation of agreements, and Stein (1993) reviews those centred on the Arkansas River (Colorado and Kansas), the Vermejo River (Colorado and New Mexico) and North Platte River (Nebraska and Wyoming). Stein (1993) concludes that inter-state compacts need to include flexibility, and indeed the North Platte ruling stated that should conditions "substantially change, the decree can be adjusted to meet the new conditions" (of course, it would be necessary to define what is meant by "substantially change", which is very difficult in the context of climate change).

Alterations in water management policy

Water management agencies are continually adapting policies and procedures as more information becomes available and management objectives change. This progressive adaptation will allow agencies to respond to global warming. However, there are two possible ways in which global warming might alter water policy.

The first is through a recognition that the past is not necessarily a good guide to the future; medium and long-term planning would then need to consider explicitly non-stationary behaviour in hydrological regimes in general and the effects of global warming in particular (Riebsame, 1988). In practice, different agencies would have different time horizons. Some might feel able to respond quickly to any change in climate - perhaps because additional supplies can be easily introduced - and so would not need to consider long-term non-stationarity.

The second trend is likely to be towards the use of non-structural water management measures such as demand management, increased use of treated wastewater, the introduction of tradeable water rights and pricing mechanisms. Climate change is not the only factor encouraging such approaches: both a general move away from engineering structures on environmental grounds and an increasing interest in economic instruments for resource management and allocation also contribute.

The response of the Denver Water Department (DWD) in the US to the veto on environmental grounds of a proposed new supply reservoir provides an analogy for the response of water management institutions to shortages triggered by global warming (Rhodes *et al*, 1992). After the Two Forks reservoir scheme was vetoed, the DWD developed a water management policy which included demand-based options - for example incentive pricing, limits on appliance water use and land use zoning - and supply augmentation options such as purchase and transfer of agricultural water rights and water rights exchanges.

The type and degree of change in water management philosophy will vary with institutional and legal framework, and will also be influenced by changes triggered by factors other than global warming. Rhodes *et al* (1992) emphasised that the Denver response was highly conditioned by the legal framework in place, environmental legislation and awareness, and historical factors in the development of the DWD.

FLOODS

There have been far fewer studies on potential changes in flood risk than on changes in the water balance and water supply, for good reason: it is currently very difficult to develop credible scenarios at the catchment scale for changes in flood-generating rainfall and snowmelt events (Beran & Arnell, 1994). There is some evidence that the paths and intensities of depressions and storms might change (eg Stephens & Held, 1993), and it is probable that there will be an increase in convective activity and hence intense rainfalls. Wetter conditions in winter - as is possible in mid-latitudes - would tend to increase the risk of flooding. An increase in air temperature would initially be expected to result in fewer and smaller snow-melt floods, but increased precipitation might compensate for the higher temperatures, resulting in greater snowfall during precipitation events. Also, a reduction in winter snow cover might lead to a greater risk of rain-generated floods during winter; the flood season would be brought forward. Lettenmaier & Gan (1990), for example, found a reduction in snow-melt floods in the Sacremento-San Jaoquin basin in California, but an increase in rain-generated floods: the risk of occurrence of a particular discharge was therefore increased.

An increase in the frequency of floods would have three major implications. The first is an increase in flood damages and effects, although this must be set against other factors - such as population growth, economic development and expansion into floodplains - that are also contributing to rising flood losses. The second is an increase in the cost of protecting against and responding to floods. Most obviously, these include the cost of providing structural flood defences or improving defences to maintain current standards of protection. More subtly, costs to society may be increased if larger areas of potentially-productive land are excluded from development by floodplain zoning policies. Third, an increase in flood losses may place significant strains on public finances - if damages are covered by public funds - or the insurance industry. The amount of cover provided by the insurance industry varies from country to country, and in countries where cover is restricted the direct impact will obviously be limited. However, the international insurance industry may be very seriously threatened by the occurrence of a few large, closely-spaced storm events: at the worst case, reinsurance capacity may be devastated, with consequent implications for the direct insurance market and the financial sector in general.

HEALTH

An estimated 250 million additional cases of water-related disease occur each year, resulting in an annual total of approximately 10 million deaths (Nash, 1993). Water-related diseases are conventionally divided into four classes. The first is *fecal-oral*. Water is the passive agent for the transmission of pathogens from excreta to humans. This category includes diarrheas and dysentries (including cholera) as well as enteric fevers such as typhoid. *Water-washed* diseases largely consist of infectious skin and eye diseases, and *water-based* diseases are transmitted by means other than ingestion by hosts that live in water. Schistosomaisis and guinea worm are the two most widespread examples. *Water-related insect vector* diseases are spread by insects that either breed or bite near water; malaria, river blindness (onchocerciasis) and yellow fever are examples. The first three classes are essentially determined by the quality of water, whilst water-vector diseases are determined by water quantity, and particularly by variations in water availability through the year.

To these microbiological diseases can be added those caused by chemicals, arising as a result of pollution. Examples include nitrates, heavy metals and organic pollutants, all of which are associated with public health concerns.

There are many factors triggering changes in water-related health (Nash, 1993). The most important is the combined effect of population growth, urbanisation and poverty, which together mean increasing numbers of people have access only to poor quality water. Industrial expansion and agricultural pollution affect the incidence of chemical-based illnesses, and large-scale water developments can provide increased opportunities for water-based and water-vector diseases. The additional effects of climate change have not yet been investigated. Potential areas of concern include the effect of higher temperatures on the geographic distribution of disease-causing organisms and, for water-vector diseases, changes in the seasonal distribution of water.

POWER GENERATION

A change in hydrological regimes has an obvious potential impact on hydropower production, but may also affect thermal (fossil fuel, geothermal and nuclear) power generation.

Mimikou *et al* (1991) simulated the behaviour of a combined water supply and hydropower system in central Greece. One of the four reservoirs was designed for power generation alone, and Figure 5 shows the change in the risk of being unable to generate the design power under a range of scenarios: the possible change is large. Under a 20% reduction of precipitation, the risk of system failure (being unable to generate design power) rose from less than 1% to between 20 and 25%. In Norway, Saelthun *et al* (1990) found under the change scenario considered a slight increase in production potential, due largely to a shift in the timing of inflows as a result of reduced snowfall. The distribution of power generation potential also changed through the year (Figure 6), with the current "waste" of power in spring being much reduced. Nash & Gleick (1993) simulated a 26% reduction in annual power generation in the Upper Colorado basin, following a 10% reduction in runoff. A fourth example is given by Hanemann & McCann (1993), who examined changes in hydropower production in northern California. Under the change scenario considered, annual power production would decrease by 3.8%, but would decrease by as much as 20% during the peak demand season. If all the lost power were made up by thermal power stations using natural gas, power generation costs would increase by $145 million per year. In practice other ways of coping with the shortfall, such as increased energy efficiency, might reduce the apparent costs of global warming for power production.

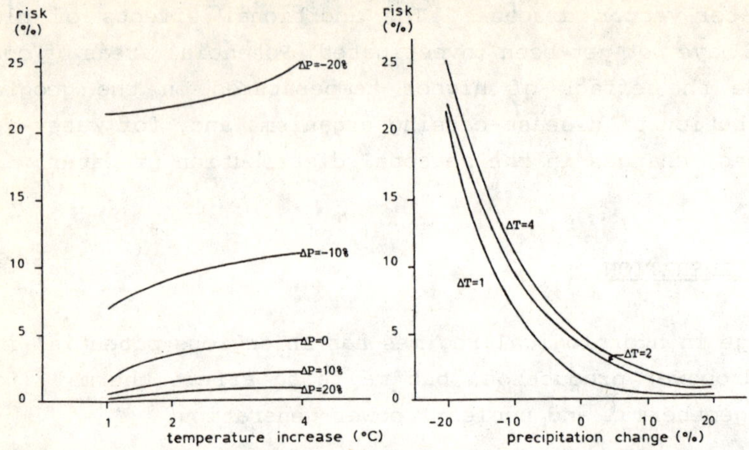

Figure 5 Risk of being unable to generate design power, under changes in temperature and rainfall: Mesohora Reservoir, Greece (Mimikou *et al*, 1991)

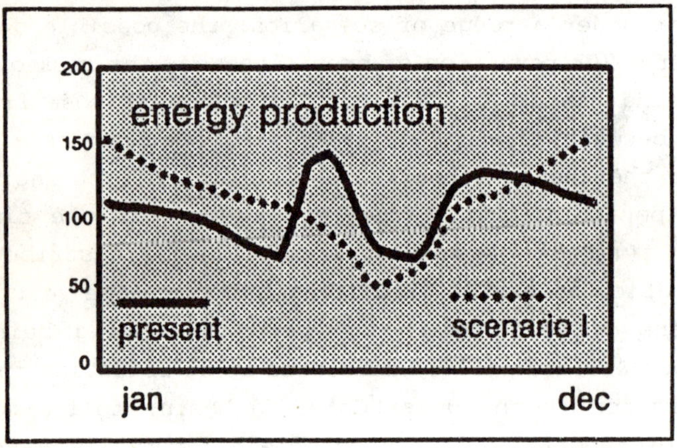

Figure 6 Distribution of power generation potential through the year, for a reservoir in Norway (Saelthun *et al*, 1990)

Thermal and nuclear power generation requires cooling water. In once-through cooling systems large quantities of water are abstracted from a river or lake, fed through the power plant, and returned to the watercourse at a higher temperature. Closed-cycle cooling systems use cooling towers to recycle cooling water; withdrawals are lower than for once-through systems, but consumptive losses are far higher. A reduction in river flow would restrict power generation by reducing the volume of water available for cooling: several French nuclear power plants were forced to close down or operate well below design capacity during the drought of 1991. Higher water temperature will also reduce power production. Miller *et al* (1993) showed for the Tennessee Valley Authority power supply system that power generation would be reduced due to plant deratings and plant shutdowns to avoid non-compliance with regulatory constraints (principally concerning the temperature of downstream water), under global warming. Plant efficiency would also be reduced, and cooling towers would be required more frequently. The effects on annual power production would be small, but the system would be under greater pressure during extreme periods.

AQUATIC ECOSYSTEMS AND FISHERIES

Aquatic ecosystems are determined by hydrological regimes and their variability through the year, and water quality; a change in regimes and quality will therefore have an effect on such ecosystems. This section considers instream ecosystems, wetlands and exploitable fisheries.

Instream ecosystems

The fish population in a river or lake will be influenced by five potential changes. Fish physiology - such as spawning, survival and growth - may be affected by temperature. Each species has upper and lower temperature limits, which vary through life stage. The effect of a temperature rise on a given fish species in a watercourse would therefore depend on how close that species was to its thermal limits; in some cases the reduction in the risk of extreme low temperatures would improve survival rates, but in others temperatures may pass lethal limits. Higher lake temperatures in the UK, for example, would threaten the survival of a number of lake fish (such as the whitefish and charr) and higher river temperatures would slow the growth rate of brown trout (Arnell *et al*, 1994). Table 3 gives temperature limits for a number of common freshwater fish. Higher water temperatures might also result in a change in the incidence of fish diseases and fungal infections (higher temperatures might reduce fungal infection).

Fish habitats will be affected by changes in flow regimes and water quality; again different fish have different requirements at different stages in the life cycle. Important dimensions of water quality are dissolved oxygen content, salinity, nutrient contents and acidity. Fish food sources might also change, and changes in both river flows and ocean circulation patterns might affect migration patterns. Finally, higher temperatures might mean that "exotic" species become established and compete successfully with current populations.

Although there has been considerable biological research into fish and their habitats, this has not yet been applied to the global warming issue.

Table 3 Temperature limits for eggs and adults of a number of freshwater fish found in the UK (from Arnell *et al*, 1994): values to the nearest °C.

		Temperature limits for eggs		Temperature limits for adults	
Species	Common name	Spawning range	Lethal	Optimum	Upper critical
Coregonus lavaretus	Whitefish	0-4	>8	8-15	20-25
Salvelinus alpinus	Charr	3-15	>8	5-16	22-27
Thymallus thymallus	Grayling	6-10	>14	4-18	18-24
Salmo trutta	Brown trout	1-10	>13	4-19	19-30
Esox lucius	Pike	4-17	>23	9-25	30-34
Cyprinus carpio	Carp	12-30	>26	15-32	30-41

Wetlands

Freshwater wetlands are very important and rich ecosystems. Plant and animal communities are adapted in a dynamic way to variations in climate, river inflows, river water levels and soil and groundwater levels (Covich, 1993). Many of these ecosystems are very sensitive to change, as has been demonstrated repeatedly following human-induced changes in flow regimes or over-abstraction of groundwater. Global warming can therefore be expected to have major effects on freshwater wetlands, with the details depending on the type of wetland community and the degree of change. Poiani & Johnson (1991), for example, found that under a range of feasible climate change scenarios, seasonal fluctuations in the volume of water in North American prairie wetlands may increase and lead to reductions in important waterfowl habitats. In arid regions, salinity patterns may also change (Covich, 1993). Changed river flow regimes might in some circumstances increase the richness of wetland ecosystems, through for example an increase in the rate of flooding.

Freshwater wetland ecosystems are very complicated, and are characterised by strong feedbacks and inter-relationships between atmosphere, water and plant. Wetland systems are currently poorly-understood, and it is therefore very difficult to model the potential impacts of global warming. However, analogies for the potential degree of change can be found in effects of river regulation, reservoir impoundment and groundwater pumping.

Exploitable fisheries

One tenth of the world's commercial fish yield is taken from inland waters (Covich, 1993), and subsistence fishing is also important in some areas. Fish yield will be influenced by changes in fish populations (as outlined above), and also by intensity of fishing.

Minns & Moore (1993) examined possible changes in the yield of three fish species - lake whitefish, northern pike and walleye - in lakes in eastern Canada. The dominant climatic control is lake water temperature, but yield is also determined by lake area, depth, sediment concentration and intensity of fishing. Minns & Moore (1993) used regression equations to predict changes in the presence or absence of fish in eastern Canadian lakes, together with changes in estimated yields. Their results showed a general northwards shift in fish yields, and areas with high yields at present would under a warmer climate suffer significant reductions. Figure 7 shows the frequency distribution of fish yields in eastern Canadian lakes, for the three species considered. Hanemann & Dumas (1993) simulated a 50% reduction in chinook salmon populations in the Sacremento River, California, due largely to a reduced spawning habitat in autumn, increased scouring in winter, and increased mortality in the Sacremento Delta. Loomis & Lee (1993) converted this decline in salmon into a reduction in the recreational value of the Sacremento River of $35 million per year, based on the estimated fall in the number of fishing trips.

NAVIGATION

Rivers and lakes can constitute important transport routes. The Rhine and Danube rivers in Europe, for example, and the Mississippi River in the United States play major roles in the transport of bulk goods. Changes in flow regimes and lake levels can be expected to impact upon navigation potential, but there have been very few studies.

A reduction in river flows would restrict both the maximum draught of a vessel (and hence load) and the period over which navigation was feasible. The low flows in the Mississippi during the drought of 1988 severely restricted navigation, and the resulting difficulties in transporting the grain harvest had knock-on effects in the agricultural sector. High flows too may restrict navigation (increased energy costs going upstream, flooding of docking facilities, etc), and the net effect of climate change on navigation will depend on how the "navigable season" changes; quantitative studies are needed. Water management practices and structures, however, may serve to buffer the effect of climate change on navigation. Many navigable rivers are highly regulated, and the impact of climate change will be on the costs and ability to maintain navigable conditions under changed hydrological circumstances. Miller & Brock (1989), for example, concluded that a reduction in runoff in the Tennessee Valley would have little effect on navigation potential, because the reservoirs that sustain navigation would still be capable of doing so in the future.

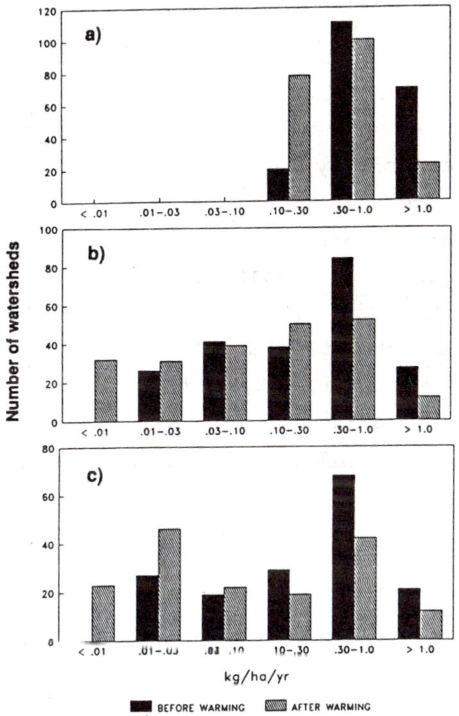

Figure 7 Yield of freshwater fish in eastern Canadian lakes, under current and warmer conditions (Minns & Moore, 1993): a) lake whitefish, b) northern pike, c) walleye

Marchand *et al* (1988) made a quantitative study of the effects of changes in Great Lakes levels on shipping. Climate change would affect shipping through changes in the duration of ice cover and changes in lake levels: the depth of key navigation channels is an important constraint on ship loading. Marchand *et al* (1988) examined four bulk cargoes - iron ore, grain, coal and limestone - and determined the number of trips necessary to transport defined quantities under different scenarios. The minimum, average and maximum annual navigation costs for some of these scenarios are summarised in Table 4, together with details of the scenarios. Table 4 shows that the increased costs, under the change scenarios assumed, could be very substantial.

RECREATION

Considerable investment has been made in developed countries in water-based recreation. Climate change has several possible effects, but all are very difficult to quantify.
Many reservoirs, especially in the United States, have important recreation roles, and changes in the volume of water stored (and hence reservoir surface area and depth) will affect the duration over which recreation is feasible, the attractiveness of the reservoir, and also the use of lake-side facilities. This potential impact of climate change has been widely noted (e.g. Miller & Brock, 1989; Frederick, 1993).

Table 4 Annual navigation costs ($M) under various change scenarios (Marchand *et al*, 1988)

Scenario	Annual costs ($M)		
	Minimum	Mean	Maximum
Baseline (1900-1976)	194	198	207
Reduction in lake levels	197	202	218
Reduction in lake levels plus economic growth	203	217	244
Reduction in lake levels, economic growth, reduced ice cover and increased withdrawals from lakes for consumptive use	205	226	258

Recreational use of freshwaters might also be affected by changes in water quality. At the extreme water quality may exceed legal or safe limits for specific activities; in a more subtle way, changes in water quality may impair the aesthetic attractiveness of a watercourse.

CONCLUSIONS

This review has shown that changes in hydrological regimes due to global warming may have a wide range of impacts in many different sectors. Many case studies have been presented, indicating potential changes in quantitative and qualitative ways. The studies are not necessarily directly comparable because they use different methodologies and different change scenarios. They show, however, that sensitivity varies considerably between systems and environments, and that it is very difficult to generalise socio-economic impacts from individual case studies. There are several general points to draw in conclusion.

Virtually all studies ignore incremental adaptation to changed conditions; the actual impacts of global warming on managed systems will be less than estimated by most studies, simply because water managers adapt continually to changed circumstances. These incremental adaptations may, of course, have economic costs. Unmanaged systems - such as aquatic ecosystems - may be more vulnerable to change.

The review has proceeded through several water use sectors, and treated them independently. In practice, of course, different water use sectors are linked, and in many cases are competing for scarce resources. Actions taken in one sector (such as water supply) affect another (such as health) and these inter-sector linkages may complicate the impacts of global warming. The vulnerability of a region is a function of the sum of inter-connected implications in the region. Also, altered pressures caused by global warming may increase competition between different water-use sectors; Miller (1989) and Frederick (1993) both discuss the potential for increasing conflict between water users under conditions of increased scarcity.

Global warming is not the only change which will affect the water sector over the next few decades, and is not necessarily always the most important. Changes in demand for water may have a greater impact on supply reliability in some catchments, and climate change effects on health may be trivial alongside the effects of increasing urbanisation and poverty. When assessing the impacts of global warming, it is important to make comparisons with the correct baseline.

The estimated impacts of global warming on water resources and water management are very uncertain, due largely to differences between different climate change scenarios. This uncertainty is not necessarily much of a practical problem for short and medium-term water management, because this is affected by other uncertainties - economic and legislative, for example - but does make it difficult to determine the significance of global warming. Physical scientists can contribute to reducing this uncertainty by improving the simulation of global and regional climates, and hydrologists can translate improved climate change scenarios into potential changes in hydrological regimes. However, it is also necessary to improve the understanding of the quantitative links between "water" and "impact"; credible assessments of the potential impact of global warming change must be based on credible, validated impact models.

References

Allen, R.G., Gichuki, F.N. & Rosenzweig, C. (1991) CO_2-induced climatic changes in irrigation-water requirements. J. Water Res. Planning and Management 117, 157-178.

Arnell, N.W. & Reynard, N.S. (1993) Impact of climate change on river flow regimes in the United Kingdom. Institute of Hydrology. Report to Department of the Environment Water Directorate. 130pp

Arnell, N.W., Jenkins, A. & George, D.G. (1994) The implications of climate change for the National Rivers Authority. NRA R&D Report 12. NRA: Bristol. 94pp

Beran, M.A. & Arnell, N.W. (1994) Climate change and hydrological disasters. in V.P. Singh (ed.) Hydrology of disasters. Kluwer: Dordrecht *in press*.

Cohen, S.J. (1986) Impact of CO_2-induced climatic change on water resources in the Great Lakes Basin. Climatic Change 8, 135-153.

Cohen, S.J. (1987) Projected increases in municipal water use in the Great Lakes due to CO_2-induced climatic change. Water Resources Bulletin 23, 81-101.

Cohen, S.J. (1991) Possible impacts of climatic warming scenarios on water resources in the Saskatchewan River sub-basin, Canada. Climatic Change 19, 291-317.

Cohen, S.J. (1994) Potential changes in hydrological systems. (this volume)

Cole, J.A., Slade, S., Jones, P.D. & Gregory, J.M. (1991) Reliable yield of reservoirs and possible effects of climatic change. Hydrol. Sci. J. 36, 579-597.

Covich, A.P. (1993) Water and ecosystems. in P.H. Gleick (ed.) Water in crisis: a guide to the world's freshwater resources. Oxford University Press: New York. 40-55.

Croley, T.E. (1993) Climate change impacts on Great Lakes water supplies. in R. Herrmann (ed.) Managing water resources during global change. Proc. Int. Symp. American Water Resources Association, Bethesda MD. 241-250.

Frederick, K.D. (1993) Climate change impacts on water resources and possible responses in the MINK region. Climatic Change 24, 83-115.

Gleick, P.H. (1988) The effects of future climatic changes on international water resources:the Colorado River, the United States and Mexico. Policy Sciences 21, 23-39.

Gleick, P.H. (1992) Effects of climate change on shared freshwater resources. in. Mintzer, I.M. (ed.) Confronting climate change: risks, implications and responses. Cambridge University Press: Cambridge. 127-140.

Gleick, P.H. (1993) Water in the 21st century. in P.H. Gleick (ed.) Water in crisis: a guide to the world's freshwater resources. Oxford University Press: New York. 105-113

Hanemann, W.M. & Dumas, C.F. (1993) Simulating impacts on Sacremento River fall run chinook salmon. in: Integrated modelling of drought and global warming: impacts on selected California resources. Dracup, J.A., Pelmuder, S.D., Howitt, R. *et al* (eds) National Institute for Global Environmental Change, University of California, Davis. Report to Environmental Protection Agency. 69-95.

Hanemann, W.M. & McCann, R. (1993) Economic impacts on the Northern California hydropower system. in: Integrated modelling of drought and global warming: impacts on selected California resources. Dracup, J.A., Pelmuder, S.D., Howitt, R. *et al* (eds) National Institute for Global Environmental Change, University of California, Davis. Report to Environmental Protection Agency. 55-68.

Herrington, P. & Hoschatt, M. (1993) Climate change and the demand for water. University of Leicester, Department of Economics. Report to Department of the Environment Water Directorate.

Hewett, B.A.O, Harries, C.D. & Fenn, C.R. (1993) Water resource planning in the uncertainty of climatic change: a water company perspective. in R. White (ed.) Engineering for climatic change. Thomas Telford: London. 38-53.

Jenkins, A., McCartney, M. & Sefton, C. (1993) Impact of climate change on river water quality in the United Kingdom. Institute of Hydrology. Report to Department of the Environment Water Directorate.

Lettenmaier, D.P. & Gan, T.Y. (1990) Hydrologic sensitivities of the Sacremento-San Jaoquin river basin, California, to global warming. Water Resources Research 26, 69-86.

Loomis, J. & Lee, S. (1993) Net economic value of recreational fishing on the Sacremento River in 1980. in: Integrated modelling of drought and global warming: impacts on selected California resources. Dracup, J.A., Pelmuder, S.D., Howitt, R. *et al* (eds) National Institute for Global Environmental Change, University of California, Davis. Report to Environmental Protection Agency. Contract CR818756-01-0, 97-106.

Marchand, D., Sanderson, M., Howe, D. & Alpaugh, C. (1988) Climatic change and Great Lakes levels: the impact on shipping. Climatic Change 12, 107-133.

McMahon, T.A., Nathan, R.J., Finlayson, B.L. & Haines, A.T. (1989) Reservoir system performance and climatic change. G.C. Dandy & A.R. Simpson (eds.) Proc. National Workshop on Planning and Management of Water Resource Systems: Risk and Reliability. Australian Government Publishing Service: Canberra. 106-123.

Miller, B.A. & Brock, W.G. (1989) Potential impacts of climate change on the Tennessee Valley Authority reservoir system. in J.B. Smith & D.A. Tirpak (eds.) The potential effects of global climate change on the United States. US Environmental Protection Agency, Washington D.C. Appendix A - Water resources. 9.1-9.46.

Miller, B.A., Alavian, V., Bender, M.D. *et al* (1993) Impacts of changes in air and water temperature on thermal power generation. in R. Herrmann (ed.) Managing water resources during global change. Proc. Int. Symp. American Water Resources Association, Bethesda MD. 439-448.

Miller, K.A. (1989) Hydropower, water institutions and climate change. Water Resources Development 5, 71-83.

Mimikou, M., Hadjisavva, P.S., Kouvopoulos, Y.S. & Afrateos, H. (1991) Regional climate change impacts: II. Impacts on water management works. Hydrol. Sci. J. 36, 259-270.

Minns, C.K. & Moore, J.E. (1993) Predicting the impact of climate change on the spatial pattern of freshwater fish yield capability in eastern Canadian lakes. Climatic Change 22, 327-346.

Nash, L. (1993) Water quality and health. in P.H. Gleick (ed.) Water in crisis: a guide to the world's freshwater resources. Oxford University Press: New York. 25-39.

Nash, L. & Gleick, P.H. (1993) The Colorado River basin and climatic change. The sensitivity of streamflow and water supply to variations in temperature and precipitation. Pacific Institute for Studies in Development, Environment and Security, California. Report to US Environmental Protection Agency EPA 230-R-93-009.

Nemec, J. & Schaake, J.C. (1982) Sensitivity of water resource systems to climate variation. Hydrol. Sci. J. 27, 327-343.

Poiani, K.A. & Johnson, W.C. (1991) Global warming and prairie wetlands. Bioscience 41, 611-618.

Rhodes, S.L., Miller, K.A. & MacDonnell, L.J. (1992) Institutional response to climate change: water provider organizations in the Denver metropolitan region. Water Resources Research 28, 11-18.

Riebsame, W.E. (1991) Adjusting water resources management to climate change. Climatic Change 13, 69-97.

Saelthun, N.R., Bogen, J., Flood, M.H. *et al* (1990) Climate change impact on Norwegian water resources. Norwegian Water Resources and Energy Administration Publication NR V42.

Sanderson, M. (1978) Implications of climatic change for navigation and power generation in the Great Lakes. Climate Change Digest CCD 87-03. Environment Canada.

Stein, J.F. (1993) The impact of climatic change on interstate apportionment. in R. Herrmann (ed.) Managing water resources during global change. Proc. Int. Symp. American Water Resources Association, Bethesda MD. 619-626.

Stephens, D.B. & Held, I.M. (1993) GCM response of northern winter stationary waves and storm tracks to increasing amounts of carbon dioxide. J. Climate 6, 1859-1870.

Thomas, C. & Howlett, D. (1993) Resource politics: freshwater and regional relations. Open University Press: Buckingham.

Tickell, C. (1993) Global warming and its effects. in R. White (ed.) Engineering for climatic change. Thomas Telford: London. 1-8.

Wolock, D.M., McCabe, G.J., Tasker, G.D. & Moss, M.E. (1993) Effects of climate change on water resources in the Delaware River basin. Water Resources Bulletin 29, 475-486.

Appendix

Poster Displays

Students were invited to bring posters on their research work and these were displayed in three different groups each chosen to cover topic areas similar to the lectures being given at the time. The titles of the posters together with the names and addresses of the students at the time of the ASI are given below in display order. In some cases others not attending the ASI were also involved in the work described.
The standard of the posters was generally very high and prizes were awarded to the three marked (*) with others also being specially commended.

Parameterization of soil hydrology and river transport for GCM's
A Shmakin, Institute of Geography, Moscow, Russia

Effect of land surface formulation on climate variability within a GCM (*)
R Scott, MIT, Cambridge, USA

Global change; sensitivity of the response to the oceanic forcing
B Timbal, CNRM, Toulouse, France

Feedback mechanisms of thermohaline circulation in a coupled ocean circulation model and atmosphere EBM
G Lohmann, AWI, Bremerhaven, Germany

Tropical precipitation climatologies from satellite retrievals and GCM simulations
C Jones, Climatic Research Unit, Norwich, UK

Influence of climate changes on groundwater quality of crystalline rocks in Alentezio region (S Portugal)
A Chambel & J Duque, University of Evora, Portugal & Amadora, Portugal

Modelling the nitrate flow in soils and groundwaters of the F R Germany

F Wendland, Research Centre Jülich GmbH, Germany

Experimental evaluation of nonisothermal unsaturated water flow in a sandy soil

A Tuli, University of Cukurova, Adana, Turkey

Sensitivity of the LMD-GCM to subgrid scale variability of field capacity (*)

A Ducharne, CNRS, Paris, France

Application of the transilient turbulence theory to forest canopies

M Inclan, DLR, Wessling, Germany

Modelling dynamics of soil moisture under a plant canopy on a local scale (*)

O Nasonova, Institute of Water Problems, Moscow, Russia

Parameterization of albedo for a melt pond

I Podgorny, Arctic and Antarctic Research Institute, St Petersburg, Russia

Modelling fallow savannah evaporation rates

C Huntingford, Institute of Hydrology, Wallingford, UK

Transport of sensible and latent heat in the upper Rhine valley

C Strodl, DLR, Wessling, Germany

Electronic atlas of temporally/spatially heterogeneous land-atmosphere interactions

A Shmakin, Institute of Geography, Moscow, Russia

Modelling of coupled land-atmosphere hydrologic processes

Z Otles, Atmospheric Science Dept, Ames, USA

Coupled hydrospheric-atmospheric research model

B Lofgren, Great Lakes Environmental Research Labs, Ann Arbor, USA

Modelling the global snow cover

B Loth, MPI, Hamburg, Germany

Italian glaciological committee

S Paludi, San Giusto Canavese, Italy

Scale effects in calculating net solar radiation for snowmelt runoff modelling

R Galarraga-Sanchez, University of Arizona, Tucson, USA

Modelling of albedo variation in the ablation zone of West Greenland

Z Zuo, Institute for Marine and Atmospheric Research, Utrecht, Netherlands

An influence of clouds on climate stability

K Szilder, Dept of Geography, University of Alberta, Canada

Transformation of measured flow data to grid points - first results within the WCP water project 3.3

T Lullwitz, Global Runoff Data Centre, Koblenz, Germany

The global runoff data centre - its potential and role in water resources assessment and planning

T Lullwitz, Global Runoff Data Centre, Koblenz, Germany

Index

The ASI Series Books Published as a Result of Activities of the Special Programme on Global Environmental Change

This book contains the proceedings of a NATO Advanced Research Workshop held within the activities of the NATO Special Programme on Global Environmental Change, which started in 1991 under the auspices of the NATO Science Committee.

The volumes published as a result of the activities of the Special Programme are:

Vol. 1: **Global Environmental Change.** Edited by R. W. Corell and P. A. Anderson. 1991.
Vol. 2: **The Last Deglaciation: Absolute and Radiocarbon Chronologies.** Edited by E. Bard and W. S. Broecker. 1992.
Vol. 3: **Start of a Glacial.** Edited by G. J. Kukla and E. Went. 1992.
Vol. 4: **Interactions of C, N, P and S Biogeochemical Cycles and Global Change.** Edited by R. Wollast, F. T. Mackenzie and L. Chou. 1993.
Vol. 5: **Energy and Water Cycles in the Climate System.** Edited by E. Raschke and D. Jacob. 1993.
Vol. 6: **Prediction of Interannual Climate Variations.** Edited by J. Shukla. 1993.
Vol. 7: **The Tropospheric Chemistry of Ozone in the Polar Regions.** Edited by H. Niki and K. H. Becker. 1993.
Vol. 8: **The Role of the Stratosphere in Global Change.** Edited by M.-L. Chanin. 1993.
Vol. 9: **High Spectral Resolution Infrared Remote Sensing for Earth's Weather and Climate Studies.** Edited by A. Chedin, M.T. Chahine and N.A. Scott. 1993.
Vol. 10: **Towards a Model of Ocean Biogeochemical Processes.** Edited by G. T. Evans and M.J.R. Fasham. 1993.
Vol. 11: **Modelling Oceanic Climate Interactions.** Edited by J. Willebrand and D.L.T. Anderson. 1993.
Vol. 12: **Ice in the Climate System.** Edited by W. Richard Peltier. 1993.
Vol. 13: **Atmospheric Methane: Sources, Sinks, and Role in Global Change.** Edited by M. A. K. Khalil. 1993.
Vol. 14: **The Role of Regional Organizations in the Context of Climate Change.** Edited by M. H. Glantz. 1993.
Vol. 15: **The Global Carbon Cycle.** Edited by M. Heimann. 1993.
Vol. 16: **Interacting Stresses on Plants in a Changing Climate.** Edited by M. B. Jackson and C. R. Black. 1993.
Vol. 17: **Carbon Cycling in the Glacial Ocean: Constraints on the Ocean's Role in Global Change.** Edited by R. Zahn, T. F. Pedersen, M. A. Kaminski and L. Labeyrie. 1994.
Vol. 18: **Stratospheric Ozone Depletion/UV-B Radiation in the Biosphere.** Edited by R. H. Biggs and M. E. B. Joyner. 1994.
Vol. 19: **Data Assimilation: Tools for Modelling the Ocean in a Global Change Perspective.** Edited by P. O. Brasseur and J. Nihoul. 1994.

Vol. 20: **Biodiversity, Temperate Ecosystems, and Global Change.**
Edited by T. J. B. Boyle and C. E. B. Boyle. 1994.

Vol. 21: **Low-Temperature Chemistry of the Atmosphere.**
Edited by G. K. Moortgat, A. J. Barnes, G. Le Bras and J. R. Sodeau. 1994.

Vol. 22: **Long-Term Climatic Variations – Data and Modelling.**
Edited by J.-C. Duplessy and M.-T. Spyridakis. 1994.

Vol. 23: **Soil Responses to Climate Change.**
Edited by M. D. A. Rounsevell and P. J. Loveland. 1994.

Vol. 24: **Remote Sensing and Global Climate Change.**
Edited by R. A. Vaughan and A. P. Cracknell. 1994.

Vol. 25: **The Solar Engine and Its Influence on Terrestrial Atmosphere and Climate.**
Edited by E. Nesme-Ribes. 1994.

Vol. 26: **Global Precipitations and Climate Change.**
Edited by M. Desbois and F. Désalmand. 1994.

Vol. 27: **Cenozoic Plants and Climates of the Arctic.**
Edited by M. C. Boulter and H. C. Fisher. 1994.

Vol. 28: **Evaluating and Monitoring the Health of Large-Scale Ecosystems.**
Edited by D. J. Rapport, C. L. Gaudet and P. Calow. 1995.

Vol. 29: **Global Environmental Change Science: Education and Training.**
Edited by D. J. Waddington. 1995.

Vol. 30: **Ice Core Studies of Global Biogeochemical Cycles.**
Edited by R. J. Delmas. 1995.

Vol. 31: **The Role of Water and the Hydrological Cycle in Global Change.**
Edited by H. R. Oliver and S. A. Oliver. 1995